Manual of Concrete Inspection

ACI 311.1R-07 with Selected References

Reported by ACI Committee 311

PUBLICATION MNL-2(19)
American Concrete Institute
Farmington Hills, MI

ELEVENTH EDITION

First Printing, February 2019

Printed in the United States of America.

Copyright © 2019
American Concrete Institute
38800 Country Club Drive
Farmington Hills, MI
48331-3439 USA
+1.248.848.3700
www.concrete.org

All rights reserved including rights of reproduction and use in any form or by any means, including the making of copies by any photo process, or by any electronic or mechanical device, printed, written, or oral, or recording for sound or visual reproduction or for use in any knowledge or retrieval system or device, unless permission in writing is obtained from the copyright proprietors.

ISBN: 978-1-64195-052-7

CONTENTS

311.1R-07   Manual of Concrete Inspection

311.4R-05   Guide for Concrete Inspection

311.6-18   Specification for Testing Ready Mixed Concrete

311.7-18   Specification for Inspection of Concrete Construction

ACKNOWLEDGMENTS

The committee wishes to thank Anne Balogh for her extensive work in redrafting and unifying the previous text in preparation for this edition.

Additionally, the committee wishes to express its thanks to Portland Cement Association (PCA) for generously providing many of the new photos contained in the manual and to committee member Michelle L. Wilson who organized and coordinated the selection of photos and graphics throughout the document.

Finally, the committee wishes to thank Chair George R. Wargo for his efforts in coordinating the work of all contributors.

ABOUT THIS BOOK

This manual is based on information from many sources, organizations, and individuals whose contributions are gratefully acknowledged. Published references are listed at the end of the text. References to standard specifications and methods of testing are listed separately.

The original manuscript was prepared by Joe W. Kelly, Chair of ACI Committee 311, and revised over a period of years to achieve a first edition in 1941. The second edition, also in 1941, included a number of corrections and minor revisions.

The third edition, in 1955, incorporated many constructive suggestions from users. The fourth edition, in 1957, brought several sections up to date and contained editorial corrections.

The fifth edition provided new information on settlement of concrete, shoring and forming, strength requirements, cold-weather concreting, and shotcrete. The sixth edition primarily provided updated information in all chapters and included editorial and substantive changes throughout.

The seventh edition presented a complete revision of the manual by eliminating sections of the previous edition covering concrete methods no longer in use. Chapters 2, 11, 12, 13, 14, 15 (partial), 16, 17, and 18 covered material that was included in the manual for the first time. The eighth and ninth editions were revised to reflect changes in technology and construction practices.

The tenth edition presented an extensive revision and update to the text along with new photos, charts, and forms. This eleventh edition substitutes *311.5-04 Guide for Concrete Plant Inspection and Testing of Ready-Mixed Concrete* with *311.6-18 Specification for Testing Ready Mixed Concrete* and *311.7-18 Specification for Inspection of Concrete Construction*.

ON THE COVER

2018 ACI Excellence in Concrete Construction Awards Overall Winner

Viaduct Over River Almonte: Cáceres, Extremadura, España

Nominator: Asociación Española de Ingeniería Estructural
Owner: Arenas & Asociados
Architectural & Engineering Firm: Arenas & Asociados - Idom
General & Concrete Contractor: FCC Construcción - Conduril
Concrete Supplier: CG Hormigones

The Viaduct Over the Almonte River is located in the Alcántara-Garrovillas reservoir section stretching 6265 meters. It forms part of the Madrid-Extremadura High-Speed Railway Line that runs through the municipalities of Garrovillas de Alconétar and Santiago del Campo, in Cáceres. The multi-criteria analyses highlighted the concrete arch solution as the most economical, the best in terms of durability and maintenance, and the one that would perform best in resisting dynamic load effects and wind. This focus on service life prevailed during design and construction with the aim of creating a bridge that would resist the passage of time with minimum maintenance.

Visit *www.ACIExcellence.org* for specific details on the winning projects.

ACI 311.1R-07

ACI Manual of Concrete Inspection

Reported by ACI Committee 311

George R. Wargo
Chair

Michael T. Russell
Secretary

Gordon A. Anderson	John V. Gruber	Venkatesh S. Iyer	Woodward L. Vogt
Joseph F. Artuso	Jimmie D. Hannaman, Jr.	Claude E. Jaycox	Bertold E. Weinberg
Mario R. Diaz	Robert L. Henry	Robert S. Jenkins	Michelle L. Wilson
Donald E. Dixon	Charles J. Hookham	Roger D. Tate	Roger E. Wilson

PREFACE

This manual is intended to guide, assist, and instruct concrete inspectors and others engaged in concrete construction and testing, including field engineers, construction superintendents, supervisors, laboratory and field technicians, and workers. Designers may also find the manual to be a valuable reference by using the information to better adapt their designs to the realities of field construction. Because of the diverse possible uses of the manual and the varied backgrounds of the readers, it includes the reasoning behind the technical instructions.

The field of concrete construction has expanded dramatically over the years to reflect the many advances that have taken place in the concrete industry. Although many of the fundamentals presented in previous editions of this manual remain relevant and technically correct, this tenth edition incorporates new material to address these advances in technology. A list of only a few of the recent developments in materials, equipment, and processes includes:

- Shrinkage-compensating cement;
- Increased use of supplementary cementitious materials (SCMs);
- Polymer-modified mixtures;
- Self-consolidating concretes;
- New and refined admixtures;
- Fiber-reinforced concrete;
- Epoxy resins;
- High-capacity and automated concrete production equipment;
- High-performance and high-strength concrete; and
- Epoxy-coated and stainless steel-clad reinforcement.

The need to cover new issues affecting inspection is the reason ACI Committee 311 continues to revise the ACI Manual of Concrete Inspection. In preparing this edition of the manual, as with previous editions, the committee's task was to interpret the policies set forth by other authorized bodies rather than to make policy on construction practices. The main emphasis of the manual is on the technical aspects of inspection and construction. For further information about construction practices, readers are encouraged to refer to the ACI Manual of Concrete Practice.

Because the content of this manual is general and broad in nature, no part of the manual should be included by reference in contract documents. Applicable inspection requirements for each project should be determined and included in the specifications.

Chapter 1—Inspection and the inspector, p. 6
    1.1—Inspection processes
        1.1.1—Why inspection is needed
        1.1.2—Purposes of inspection
        1.1.3—Owner and contractor inspections
    1.2—Inspector
        1.2.1—Duties
        1.2.2—Education and certification
        1.2.3—Authority
        1.2.4—Relations with contractors, supervisors, and workers
        1.2.5—Safety
    1.3—Importance of clear specifications

Chapter 2—Statistical concepts for quality assurance, p. 10
    2.1—Quality-control and quality-acceptance inspections
    2.2—Traditional quality assurance
    2.3—Statistical concepts and procedures
    2.4—Basic statistical concepts
        2.4.1—Definitions
        2.4.2—Normal distribution curves
        2.4.3—Applying normal distribution curves to concrete compressive strength
    2.5—Statistical tools
        2.5.1—Frequency distributions
        2.5.2—Control charts

ACI Committee Reports, Guides, Manuals, Standard Practices, and Commentaries are intended for guidance in planning, designing, executing, and inspecting construction. This document is intended for the use of individuals who are competent to evaluate the significance and limitations of its content and recommendations and who will accept responsibility for the application of the material it contains. The American Concrete Institute disclaims any and all responsibility for the stated principles. The Institute shall not be liable for any loss or damage arising therefrom.

Reference to this document shall not be made in contract documents. If items found in this document are desired by the Architect/Engineer to be a part of the contract documents, they shall be restated in mandatory language for incorporation by the Architect/Engineer.

ACI 311.1R-07 supersedes ACI 311.1R-99 and was adopted December 2007 and published March 2008.
Copyright © 2007, American Concrete Institute.
All rights reserved including rights of reproduction and use in any form or by any means, including the making of copies by any photo process, or by electronic or mechanical device, printed, written, or oral, or recording for sound or visual reproduction or for use in any knowledge or retrieval system or device, unless permission in writing is obtained from the copyright proprietors.

Appendix 1—Sampling by random numbers, p. 17
A1.1—Example 1: Sampling by time sequence
A1.2—Example 2: Sampling by material weight
A1.3—Example 3: Sampling by depth of concrete pavement

Appendix 2—Normal distribution curves, p. 20

Appendix 3—Computing standard deviation and required average concrete strength, p. 21
A3.1—Calculating standard deviation s
A3.2—Calculating required average strength f'_{cr}

Appendix 4—Control charts on concrete materials, p. 22
A4.1—Example 1: Calculations to determine moving averages for sand-equivalent test
A4.2—Example 2: Calculations to determine moving averages for 1-1/2 x 3/4 in. concrete aggregate (maximum variation of percentage of material passing 1 in. sieve)

Chapter 3—Inspection and testing of materials, p. 23
3.1—Cement
   3.1.1—Standard types
   3.1.2—Blended cements and other hydraulic cements
   3.1.3—Optional requirements
   3.1.4—Sampling and testing procedures
3.2—Aggregates
   3.2.1—Specifications
   3.2.2—Sampling procedures
3.3—Water
3.4—Admixtures
3.5—Steel reinforcement
3.6—Curing compounds
3.7—Joint materials

Appendix 5—Aggregate sampling and testing, p. 33
A5.1—Sieve or screen analysis of fine and coarse aggregate: ASTM C136
A5.2—Sampling aggregates: ASTM D75
A5.3—Materials finer than No. 200 (75 μm) sieve: ASTM C117
A5.4—Clay lumps and friable particles in aggregates: ASTM C142
A5.5—Organic impurities in fine aggregate: ASTM C40
A5.6—Specific gravity and absorption of coarse aggregate: ASTM C127
A5.7—Specific gravity and absorption of fine aggregate: ASTM C128
A5.8—Total moisture content of aggregate by drying: ASTM C566

Chapter 4—Handling and storage of materials, p. 36
4.1—Cement
   4.1.1—Storage and hauling of bulk cement
   4.1.2—Storage of bagged cement
4.2—Aggregates
   4.2.1—Storage in stockpiles
   4.2.2—Storage in bins
   4.2.3—Finish screening
   4.2.4—Transporting
4.3—Supplementary cementitious materials
4.4—Admixtures

Chapter 5—Fundamentals of concrete, p. 38
5.1—Nature of concrete
   5.1.1—Components
   5.1.2—Water-cementitious material ratio
5.2—Fresh concrete
   5.2.1—Workability
   5.2.2—Consolidation
   5.2.3—Hydration, setting, and hardening
   5.2.4—Heat of hydration
5.3—Hardened concrete
   5.3.1—Curing
   5.3.2—Strength
   5.3.3—Durability
   5.3.4—Chemical attack
   5.3.5—Freezing-and-thawing effects
   5.3.6—Alkali-aggregate reactivity
   5.3.7—Volume changes

Chapter 6—Proportioning and control of concrete mixtures, p. 43
6.1—Factors to consider
6.2—Methods of specifying concrete proportions
   6.2.1—Strength specifications
   6.2.2—Prescriptive specifications
6.3—Proportioning for specified strength or *w/cm*
   6.3.1—Cement types
6.4—Concrete with supplementary cementitious materials
   6.4.1—Mixture proportioning and control
   6.4.2—Water-cementitious material ratio
   6.4.3—Aggregate selection
   6.4.4—Air entrainment
   6.4.5—Quantity of paste
   6.4.6—Proportion of fine-to-coarse aggregate
6.5—Proportioning for resistance to severe exposure conditions
   6.5.1—Paste quality
   6.5.2—Required air entrainment
   6.5.3—Aggregate proportions
   6.5.4—Proportioning by absolute volume
   6.5.5—Computing absolute volume and percentage of solids
   6.5.6—Example of proportioning by absolute volume
6.6—Control of concrete proportions
   6.6.1—Laboratory batch quantities
   6.6.2—Field batch quantities
   6.6.3—Field control of selected proportions
6.7—Computations for yield
   6.7.1—Density method

Chapter 7—Batching and mixing, p. 53
7.1—Batching operations
   7.1.1—Measurement tolerances

7.1.2—Weighing equipment
7.1.3—Batching equipment
7.1.4—Measuring water
7.1.5—Measuring admixtures
7.2—Mixing operations
7.2.1—Central or site mixing
7.2.2—Ready mixed concrete
7.2.3—Volumetric batching and mixing
7.3—Plant inspection
7.3.1—Control of water content
7.3.2—Control of air content
7.3.3—Control of temperature
7.4—Placing inspection
7.4.1—Control of slump loss
7.4.2—Control of consistency
7.4.3—Measuring concrete quantity

Chapter 8—Inspection before concreting, p. 64
8.1—Preliminary study
8.2—Stages of preparatory work
8.3—Excavations and foundations
8.3.1—Building slabs-on-ground
8.3.2—Building foundations
8.3.3—Underwater placements
8.3.4—Pile foundations
8.4—Forms for buildings
8.4.1—Form tightness and alignment
8.4.2—Shoring
8.4.3—Preventing bulging and settlement
8.4.4—Coating for release
8.4.5—Form reuse and maintenance
8.5—Reinforcement
8.5.1—Cutting and bending
8.5.2—Storage and handling
8.6—Installation
8.6.1—Cover depth
8.6.2—Splicing, welding, and anchoring
8.6.3—Congestion
8.6.4—Support
8.7—Embedded fixtures
8.8—Joints
8.9—Final inspection before placing
8.10—Checklist

Chapter 9—Concreting operations, p. 71
9.1—Placing conditions
9.2—Handling of concrete
9.2.1—Conveying
9.2.2—Placing
9.3—Consolidation
9.3.1—Vibration
9.4—Finishing
9.4.1—Unformed surfaces
9.4.2—Formed surfaces
9.5—Construction joints
9.5.1—Planned
9.5.2—Unplanned

Chapter 10—Form removal, reshoring, curing, and protection, p. 81
10.1—Removal of forms and supports
10.1.1—Time of removal
10.1.2—Multistory work
10.2—Protection from damage
10.2.1—Backfilling
10.3—Curing
10.3.1—Moist curing
10.3.2—Membrane curing
10.3.3—Impermeable sheets
10.3.4—Accelerated curing
10.4—Curing and protection during weather extremes
10.4.1—Cold weather
10.4.2—Hot weather

Chapter 11—Postconstruction inspection of concrete, p. 86
11.1—Acceptance inspection
11.2—Visual inspection (condition survey)
11.3—Other roles and responsibilities
11.3.1—Nondestructive evaluation (NDE)
11.3.2—Destructive testing
11.3.3—Summary
11.4—Observations leading to repair/rehabilitation
11.4.1—Minor defects
11.4.2—Structural defects

Chapter 12—Slabs for buildings, p. 92
12.1—Positioning of reinforcement
12.2—Mixture requirements
12.3—Slabs-on-ground
12.3.1—Subgrade and subbase
12.3.2—Placing and consolidation of concrete
12.3.3—Finishing
12.3.4—Hardened surfaces
12.3.5—Two-course construction and special toppings
12.3.6—Curing and protection
12.4—Structural slabs
12.5—Joint construction

Chapter 13—Pavement slabs and bridge decks, p. 96
13.1—Foundation (subgrade and subbase course)
13.1.1—Fine grading
13.1.2—Stabilized base
13.2—Forms
13.2.1—Keyway forms
13.3—Steel reinforcement
13.3.1—Storage
13.3.2—Installation methods
13.4—Concrete
13.4.1—Materials
13.4.2—Mixture proportioning
13.4.3—Batching and mixing
13.5—Paving operations
13.5.1—Concrete placement
13.5.2—Vibration
13.5.3—Slipform paving

13.5.4—Fixed-form paving
13.5.5—Finishing
13.5.6—Texturing
13.5.7—Curing and protection
13.6—Final acceptance
13.7—Joints
13.7.1—Transverse contraction joints
13.7.2—Transverse construction joints
13.7.3—Longitudinal contraction joints
13.7.4—Longitudinal construction joints
13.7.5—Expansion joints
13.7.6—Joint sealing
13.8—Bridge decks
13.8.1—Concrete placement
13.8.2—Finishing

Chapter 14—Architectural concrete, p. 107
14.1—Determining requirements for acceptability
14.1.1—Preconstruction samples
14.2—Importance of uniformity
14.3—Forms
14.3.1—Form sheathing or lining
14.3.2—Textures and patterns
14.3.3—Form joints
14.3.4—Form sealers and release agents
14.3.5—Form ties
14.3.6—Form removal
14.4—Reinforcement
14.5—Concrete materials and mixture proportions
14.5.1—Cement
14.5.2—Aggregates
14.5.3—Admixtures
14.6—Batching, mixing, and transporting
14.7—Placing and consolidation
14.8—Surface treatments
14.8.1—Sandblasting
14.8.2—Bush hammering
14.8.3—Grinding
14.8.4—Manual treatment
14.8.5—Exposed-aggregate finishes
14.9—Curing and protection
14.10—Repairs
14.11—Precast members
14.11.1—Storage and transportation
14.11.2—Erection
14.12—Final acceptance
14.12.1—Bug holes
14.12.2—Color variations

Chapter 15—Special concreting methods, p. 113
15.1—Slipforming vertical structures
15.1.1—Formwork
15.1.2—Reinforcing steel
15.1.3—Control of concrete placement
15.1.4—Finishing and curing
15.1.5—Mixture requirements
15.2—Slipforming cast-in-place pipe
15.2.1—Forms
15.2.2—Control of concrete placement
15.3—Tilt-up construction
15.3.1—Casting platform
15.3.2—Forms
15.3.3—Bond prevention
15.3.4—Concrete placement
15.3.5—Panel erection
15.4—Lift-slab construction
15.4.1—Forms
15.4.2—Bond prevention
15.4.3—Slab erection
15.5—Preplaced-aggregate concrete
15.5.1—Aggregate placement
15.5.2—Grout materials and mixing
15.5.3—Grouting operations
15.6—Underwater concrete construction
15.6.1—Equipment and methods
15.6.2—Mixture requirements for tremie concrete
15.7—Vacuum dewatering of concrete
15.7.1—Equipment and methods
15.7.2—Reduction in slab thickness
15.8—Pumped concrete
15.8.1—Types of equipment
15.8.2—Mixture requirements
15.8.3—Control of placement
15.8.4—Taking concrete samples
15.9—Shotcrete
15.9.1—Shotcreting processes
15.9.2—Mixture proportions
15.9.3—Safety
15.9.4—Forms and ground wires
15.9.5—Surface finishing
15.9.6—Curing and protection
15.9.7—Control testing

Chapter 16—Special types of concrete, p. 123
16.1—Structural lightweight-aggregate concrete
16.1.1—Aggregates
16.1.2—Mixture proportioning and control
16.1.3—Testing
16.1.4—Batching and mixing
16.1.5—Placing, consolidation, and finishing
16.1.6—Curing and protection
16.2—Lightweight fill concrete
16.2.1—Aggregates
16.2.2—Mixture proportioning and testing
16.3—Low-density concrete
16.3.1—Aggregates
16.3.2—Foams for cellular concrete
16.3.3—Mixture proportioning and control
16.3.4—Testing
16.3.5—Batching and mixing
16.3.6—Placing and consolidation
16.3.7—Curing and protection
16.4—Heavyweight concrete
16.4.1—Aggregates
16.4.2—Mixture proportions and control

16.4.3—Batching and mixing
16.4.4—Placing, consolidation, and finishing
16.4.5—Curing and protection
16.5—Mass concrete for dams
16.5.1—Mixture proportioning
16.5.2—Testing
16.5.3—Temperature control
16.5.4—Special equipment and procedures
16.6—Structural mass concrete
16.7—Shrinkage-compensating concrete
16.7.1—Materials
16.7.2—Mixture proportioning and control
16.7.3—Production, placing, and finishing
16.7.4—Curing and protection
16.8—High-performance concrete

Chapter 17—Precast and prestressed concrete, p. 132
17.1—Precast concrete
17.1.1—Scope of inspection
17.1.2—Inspecting for quality control and quality assurance
17.1.3—Record keeping and test reports
17.1.4—Formwork
17.1.5—Embedded items
17.1.6—Bar and wire reinforcement
17.1.7—Curing
17.1.8—Handling, storage, and transportation
17.1.9—Erection
17.1.10—Repairs
17.2—Precast prestressed concrete
17.2.1—Concrete materials
17.2.2—Tendons for pretensioning
17.2.3—Tendon handling and storage
17.2.4—Attachments for tendons
17.2.5—Deflection devices
17.2.6—Tensioning of tendons
17.2.7—Wire failure in tendons
17.2.8—Detensioning
17.3—Cast-in-place prestressed concrete
17.3.1—Concrete materials
17.3.2—Post-tensioning tendons
17.3.3—Anchorages and tensioning
17.3.4—Grouting procedures
17.3.5—Postconstruction inspection

Chapter 18—Grout, mortar, and stucco, p. 140
18.1—Pressure grouting
18.2—Grouting under base plates and machine bases
18.2.1—Damp-pack mortar
18.2.2—Gas-forming grouts
18.2.3—Catalyzed metallic grouts
18.2.4—Cementitious systems
18.2.5—Nonshrink grouts
18.3—Mortar and stucco

Chapter 19—Testing of concrete, p. 144
19.1—Sampling
19.2—Tests of freshly mixed concrete
19.2.1—Consistency
19.2.2—Air content
19.2.3—Density of freshly mixed concrete: ASTM C138/C138M
19.2.4—Temperature
19.3—Strength tests
19.3.1—Compressive strength: ASTM C31/C31M, C192/C192M, and C39/C39M
19.3.2—Capping cylindrical concrete specimens for compressive strength tests: ASTM C617
19.3.3—Use of unbonded caps for compressive strength tests: ASTM C1231/C1231M
19.3.4—Testing concrete cylinders: ASTM C39/C39M
19.3.5—Flexural strength of concrete: ASTM C31/C31M, C192/C192M, C78, and C293
19.3.6—Molding flexural specimens
19.3.7—Curing flexural specimens
19.3.8—Testing beams for flexural strength
19.3.9—Splitting tensile strength of cylindrical concrete specimens: ASTM C496/C496M
19.3.10—Compressive strength of lightweight insulating concrete: ASTM C495
19.4—Accelerated curing of test specimens
19.5—Uniformity tests of mixers
19.5.1—Truck mixers
19.5.2—Stationary mixers
19.5.3—Washout test for coarse-aggregate content
19.5.4—Air-free density of concrete test
19.5.5—Air-free density of mortar test
19.6—Density of structural lightweight concrete
19.7—Tests of completed structures
19.7.1—Cores from hardened concrete
19.7.2—Load tests
19.7.3—Nondestructive tests
19.8—Shipping and handling samples

Chapter 20—Records and reports, p. 154
20.1—General information
20.2—Specific information
20.3—Maintaining records
20.4—Quality-control charts
20.4.1—Concrete delivery ticket

Chapter 21—Inspection checklist, p. 168
21.1—Materials (Chapters 3 and 4)
21.2—Proportioning of concrete mixtures (Chapter 6)
21.3—Batching and mixing (Chapter 7)
21.4—Before concreting (Chapter 8)
21.5—Concreting operations (Chapter 9)
21.6—After concreting (Chapter 10)
21.7—Special work
21.7.1—Cold weather concreting
21.7.2—Hot weather concreting
21.7.3—Air-entrained concrete
21.7.4—Two-course floors (Chapter 12)
21.7.5—Architectural concrete (Chapter 14)
21.7.6—Tilt-up construction (Chapter 15)
21.7.7—Preplaced-aggregate concrete (Chapter 15)

21.7.8—Underwater construction (Chapter 15)
21.7.9—Vacuum dewatering (Chapter 15)
21.7.10—Pumped concrete (Chapter 15)
21.7.11—Shotcrete (Chapter 15)
21.7.12—Structural lightweight-aggregate concrete (Chapter 16)
21.7.13—Low-density concrete (Chapter 16)
21.7.14—High-density concrete (Chapter 16)
21.7.15—Mass concrete (Chapter 16)
21.7.16—Shrinkage-compensating concrete (Chapter 16)
21.7.17—Prestressed concrete (Chapter 17)
21.7.18—Grouting under base plates (Chapter 18)
21.7.19—Pressure grouting (Chapter 18)
21.7.20—Mortar and stucco (Chapter 18)
21.7.21—Tests of concrete (Chapter 19)
21.7.22—Records and reports (Chapter 20)

Chapter 22—References, p. 171
22.1—Referenced standards and reports
22.2—Cited references

Index, p. 177

CHAPTER 1—INSPECTION AND THE INSPECTOR
1.1—Inspection processes
1.1.1 *Why inspection is needed*—The purpose of inspection is to verify that the requirements and intent of the contract design documents are faithfully accomplished. In concrete construction, inspection includes not only visual observations and field measurements, but also field and laboratory testing and the collection and evaluation of test data. In many instances, inspectors also act as or are assisted by the field technicians assigned to perform the testing.

One important responsibility of the concrete inspector is to assess the quality of the materials used in the concrete. It is difficult, and usually impossible, to produce specified concrete from nonconforming materials. Thus, the final materials entering the concrete mixture should be of specified quality.

An important factor in quality construction is good workmanship in all operations and processes (Fig. 1.1). Observing this aspect becomes an important responsibility of the inspector. Even when concrete is made using high-quality materials, proportioned correctly, and batched correctly, the resulting concrete structure can be unsatisfactory if construction workmanship is of poor quality.

Manual skills, technical knowledge, motivation, and pride all contribute to good workmanship. Most individuals involved with concrete design and construction take pride in their efforts and strive to attain superior quality (Fig. 1.2). Not all personnel, however, will receive the necessary training to do their jobs properly. The need to meet fast-track construction schedules and stay within cost limits often places too much emphasis on production rates. If speed becomes a top priority, construction quality may not receive adequate attention if not properly executed. Ironically, cost may also suffer. Techniques that speed concrete placement can actually add to material costs or result in the need for expensive repairs.

Jacob Feld, a noted investigator of structural failures, listed examples in his book, *Lessons from Failures of Concrete Structures*, showing that a high percentage of failures of concrete structures that he had investigated were caused in significant part by poor construction—in other words, poor workmanship. He stated: "The one thing which these failures conclusively point to is that all good concrete construction should be subjected to rigid inspection... It is believed that only by this kind of inspection is it possible to guard against the failure of concrete structures" (Feld 1964).

For every monumental structural collapse, innumerable instances of small failures occur. This is particularly true when important concrete properties, such as durability and watertightness, do not conform to design requirements.

Superior concrete structures can be built at a reasonable cost if concrete producers and contractors are vigilant. As the late F. R. McMillan said in the foreword to his famous *Concrete Primer* (McMillan and Tuthill 1987): "Many who have been interested in the cause of better concrete have noted the difficulty of making any real progress until someone in authority has been convinced that good concrete can be had, that it should be had, and, having been so convinced, has sent out the word that it must be had."

1.1.2 *Purposes of inspection*—The desire for quality has led to the use of inspection personnel to monitor and document quality of concrete construction. The responsibilities and duties of inspectors have broadened over the years. Today, several inspection teams may be used on one project to represent the interests of the various parties involved. Inspectors may be employed:
- By project owners to provide quality assurance for the work;
- By government agencies and large industries to assure the quality;
- By architects and engineers to verify and document compliance with project specifications and drawings;
- By contractors to provide quality-control inspection for projects under construction. This helps provide assurance to the contractor that the finished construction will meet all requirements of the contract documents and thus will be accepted by the owner;
- By producers of concrete materials and products who need assurance that finished products will meet the requirements of the contract documents. Examples include producers of cement, aggregate, ready mixed concrete, and precast products;
- By licensing and building-permit jurisdictions charged with enforcing building codes and other regulations. In this case, the inspector will be responsible only for assuring that the finished structure conforms to requirements of the codes or regulations; or

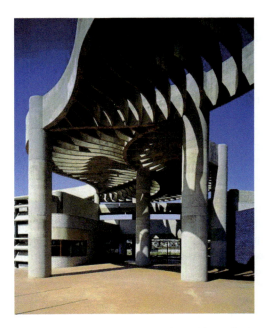

Fig. 1.1—Increasingly sophisticated building designs call for higher standards of materials and workmanship.

- By commercial laboratories designated to provide testing and inspection services.

Regardless of the function, inspection, including laboratory and field testing, may be performed by a team or, for very small projects, by just one person (Fig. 1.3).

1.1.3 *Owner and contractor inspections*

1.1.3.1 *Owner inspection*—Owner inspection provides assurance to owners that the requirements of the contract documents (drawings and specifications) are fulfilled. ACI 311.4R, Section 2.3, was prepared to guide architects, engineers, and owners in the development of effective inspection programs. It states:

> For the protection of the public and the owner, the responsibility for planning and detailing owner inspection should be vested in the A/E as a continuing function of the design responsibility. The A/E should ensure that the program for owner inspection meets all requirements of design specifications and the local building code. The inspection responsibility may be discharged directly, may be conducted by owner personnel, or may be delegated to an independent inspection organization reporting to the A/E.
>
> If the A/E is also responsible for construction, an independent inspection organization should be retained directly by the owner. When the owner provides the A/E service, the owner should also provide inspection or retain an independent inspection organization. Inspection requirements on projects supervised by a construction manager should also be detailed by the A/E and should be carried out by inspection personnel representing the owner.
>
> The fee for owner inspection and testing should be a separate and distinct item and should be paid by the owner, or by the A/E acting on behalf of the owner, directly to the inspection organization. The owner or

Fig. 1.2—Concrete's ability to be formed into any shape lets artistry and function go hand in hand.

Fig. 1.3—ACI-certified field technician performing a slump test.

> A/E should avoid the undesirable practice of arranging payment through the contractor for inspection services intended for use by the owner as a basis of acceptance. Such a practice is not in the owner's interest, and may result in a conflict of interests. Impartial service is difficult under such circumstances, and the fees for inspection are eventually paid by the owner in any case.

Under a typical construction contract, inspectors representing the owner have no responsibility or authority to manage the contractor's workforce. The owner's inspection team is responsible for determining that materials, procedures, and end products conform to the requirements of the contract documents. Because inspectors are only responsible for evaluating the contractor's work for conformance, they may

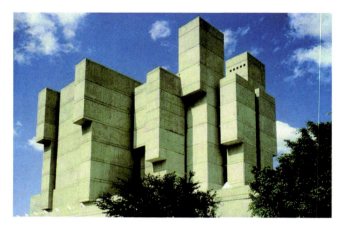

Fig. 1.4—Building design accented with horizontal joints; quality construction provides a highly functional and aesthetic workplace.

not burden the contractor by adding requirements not given in the contract documents.

1.1.3.2 *Quality-control inspection*—Personnel maintained or hired by the contractor perform inspection and testing services used to control quality during construction. In some construction contracts, particularly those with certain government agencies, the contractor is required to provide a specified amount of inspection and testing as part of a formal quality-control program. Even when not contractually required, many contractors maintain inspection and testing personnel, separate from the line of supervision, to monitor quality control. These inspectors report directly to the contractor's management. Sometimes inspection work is performed by the contractor's construction supervisors as an automatic part of construction operations.

By performing their own quality-control inspection, contractors can avoid later rejection of work that would be very costly to replace or correct. Inspection performed by or for the contractor, particularly when contractually required, should be more detailed than that performed for owner inspection, with greater attention being given to form alignment, positioning of embedments and reinforcing bars, and general placement practices. Even when contract documents require extensive quality-control inspection and testing by the contractor, however, owners should not reduce or eliminate their own inspection efforts. The owner should maintain formal oversight of the contractor to ensure that the quality-control program achieves its objectives (Fig. 1.4).

Manufacturers of concrete materials and products also use inspection and testing to maintain quality control and to ensure that contract requirements are met. These programs should parallel the contractor's quality-control efforts. Owners should independently audit and assess the effectiveness of these quality-control programs.

1.2—Inspector

1.2.1 *Duties*—Although the duties and emphasis of the different inspection teams involved on a project may vary and sometimes overlap, the basic approach is the same. The owner's inspectors generally emphasize inspection of subgrade and contact surfaces, inspection of reinforcing steel, inspection of concrete materials as they are delivered, testing of the unhardened and hardened concrete, and inspection of the finished structure. The contractor's quality-control inspectors usually emphasize inspection of materials production, setting of formwork, and placement and curing of concrete. The duties that inspectors are most frequently asked to perform include:

- Identifying and examining materials before acceptance. This includes verifying quality based on certifications and test results provided by producers and suppliers as well as sampling and testing materials delivered to the job site;
- Monitoring batching and mixture proportioning (including adjustments) and testing concrete for consistency, air content, temperature, and density;
- Examining foundations, subgrade, forms, reinforcing steel, embedded items, and other work in preparation for concreting;
- Inspecting the mixing, conveying, placing, consolidating, finishing, curing, and protection of concrete;
- Preparing required concrete specimens for laboratory tests, including monitoring for adequate field curing and proper protection of the specimens;
- Observing the equipment, working conditions, weather, and other items affecting the concrete or other related parts of the structures;
- Observing methods used for curing and protection of concrete, especially during periods of hot or cold weather;
- Evaluating test results for conformance to the specifications;
- Reporting nonconforming conditions and materials in a prompt manner;
- Verifying that unacceptable items and procedures are corrected; and
- Preparing records and reports.

To be effective, inspectors need support from management. They need to also be observant and able to evaluate the relative importance of various work items to give priority to more important matters. Above all, inspectors should be completely familiar with all acceptance criteria in the contract documents, and they should promptly document and report any nonconformance to these criteria to contractors and to their own supervisors.

1.2.2 *Education and certification*—Inspectors and technicians can receive related technical training at colleges, trade schools, and similar educational institutions. ACI Certification Program Sponsoring Groups may also provide specific training before certification exam sessions. Certification of inspectors and technicians is becoming the standard and, in many cases, is mandatory. It provides third-party assurance that the inspector or technician has at least the basic skills and knowledge to perform the job. Some states and local jurisdictions require certification, but in most cases, the requirement is placed in industry standards and then referenced in the building codes. Industry standards containing a certification requirement or recommendation for certification include ACI 301, 311.4R, 311.5, 318, and 349 and ASTM C31/C31M, C39/C39M, C42/C42M, C78, C94/C94M, C192/C192M, C685/C685M, C1077, and E329.

ACI provides certification programs for:
- Concrete Field Testing Technician—Grade I
- CSA-Based Concrete Field Testing Technician—Grade I
- Concrete Flatwork Finisher & Technician
- Specialty Commercial/Industrial Flatwork Finisher & Technician
- Concrete Strength Testing Technician
- Concrete Laboratory Testing Technician—Level 1 & 2
- Aggregate Base Testing Technician
- Aggregate Testing Technician—Level 1 & 2
- Concrete Construction Special Inspector & Associate
- CSA-Based Concrete Construction Special Inspector
- Concrete Transportation Construction Inspector
- Associate Concrete Transportation Construction Inspector
- Tilt-Up Supervisor & Technician
- Shotcrete Nozzleman (Dry Mix Process)
- Shotcrete Nozzleman (Wet Mix Process)

ACI continues to develop new certification programs in response to expanding industry needs.

In addition to having a technical understanding of the principles involved in the assigned construction, inspectors must have practical experience. They should know how and why the work is to be done in a particular way. Inexperienced but technically trained inspectors should undergo on-the-job training under the supervision of more experienced individuals before working alone. Employers also should encourage inspectors to keep their skills up to date by continuing their technical training. Employers can assist in this effort by providing periodic training courses.

1.2.3 *Authority*—The quality-assurance/quality-control duties and responsibilities of the owner, engineer, contractor, and supplier should be clearly detailed in the contract documents and thoroughly understood by all parties. These duties and responsibilities should be reviewed at a preconstruction or concrete preplacement meeting. Clearly defining each party's authority, responsibilities, and lines of communication before concrete is placed will have a positive impact on the concrete placement.

At the start of each job, the inspector's supervisor should clearly explain the authority that the inspector has and the actions to be taken in various situations that may be encountered. These duties and responsibilities should be provided to the inspector in writing. The inspector and his firm should contractually be a legal agent of the owner or contractor to have the authority to stop work. An inspector may be authorized to:
- Stop work until preliminary conditions (such as preparation of forms, contact surfaces, subgrades, construction joints, and the placing of reinforcement) are satisfactory and accepted and until inspection personnel are available to observe concreting operations; and
- Stop work if the materials, equipment, or workmanship do not conform to the contract documents.

In both cases, the inspector is usually authorized to take direct action in concert with the contractor's supervisors. Inspectors should stop work only as a last resort, when it is evident that nonconforming concrete will result from continuing operations, and only if authority to do so has been established. In cases where only construction personnel may

Fig. 1.5—A preconstruction meeting between certified inspector and contractor will help prevent problems during concreting operations.

stop work, inspectors should promptly advise responsible construction supervisors and owners or owner representatives of all nonconforming conditions as soon as they are identified. Actions taken by the contractor in response to this notification should be documented. Matters of general policy or major points not covered or clearly conveyed in the specifications should be discussed with the inspector's supervisor and then with the engineer, usually via request for information (RFI) communication (Fig. 1.5).

1.2.4 *Relations with contractors, supervisors, and workers*—Inspectors representing the owner should cooperate with the contractor consistent with the owner's interests to help reduce construction costs and improve schedules. Inspections should be made promptly when requested, and conditions that will lead to unsatisfactory work should be pointed out to the contractor immediately to avoid the waste of materials, time, and labor. Inspectors should be on the job during reinforcement placement before concreting and whenever concrete is being placed, finished, cured, or repaired.

The inspector should not delay the contractor unnecessarily nor interfere with the contractor's methods unless it is evident that the work will be unacceptable. Demands should never be made of the contractor that are not in accordance with the contract documents. If the contract documents permit a choice of methods, the inspector may suggest, but not demand, which of the methods should be used. At no time is the inspector permitted to direct the work of the contractor.

Inspectors should maintain an impersonal, agreeable, and helpful attitude toward contractors and their employees. They should never accept personal favors nor criticize the contractor's organization or workers. By dealing fairly and recognizing and commending good work, inspectors usually can gain the respect and cooperation of the contractor's supervisors and workers.

Establishing a clear line of communication also is important. Formal communications should be conducted only with the authorized representatives of the contractor, preferably in the form of advisements relating to the quality of the work in progress. Matters involving a potential change in project costs, completion time, or other important factors should be referred to the owner or owner's authorized representative.

The inspector usually deals directly with the subcontractors' supervisors. If subcontractor response to nonconforming conditions is unsatisfactory, the matters requiring correction should be immediately referred to the general contractor who is legally responsible for the work.

The same basic approach to inspection and the inspector's relations with work crews applies to inspectors employed by the contractor. In many instances, however, the working relationships with supervisors and work crews are affected by the fact that everyone is employed by the contractor. Inspectors should receive instructions from the contractor's management detailing the guidelines for these relationships.

1.2.5 *Safety*—Inspectors should follow all rules for jobsite safety established by the owner or contractor. Job-site safety is everyone's responsibility. Any time inspectors observe unsafe conditions or practices, they should immediately advise the responsible contractor, and then document the safety advisement as well as any actions taken by the contractor in response. Inspectors should contact their supervisors for additional guidance on reporting safety issues.

1.3—Importance of clear specifications

This manual describes work practices and procedures necessary to ensure satisfactory concrete construction. It cannot be emphasized too strongly, however, that inspectors are governed strictly by the requirements of the contract documents defining the work. This manual and other references should be used only as information sources and to provide additional guidance on items not covered by the contract documents. In some situations, inspectors will be governed by performance criteria supplied by management or provisions of applicable building codes or regulations that state criteria in terms of the desired results but without stating the methods for achieving the desired results. Likewise, inspectors may be governed by prescriptive criteria that define the exact materials or processes to be executed.

Because the contract documents are the main criteria governing the decisions and actions of an inspector, clear specifications and drawings are essential. Much of the friction in construction arises from differing interpretations of vague or incomplete contract documents. The contract documents should be specific and carefully written. Although the inspector has no control over the contract documents for the project under construction, job-site inspectors should provide feedback to the designer by suggesting changes to their supervisors that may be appropriate to established requirements.

A common and erroneous assumption is that specified tolerances for lines, grades, dimensions, and surface finish apply to the setting of forms, screeds, and grade strips. This, however, is not the case. Tolerances apply only to the completed concrete. Forms, screeds, and grade strips should be set at the exact position indicated on the drawings, insofar as possible, so the finished concrete will conform to the required measurements.

Designers usually determine the governing measurements of line, grade, and dimensions. When required by their duties and responsibilities, inspectors check the alignment of forms and screeds and the positioning of reinforcement and embedded items. They also measure lengths, volumes, and weights as required to ensure that the quantities of materials and finished work meet all requirements. Recognizing that even the most careful measurement can never be exact, inspectors should exercise judgment as to the tolerance to be permitted in specific cases if the contract documents do not state limiting values or permissible tolerances. Because it is impossible to align forms and reinforcing steel to the nearest hundredth of an inch, permissible deviation from exact value should be governed by the effect the deviation would have on the performance or appearance of the structure. For example, a displacement of reinforcing bars of 1/2 in. (12 mm) might be of no consequence in a foundation, but it could seriously weaken a thin slab or impair protection of the bars from corrosion. Standards for tolerances not covered by the contract documents should be established early in the construction period. The most widely referenced source for concrete construction tolerances is ACI 117.

REFERENCE LIBRARY

In addition to standards and publications referenced in the contract documents, concrete inspectors should keep a reference library of the latest editions of other useful resources including, at a minimum:

- *Field Reference Manual: Standard Specifications for Structural Concrete for Building with Selected ACI and ASTM References* (ACI SP-15) (American Concrete Institute 1999) (compilation publication containing ACI 301 and related ACI and ASTM documents)
- *Placing Reinforcing Bars* (Concrete Reinforcing Steel Institute 1997a)
- "Building Code Requirements for Structural Concrete (ACI 318)"
- *Concrete Manual* (U.S. Bureau of Reclamation 1988)
- *Design and Control of Concrete Mixtures* (Portland Cement Association 2002)
- "Guide for Concrete Inspection (ACI 311.4R)" and "Batch Plant Inspection and Field Testing of Ready-Mixed Concrete (ACI 311.5)" (both of which are reprinted in Appendixes A and B, respectively, of this manual).
- Chapter 21 lists additional references and standards that may be helpful for specific jobs.
- Chapter 22 provides titles for industry references.

CHAPTER 2—STATISTICAL CONCEPTS FOR QUALITY ASSURANCE

2.1—Quality-control and quality-acceptance inspections

Quality assurance refers to all of the programs and functions involved in obtaining quality concrete that will provide satisfactory service in the finished concrete structure. It includes design, production, sampling, testing, and decision criteria. The two types of inspection employed in quality assurance are quality-control inspection, primarily a function of the contractor, and owner-acceptance inspection, primarily a function of the owner or owner's representative.

Legal requirements for documentary evidence of satisfactory compliance have led to greater use of specifications based on statistical concepts. Using statistics permits decisions about quality to be made with an established degree of confidence. This chapter gives background on the underlying statistical concepts for use in concrete construction quality-control and owner-acceptance programs, along with examples of the procedures inspectors are likely to encounter. More detailed information on the use of statistics in concrete can be found in the references noted in this chapter.

2.2—Traditional quality assurance

Many specifications that have traditionally been used for concrete are prescriptive rather than performance-based specifications. Prescriptive specifications define the exact materials and detail the operations of the contractor and the equipment to be used to produce the concrete. Such specifications were developed because adequate definitions and test methods for evaluating the performance of the end product were lacking. The values used to define required performance usually were subjective, based on experience and judgment, rather than rational concepts.

Though prescriptive specifications—combined with the skills of experienced designers and the cooperation of experienced contractors—have produced good concrete structures, these structures have not always met the desired quality. Under these specifications, a random, supposedly representative, sample is taken. This sample is tested and the result is compared with the specified value of the particular characteristic. If the test result is within the specified tolerances, the material passes the criteria and is accepted. If the material fails to pass the criteria, the design engineer applies engineering judgment and a decision is made as to whether the material substantially complies and should be accepted, whether the material fails and should be rejected, or whether the material should be retested. Because "substantial compliance" is subjective, it varies from person to person and job to job, creating confusion and disputes. Research has shown that as much as 30% of some construction controlled by traditional methods has been outside the stated limits when closely examined by statistical methods using random sampling, even though it was considered completely acceptable under the control practices used (California Department of Transportation 1968).

When a failing test is encountered, retesting without obtaining a new sample is not appropriate (unless the original test has been performed improperly, in which case the entire test should be voided). Even if the results of the two tests (original test and retest) are averaged, there is a built-in bias because the second test is taken only if the first test fails. For example, consider a material that has a 50% chance of passing the specification limit and a 50% chance of failing. For each test on a certain lot, there is always a 50% chance the result will pass and a 50% chance it will fail. When a test is made and the result fails, however, a retest of the same lot also has a 50% chance of passing and a 50% chance of failing. Because a second test is made only if the original test fails, however, this second test has biased the original test so that the overall probability actually becomes 75% pass, 25% fail. This is illustrated in Fig. 2.1.

2.3—Statistical concepts and procedures

Contract documents written using statistical concepts express quality requirements in terms of target values for

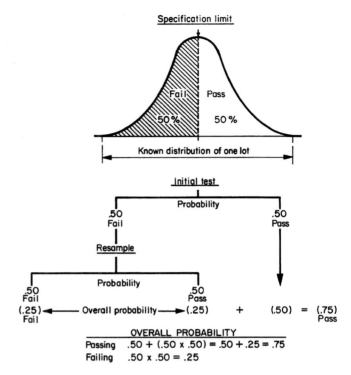

Fig. 2.1—Probability of acceptance when resampling.

contractors and express compliance requirements in terms of plus or minus tolerances from established target values. Tolerances for the target value, prescribed by design needs, can be based on statistical analyses of the variations in materials, processes, sampling, and testing existing in traditional construction practices. Tolerances derived in this manner can be both realistic and enforceable. They take into account all of the normal causes of variation and allow for the expected distribution of test results around the mean. Provisions can be made both for control to the stated level and for control of the variation from this level.

The use of statistical concepts has proved to be very effective and efficient where properly applied. In addition to indicating acceptable and unacceptable materials in construction, statistical methods can indicate gray areas where test results show that a material is not completely in compliance with requirements but can be accepted if permitted by the contract documents. Public agencies, particularly state highway departments, often find statistical concepts valuable for use on projects involving high production rates and large volumes of concrete, such as highway paving, large dams, and airfield paving.

Statistical procedures are based on the laws of probability. For these laws to function properly, the data must be collected by random sampling. A true random sample is one in which all parts of the whole have an equal chance of being chosen (ASTM D3665). Without true random samples, statistical procedures give false results.

It is possible to adapt the use of random numbers to the laboratory, to the field, and to the factory. Random selection is not merely haphazard. Numbers can be selected from a standard table of random numbers or by use of mechanical

Table 2.1—Types of sampling

	Protects against known defects	Protects against unknown defects	Protects against cycles and patterns	Low inherent risk	Low risk in unknown situations	Simple organization	High reliability	Does not require knowledgeable sampler
JUDGMENT SAMPLING (based solely on judgment of sampler)	√	X	?	?	X	√	?	X
QUOTA SAMPLING (making judgment distribution of sample by time of day, location, etc., according to distribution of facts)	√	X	√	?	X	X	?	X
SYSTEMATIC SAMPLING (selecting successive observations at constant intervals in a sequence)	√	√	X	√	?	√	√	√
STRATIFIED SAMPLING (selecting each of two or more parts independently from a corresponding part)	√	√	√	√	?	?	√	√
RANDOM SAMPLING (selecting sample in such a manner that each individual has the same chance of being chosen)	√	√	√	√	√	X	√	√

Note: √ = satisfies column heading requirement; X = does not satisfy column heading requirement; and ? = may or may not conform to heading, according to situation.

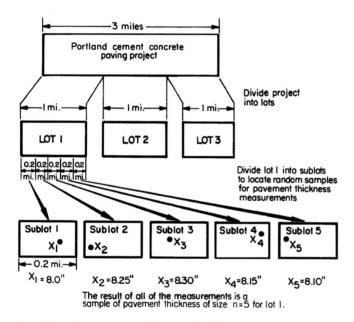

Fig. 2.2—Designation of lots and sublots.

randomizing devices (such as dice or a spinning wheel). They also can be generated by computer. No device, however, is acceptable as random unless it passes certain statistical tests.

ASTM E105 discusses the preparation of sampling plans. Sampling programs commonly used are described in the following paragraphs. Table 2.1 shows the advantages and disadvantages of these programs (McMillan and Tuthill 1987). Normally, no one sampling type is used alone.

1. *Judgment sampling* is based solely on the judgment of the sampler, with no other restrictions. The sampler decides when and where a sample should be taken;

2. *Quota sampling* is a type of judgment sampling based on time of day, geographic area, or some other known distribution of facts;

3. *Systematic sampling* involves selecting successive observations in a sequence, such as of time or area, at constant intervals;

4. *Stratified sampling* involves dividing a given quantity of material into independent parts, each to be sampled separately. Stratified sampling is inherent in any acceptance sampling based on use of sublots; and

5. *Random sampling* involves selecting a sample in such a manner that each increment comprising the lot has the same chance of being chosen for the sample (ASTM D3665; California Department of Transportation 1968).

Examples of sampling by random numbers, using the typical table of random numbers, are given in Appendix 1 (Pennsylvania State University 1974).

In addition to randomness, the concept of lots also is essential to the proper application of statistics to quality-assurance sampling of construction. A lot is a prescribed and defined quantity of material (such as volume, area, tonnage, time of production, or units) produced from the same process for the same purpose. This is the quantity offered for acceptance as a unit, and all sampling and testing requirements are defined in relation to and applied to that quantity. Only by establishing the lot size can the proper sampling location and frequency be selected to determine the quantity of material within the specified limits. Under the concept of lot-by-lot sampling and testing, the process of concrete construction may be thought of as the production of a succession of lots, each presented to the designer for acceptance or rejection. This is illustrated for a concrete pavement in Fig. 2.2. To implement the acceptance plan, each lot is separated into subdivisions of equal sizes, called sublots. Sampling locations are randomized within the boundaries of each sublot. Stratifying or separating the lot into smaller components ensures that there will be no excessive periods without any sampling and testing. It also allows sampling and testing crews to space out their work.

A sample, used in a statistical sense, is that portion of the lot taken to represent the whole. The term is not to be confused with the individual test portions or sample increments that make up the sample. Figure 2.3 illustrates the relationships between test portions, sample increments, and the sample. (Note: In discussions of sampling and testing of materials or concrete in other chapters, the term "sample" generally refers to each individual portion of material taken for testing.)

Normally, it is desirable for each sample to be made up of four or five sample increments (or four or five sublots per lot). The average of the test values of sublots in a lot, however, will have a range smaller than the range of the values of the individual test portions in the entire lot.

2.4—Basic statistical concepts

2.4.1 *Definitions*—The following symbols and terms, primarily from ACI 214R, are commonly used in statistical quality-assurance programs for concrete.

n = number of tests in a sample. The total number of test results or values under consideration.

X = an individual test result. Indicates separate test results (may be written X_1, X_2, X_3, and so on).

$\bar{X}$ = sample average (called "bar X").

The arithmetic mean of all test results; the sum of all test result values divided by n, the number of tests

$$\bar{X} = \frac{X_1 + X_2 + X_3 + \ldots + X_n}{n}$$

s = sample standard deviation.

The square root of the average obtained by dividing the sum of the squares of the numerical differences of each test result from the sample average by one less than the number of tests

$$s = \sqrt{\frac{(X_1 - \bar{X})^2 + (X_2 - \bar{X})^2 \ldots + (X_n - \bar{X})^2}{n - 1}}$$

A simpler and more adaptable form for desk calculators is

$$s = \sqrt{\frac{1}{n-1}\left(\Sigma(X_i^2) - \frac{(\Sigma X_i)^2}{n}\right)}$$

where

n = number of values;
ΣX_i = sum of n values; and
$\Sigma(X_i^2)$ = sum of squares of n values.
σ = population standard deviation.

The population consists of all possible data. The sample is a portion of the population.

V = coefficient of variation.

The sample standard deviation expressed as a percentage of the average

$$V = \frac{100s}{\bar{X}}$$

R = sample range.

The statistic found by subtracting the lowest value in a data set from the highest value in that set

$$R = X_{max} - X_{min}$$

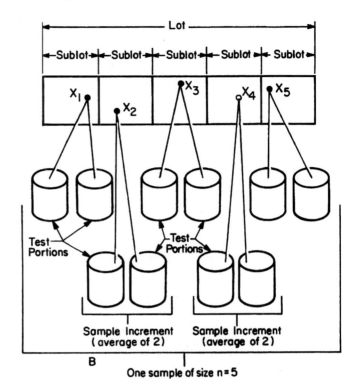

Fig. 2.3—Definition of test portion, same increment sample.

f'_c = specified compressive strength of concrete. Usually 28-day strength, but can be specified at any age.

f'_{cr} = required average compressive strength of concrete. Ensures that no more than the permissible proportion of test results will fall below the specified strength; used as the basis for selection of concrete proportions.

2.4.2 *Normal distribution curves*—Research has shown that test results of construction materials and operations form a definite pattern grouping around a central value, called the mean. The measurements generally group around the average in a symmetrical pattern, thereby allowing the use of statistics based on the familiar bell-shaped normal distribution curve. Although slight variations from the symmetrical curve can occur, especially when the number of test results is small, the error in assuming normal distribution is usually not large. The assumption of a normal distribution permits the use of established relationships of mean (average) and standard deviation to establish realistic tolerances in the contract documents for selected sample sizes. Determining these tolerances requires the use of statistical analysis and engineering judgment.

A typical distribution of results of compressive strength tests made on a specific concrete mixture (Fig. 2.4) is given in ACI 214R. Superimposed on these plotted results is a normal bell-shaped distribution curve computed from test results. This particular set of test results and distribution curve would be considered a reasonably good fit. All statistical control procedures for concrete construction, including control charts, assume that the distribution of test results

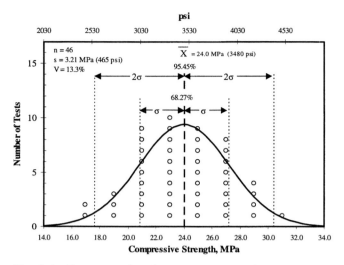

Fig. 2.4—Frequency distribution of strength data and corresponding assumed normal distribution (σ is the estimate of the population standard deviation).

Table 2.2—Required average compressive strength when data are available to establish standard deviation

Specified compressive strength f_c', psi	Required average compressive strength f_{cr}', psi
$f_c' \leq 5000$	Use the larger value computed from the following equations: $f_{cr}' = f_c' + 1.34s$ $f_{cr}' = f_c' + 2.33s - 500$
Over 5000	Use the larger value computed from the following equations: $f_{cr}' = f_c' + 1.34s$ $f_{cr}' = 0.90f_c' + 2.33s$

Note: 1 psi = 0.0069 MPa.

(regardless of concrete or material quality) will approximate such a normal curve. Appendix 2 gives additional examples of normal distribution curves.

2.4.3 *Applying normal distribution curves to concrete compressive strength*—In concrete construction, too much emphasis is often placed on the results of individual cylinder tests. Low tests are inevitable, but an occasional low test should not be a major concern if the result is not too low or the low test does not represent a portion or member of the structure deemed critical by the designer. On the other hand, a few test results above f_c' are not necessarily proof of adequate-quality concrete throughout the entire structure; it is quite possible that lower-strength concrete was placed somewhere in the structure but was not included in the tests.

What is of concern is that the number of standard cylinder test results falling below f_c' is not more than allowed by the contract documents. Typically, contract documents require that the average of all sets of three consecutive test results equals or exceeds the specified strength. If the test results do not meet this criterion, changes to the mixture proportions or improvements in quality control should be made. Furthermore, if an individual strength test result falls below f_c' by more than 500 psi (3.5 MPa) when f_c' is 5000 psi (35 MPa) or less, or by more than $0.10f_c'$ when f_c' is more than 5000 psi

(35 MPa), then the in-place concrete is in question, and appropriate actions should be taken. Refer to the section on evaluation and acceptance of concrete in ACI 318 for further information.

Tests of standard concrete control specimens during construction provide the basis for evaluating the potential strength of the concrete delivered to the job site. Placing, consolidation, and curing techniques will affect the quality of the hardened concrete in the structure. The quality of concrete can best be evaluated by analysis of at least 30 standard tests from a given mixture, as described in ACI 214R.

2.5—Statistical tools

2.5.1 *Frequency distributions*—Results of concrete compressive strength tests are plotted to form a histogram, or frequency distribution, for which the normal distribution curve can be computed. More importantly, the results are tabulated, and the average $\overline{X}$ and standard deviation s are computed to provide background data to use in selecting a required average strength f_{cr}' for future concrete construction. The required average strength is used to establish proportions that can provide actual concrete strength test results that will meet the specified concrete strength f_c' despite of production variations. This strength is statistically computed based on the $\overline{X}$ and s of the past results. Appendix 3 shows examples of how to compute s and f_{cr}' for given data.

Normally, the concrete inspector will not be concerned with plotting frequency distribution curves for concrete strength tests. The concrete producer or contractor typically determines the f_{cr}' and proportions the mixture to meet this requirement. An average strength higher than the specified strength is needed to ensure that the in-place concrete will be accepted when results of concrete control tests on the upcoming job are compared with the criteria of the contract documents. If the concrete inspector is required to check the f_{cr}' value computed by the contractor or concrete producer, the inspector should use data, particularly the standard deviation s, computed from previous testing of concrete produced by the same concrete plant or contractor.

When ACI 318 is referenced, a concrete production facility that has a previous standard deviation based on at least 30 consecutive strength tests[*] representing materials and conditions similar to those expected in the new work can be used to proportion for a required f_{cr}' representing the larger of the values computed as shown in Table 2.2, from ACI 318.

All of these required levels of overdesign are based on statistical assumption so that the concrete produced for the new construction can be expected to meet specification requirements when standard cylinders are tested during construction. These levels of overdesign are also based on the assumption that specification requirements are the same as those contained in ACI 318, which require that the average of any three consecutive strength tests equals or exceeds the required f_c' and that no individual strength test (average of two

---

[*]Allowance for using as few as 15 consecutive strength results is made by using factors that increase the sample standard deviation. These factors are: 1.16 for 15 tests, 1.08 for 20 tests, 1.03 for 25 tests, and 1.00 for 30 or more tests, as shown in Table 5.3.1.2 of ACI 318. The factors can be interpolated for an intermediate number of tests.

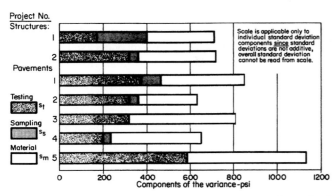

Fig. 2.5—Portland-cement concrete components of standard deviation, compressive strength.

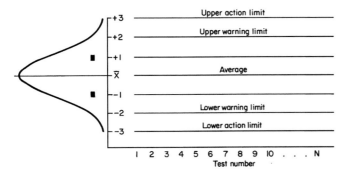

Fig. 2.6—Statistical control chart.

Table 2.3—Standards of concrete compressive strength, $f_c' \leq 5000$ psi

Overall variation					
Class of operation	Standard deviation for different control standards, psi				
	Excellent	Very good	Good	Fair	Poor
General construction testing	Below 400	400 to 500	500 to 600	600 to 700	Above 700
Laboratory trial batches	Below 200	200 to 250	250 to 300	300 to 350	Above 350
Within-test variation					
Class of operation	Coefficient of variation for different control standards, %				
	Excellent	Very good	Good	Fair	Poor
Field control testing	Below 3.0	3.0 to 4.0	4.0 to 5.0	5.0 to 6.0	Above 6.0
Laboratory trial batches	Below 2.0	2.0 to 3.0	3.0 to 4.0	4.0 to 5.0	Above 5.0

Note: 1 psi = 0.0069 MPa.

Table 2.4—Standards of concrete compressive strength, $f_c' > 5000$ psi

Overall variation					
Class of operation	Coefficient of variation for different control standards, %				
	Excellent	Very good	Good	Fair	Poor
General construction testing	Below 7.0	7.0 to 9.0	9.0 to 11.0	11.0 to 14.0	Above 14.0
Laboratory trial batches	Below 3.5	3.5 to 4.5	4.5 to 5.5	5.5 to 7.0	Above 7.0
Within-test variation					
Class of operation	Coefficient of variation for different control standards, %				
	Excellent	Very good	Good	Fair	Poor
Field control testing	Below 3.0	3.0 to 4.0	4.0 to 5.0	5.0 to 6.0	Above 6.0
Laboratory trial batches	Below 2.0	2.0 to 3.0	3.0 to 4.0	4.0 to 5.0	Above 5.0

Note: 1 psi = 0.0069 MPa.

cylinders) falls below the required f_c' by more than 500 psi (3.5 MPa) when f_c' is 5000 psi (35 MPa) or less, or by more than $0.10 f_c'$ when f_c' is more than 5000 psi (35 MPa).

The standard deviation of results for compressive strength or any other test is made up of components that are standard deviations of several sub-items. The main components are deviations due to sampling errors, testing errors, and actual variations in the material itself. Figure 2.5 shows the relationships among these sub-items on several different concrete projects. Note that these items are not additive; instead, the relationship consists of

$$s_o^2 = s_t^2 + s_m^2$$

where s_o is the overall standard deviation, s_t is the standard deviation of testing, and s_m is the standard deviation of the actual material properties.

Tables 2.3 and 2.4, from ACI 214R, show typical values for standard deviation of concrete strength tests for various standards of control and for type of testing.

2.5.2 *Control charts*—Although frequency distributions are used primarily to establish a proposed average compressive strength of concrete before work begins, control charts are the primary statistical tools used for evaluating results of tests on the concrete and concrete materials during construction.

Control charts are basically horizontal line charts. The horizontal lines for an acceptance control chart generally consist of a central line at the specified average, an upper line at the specified upper acceptance limit, and a lower line at the lower acceptance limit (if both apply). A true control chart (one used for actual process control) will usually have the center average line plus two lower and two upper lines. The upper and lower lines closest to the average are called warning limits, and the next lines are called action limits. These warning and action limit lines are placed at some multiple of s, the sample standard deviation, above and below the average value.

Figure 2.6 shows a blank sample of a typical actual process control chart (Pennsylvania State University 1974). Note how the control lines are related to a standard distribution curve (shown lying on its side to the left).

Generally, procedures for using actual process-control charts require that when a point (test result) falls on or outside either warning line, the producer should examine its operation to determine what has caused this variation and attempt to correct it. When a point falls on or outside the action line, the producer should stop its operation and make the adjustments necessary to bring the operation under control. Control charts can only indicate that a problem exists, not where it is located.

A horizontal-line control chart can be used to plot both single test results and running averages of a specified number of consecutive test results. Figure 2.7 shows typical charts of this type (Pennsylvania State University 1974).

In addition to control charts for single test results and averages, it is common to use them to record the variability, or range, of test results, either from a lot or over some specified period. These, too, are horizontal line charts, but they normally have only lower and upper control limits.

Figure 2.8 (ACI 214R) shows typical control charts for results of concrete compressive strength tests. Control charts also are commonly required for results of air-content tests. A typical example is shown in Fig. 2.9 (Pennsylvania State University 1974). Control charts are sometimes maintained for results of slump tests, as shown in Fig. 2.10 (California Department of Transportation 1968).

Typically, the only concrete material quality for which control charts are maintained is aggregate gradation, and usually for only a few of the sieve sizes. Where aggregate quality is particularly important, however, charts may be maintained on results of certain quality tests, although statistically, this may be an improper application. Control charts for aggregate gradation and quality are also shown in Fig. 2.10. Appendix 4 shows typical control charts for aggregate gradation (for individual tests and range), moving averages for sand-equivalent test results, and moving averages for grading analysis.

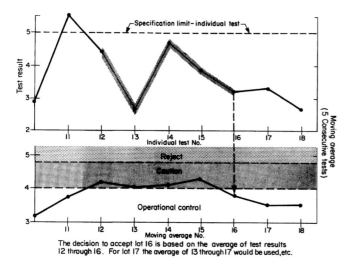

Fig. 2.7—Typical horizontal line control charts.

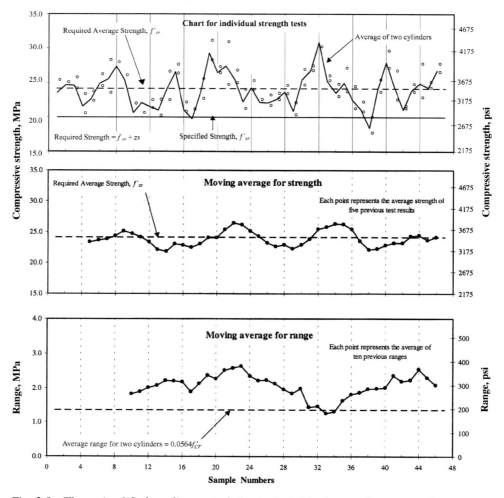

Fig. 2.8—Three simplified quality-control charts: individual strength tests, moving average of five strength tests, and range of two cylinders in each test and moving average for range.

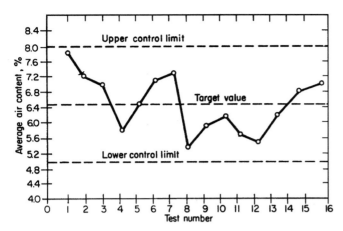

Fig. 2.9—"Bar X" control chart for air content.

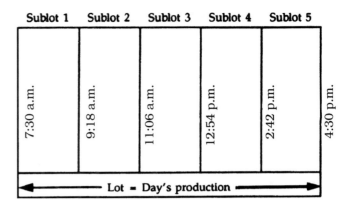

Fig. A1.1—Relationship between lot and sublots, time interval.

4:30 p.m.) to accomplish a day's production. Assume that the plant is running continuously through the lunch period.

Solution:

1. *Lot size*—The lot size is a day's production of 9 hours, starting at 7:30 a.m. and stopping at 4:30 p.m.

2. *Sublot size*—Divide the lot into five equal sublots by selecting five equal time intervals during the 9 hours that the plant is operating. The time interval for each sublot is

$$\text{sublot time interval} = \frac{(9 \text{ h per lot})(60 \text{ min per h})}{5 \text{ sublots per lot}}$$

$$= 108 \text{ min per sublot}$$

Figure A1.1 shows a diagram of the division of the 9 hours of production time into five equal sublots.

3. *Sample increments*—One sample increment per sublot is required, but the exact time within each sublot when the sample increment should be taken is unknown. These times must be selected randomly.

Use the table of random numbers (Table A1.1) to randomize the timing of sample increments. Choose consecutive random numbers from either the X or Y columns. (The inspector's supervisor may specify the particular column to use.)

For this example, use Column X to obtain the timing of five sample increments. Note that Column Y could have been used instead and that any five consecutive values could be selected by starting at any point in the table.

Choose the first five numbers from Column X (0.4721, 0.6936, 0.6112, 0.7930, and 0.0652). To randomize the sampling time within each sublot, use the time interval (108 minutes) computed in Step 2, then multiply this time interval by each of the five random numbers:

Sublot 1: 0.4721 × 108 = 51 min;
Sublot 2: 0.6936 × 108 = 75 min;
Sublot 3: 0.6112 × 108 = 66 min;
Sublot 4: 0.7930 × 108 = 86 min; and
Sublot 5: 0.0652 × 108 = 7 min.

Add the computed times to the starting times for each sublot to determine the randomized time at which each sample increment is to be obtained. The sampling times are:

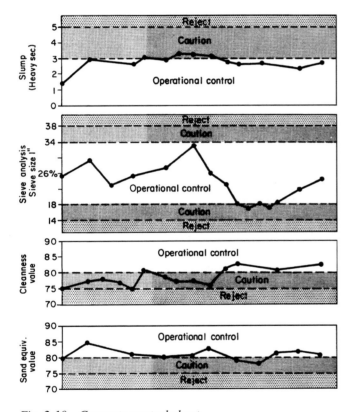

Fig. 2.10—Concrete control chart.

APPENDIX 1—SAMPLING BY RANDOM NUMBERS
A1.1—Example 1: Sampling by time sequence

The task is to sample materials, such as aggregate, concrete, or precast products, at the place of manufacture. Select the sample increments by means of a stratified random sampling plan to distribute the sampling over a half or full day, whichever is more applicable.

The contract documents will define the lot size and the number of sublots and sample increments per lot. For this example, assume the contract documents state that the lot size is a day's production, that five sublots are required from each lot, and that one sample increment per sublot must be obtained. The plant operates for 9 hours (from 7:30 a.m. to

Table A1.1—Table of random numbers

	Random positions in decimal fractions (four places)				
	X	Y		X	Y
1.	0.4721	R 0.2091	51.	0.6985	L 0.8636
2.	0.6936	L 0.3182	52.	0.3410	L 0.5636
3.	0.6112	R 0.2909	53.	0.5937	R 0.3727
4.	0.7930	R 0.8908	54.	0.6912	R 0.4545
5.	0.0652	L 0.4818	55.	0.0318	R 0.7272
6.	0.04604	L 0.2091	56.	0.1303	R 0.8090
7.	0.0167	L 0.3727	57.	0.6893	R 1.0000
8.	0.0077	R 0.6181	58.	0.3886	R 0.7817
9.	0.6777	R 0.8636	59.	0.0312	R 0.8090
10.	0.8010	L 0.8362	60.	0.0166	R 0.5909
11.	0.3027	L 0.3454	61.	0.4609	L 0.4000
12.	0.9831	L 0.2364	62.	0.0893	L 0.9726
13.	0.7159	R 0.6181	63.	0.4542	L 0.1545
14.	0.3609	R 0.6454	64.	0.9363	R 0.1000
15.	0.8915	L 0.2636	65.	0.8183	R 0.5636
16.	0.6442	R 0.3182	66.	0.9401	L 0.5091
17.	0.1904	R 0.1818	67.	0.5967	L 0.9726
18.	0.6074	R 0.8908	68.	0.7547	R 0.2636
19.	0.7522	R 0.9181	69.	0.0101	R 0.2909
20.	0.7041	L 0.8362	70.	0.2896	L 0.8362
21.	0.5102	R 0.2364	71.	0.8011	R 0.6454
22.	0.2471	L 0.3182	72.	0.6718	L 0.6454
23.	0.5693	L 0.5636	73.	0.5567	L 0.1818
24.	0.8583	R 0.4545	74.	0.0481	L 0.2636
25.	0.3093	R 0.1818	75.	0.4266	L 0.9454
26.	0.9144	R 0.9181	76.	0.3941	R 0.5636
27.	0.7944	L 0.5909	77.	0.9876	L 0.7545
28.	0.8725	R 0.2636	78.	0.6313	R 0.7272
29.	0.0135	R 0.8908	79.	0.6803	R 0.3182
30.	0.2044	R 0.7272	80.	0.7955	L 0.9726
31.	0.2517	L 0.2909	81.	0.7399	R 0.8080
32.	0.2763	L 0.8090	82.	0.9328	L 0.5909
33.	0.0314	R 0.4818	83.	0.1507	L 0.4000
34.	0.9560	L 1.0000	84.	0.3087	R 0.3182
35.	0.4622	R 0.4000	85.	0.7513	L 0.1818
36.	0.1327	L 0.7817	86.	0.6469	R 0.4818
37.	0.6922	L 0.5636	87.	0.2536	R 0.7545
38.	0.0010	L 0.1273	88.	0.1488	R 0.1818
39.	0.7609	R 0.2091	89.	0.9411	L 0.5636
40.	0.5957	L 0.1000	90.	0.0571	R 1.0000
41.	0.3115	R 0.4000	91.	0.4797	R 0.9454
42.	0.3377	R 0.8362	92.	0.0866	R 0.4272
43.	0.5651	L 0.1545	93.	0.2889	R 0.1273
44.	0.4742	R 0.6727	94.	0.4783	L 0.7000
45.	0.9483	L 0.4000	95.	0.0304	R 0.9181
46.	0.2951	R 0.6451	96.	0.8945	R 0.4515
47.	0.0441	L 0.1273	97.	0.4499	R 0.2081
48.	0.9143	L 0.1273	98.	0.9209	L 0.9454
49.	0.5723	L 0.8362	99.	0.5827	L 0.5636
50.	0.6069	R 0.4000	100.	0.4560	L 0.8908

Notes:
X = decimal fraction of total length measured along the road from starting point (or decimal fraction of other units).
Y = decimal fraction of road from outside edge toward centerline (or decimal fraction of other units).
R = measurement from right edge.
L = measurement from left edge.

Sublot no.	Sampling time		
1	7:30 a.m.	+ 51 min	8:21 a.m.
2	9:18 a.m.	+ 75 min	10:33 a.m.
3	11:06 a.m.	+ 66 min	12:12 p.m.
4	12:54 p.m.	+ 86 min	2:20 p.m.
5	2:42 p.m.	+ 7 min	2:49 p.m.

The first sample increment is obtained from Sublot 1 at 8:21 a.m. The second sample increment is not drawn until 10:33 a.m., when Sublot 2 is in production. The remaining sample increments are obtained during the time when each of the last three sublots is in production.

The five sample increments constitute a sample size of $n = 5$, taken from five sublots of a day's production (the lot). Each increment of this sample size should be tested. The data would then be used to estimate the properties of the lot. In other words, the average, the sample standard deviation s, the coefficient of variation V, or any other statistic would be computed to reflect the statistical property of the lot when it is offered by the contractor or material supplier for acceptance sampling.

A1.2—Example 2: Sampling by material weight

Assume that the contract documents specify a lot size of 3000 tons, with five sublots per lot, and that one sample increment per sublot must be obtained. In addition, assume that the total tonnage required for the project is 15,000 tons. The sampling will be done from the hauling units at the manufacturing plant.

Solution:
Follow the same basic pattern as given for the previous example. First, identify the lot size, and then determine the number of lots, sublot size, and the point at which to obtain the sample increments.

Lot size and number of lots—Because 15,000 tons are required for the project and the lot size is 3000 tons, the total number of lots is

$$\text{number of lots} = \frac{15{,}000 \text{ tons}}{3000 \text{ tons per lot}} = 5 \text{ lots}$$

2. *Sublot size*—The sublot size is

$$\text{sublot size} = \frac{3000 \text{ tons per lot}}{5 \text{ sublots per lot}} = 600 \text{ tons per sublot}$$

Figure A1.2 shows the relationship between lot and sublot size.

Sample increments—The contract documents require one sample increment per sublot, but which load that should be sampled is unknown because the sample increments have not yet been randomized. Referring to Table A1.1, choose five random numbers from Column Y, starting with No. 17. Then multiply these numbers by each of the five sublots:

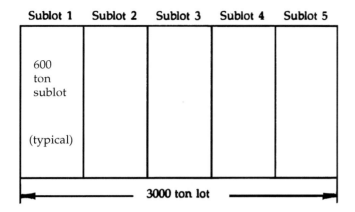

Fig. A1.2—Relationship between lot and sublots, quantity interval.

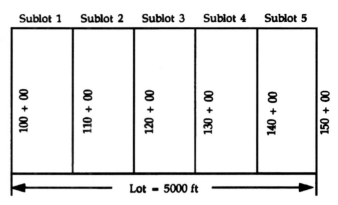

Fig. A1.3—Relationship between lots and sublots, distance interval.

Sublot no.	Random number	Sublot size, tons	Ton to be sampled
1	0.1818	600	109th
2	0.8908	600	534th
3	0.9181	600	551st
4	0.8362	600	502nd
5	0.2364	600	142nd

Obtain the first sample increment at the 109th ton of the first sublot. Then wait until the first sublot of 600 tons is completed before selecting the second sample increment at the 534th ton of the second sublot. Follow the same procedure for obtaining the remaining three sample increments.

If a cumulative log of the tonnage being produced is kept, the sampling sequence for the lot (3000 tons) would be:

Sublot 1: = 109th ton;
Sublot 2: 600 + 534 = 1134th ton;
Sublot 3: 1200 + 551 = 1751st ton;
Sublot 4: 1800 + 502 = 2302nd ton; and
Sublot 5: 2400 + 142 = 2542nd ton.

Thus, in actual practice, the hauling units containing the 109th ton, 1134th ton, 1751st ton, and so on would be sampled.

A1.3—Example 3: Sampling by depth of concrete pavement

The depth of concrete pavement in a roadway must be sampled for acceptance. Assume that the contract documents require a lot size of 5000 linear feet, with each lot divided into five sublots and one sample increment per sublot. In addition, assume that the pavement width is 12 ft and the project begins at Station 100 + 00 and ends at Station 300 + 00.

Solution:
1. *Lot size and number of lots*

$$\text{Lot size} = 5000 \text{ linear feet}$$

The distance from Station 100 + 00 to Station 300 + 00 is 20,000 ft. Thus, the number of lots is

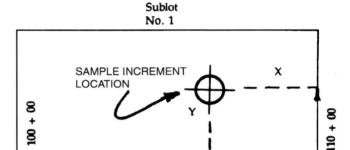

Fig. A1.4—Coordinate system for paved sublot.

$$\text{number of lots} = \frac{20{,}000 \text{ ft}}{5000 \text{ ft per lot}} = 4 \text{ lots}$$

2. *Sublot size*—The first lot begins at Station 100 + 00 and ends at Station 150 + 00. Divide the distance between these stations into five equal sublots:

$$\text{sublot size} = \frac{5000 \text{ ft per lot}}{5 \text{ sublots per lot}} = 1000 \text{ ft per sublot}$$

Figure A1.3 indicates how this lot is divided.

Note that random numbers starting somewhere else in Table A1.1, either in Column X or Y, could have been used. The inspector should select the starting point at random to avoid making the time or location of the sample increment predictable.

3. *Sample increments*—The point at which each sample increment will be obtained must be randomized in both the longitudinal (X) direction and the transverse (Y) direction. This location by X-Y coordinates is illustrated in Fig. A1.4 in which both the station and the offset are chosen in separate randomizing operations.

Referring to Table A1.1, use five consecutive random numbers from Column X and five from Column Y, starting with No. 43. Multiply the X numbers by the length of each sublot, and multiply the Y numbers by the 12 ft width of the roadway wearing surface.

Sublot No.	Longitudinal random number	Sublot size, length, ft	Sample location, ft, from start of sublot	Sample location, ft, from start of lot
1	0.5651	1000	565.1	565.1
2	0.4742	1000	474.2	1474.2
3	0.9483	1000	948.3	2948.3
4	0.2951	1000	295.1	3295.1
5	0.0441	1000	44.1	4044.1

Sublot no.	Transverse random number	Sublot size, width, ft	Location of sample, ft, from edge of roadway
1	L 0.1545	12	1.8 from left edge
2	R 0.6727	12	8.1 from right edge
3	L 0.4000	12	4.8 from left edge
4	R 0.6451	12	7.7 from right edge
5	L 0.1273	12	1.5 from left edge

Coordinates from beginning of lot and from right edge of roadway are, in feet:

Sublot 1:   X = 565.1   Y = 10.2
Sublot 2:   X = 1474.2   Y = 3.9
Sublot 3:   X = 2948.3   Y = 7.2
Sublot 4:   X = 3295.1   Y = 4.3
Sublot 5:   X = 4044.1   Y = 10.5

APPENDIX 2—NORMAL DISTRIBUTION CURVES

Figures A2.1, A2.2, A2.3, and A2.4 show examples of normal distribution curves and how they vary yet are interrelated.

The sample standard deviation s and the related coefficient of variation V are measures of the scatter or variability of data. As illustrated in the figures, when the frequency distribution is long and flat, s (or V) is large, indicating wide variation. When the variability is small, s (or V) is small, and the data are closely packed. Figure A2.4 also shows that a high value of f'_{cr} is required when the variation is wide.

Figure A2.5 illustrates the proportion of the total area under a standard normal distribution curve for each successive step of distance s from the average (mean). These proportional areas for deviations of 1s, 2s, and 3s form the basis for statistically based tolerances for test values. For a standard normal distribution, approximately 68% of all test values will fall within one standard deviation (1s) on each side of the average, approximately 95% will be within 2s of the average, and almost no test values (less than 0.3%) will be found outside the 3s limits.

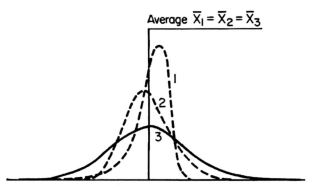

Fig. A2.2—Quite different distributions may have the same average.

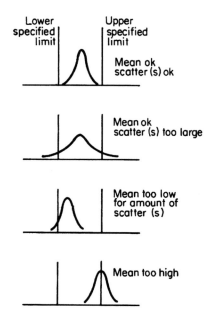

Fig. A2.3—Process control related to specification limits.

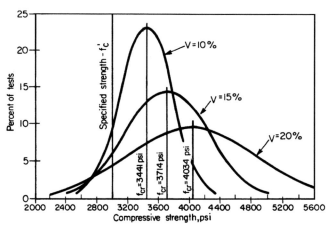

Fig. A2.4—Normal frequency curves and required f_{cr} for different coefficients of variation.

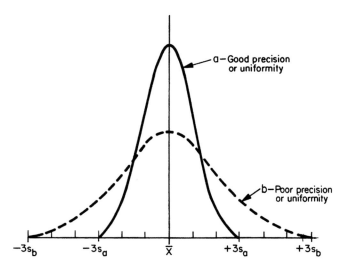

Fig. A2.1—Coordinate system for paved sublot.

APPENDIX 3—COMPUTING STANDARD DEVIATION AND REQUIRED AVERAGE CONCRETE STRENGTH

A3.1—Calculating standard deviation s

The sample standard deviation s is defined as the "root mean square deviation" of test results from their average and is computed by the following equation (described previously)

$$s = \sqrt{\frac{(X_1 - \bar{X})^2 + (X_2 - \bar{X})^2 \ldots + (X_n - \bar{X})^2}{n-1}} \quad \text{(A-1)}$$

where X_1, X_2, and X_n are individual strength test values, and n is the number of tests.

Or, in a simpler form

$$s = \sqrt{\frac{1}{n-1}\left(\Sigma(X_i^2) - \frac{(\Sigma X_i)^2}{n}\right)} \quad \text{(A-1a)}$$

where $\Sigma(X_i^2)$ is the sum of the squares of all individual tests, and $(\Sigma X_i)^2$ is the square of the sum of all individual tests. Example 1 shows how to compute s using Eq. (A-1a).

Note that $(n-1)$ is used rather than the theoretical value n, which is applicable for an unlimited number of tests. The reason is that $(n-1)$ increases the value of s and tends to compensate for the lesser reliability of a small number of tests.

Example 1— Compute s by Eq. (A-1a)

Compressive strength tests of concrete cylinders		
Test no.	X value, psi	X^2 value
1	3315	10,989,255*
2	3090	9,548,100
3	3510	12,320,100
4	2900	8,410,000
5	3690	13,616,100
6	3310	10,956,100
7	3100	9,610,000
8	3490	12,180,100
9	3295	10,857,025
10	3500	12,250,000
11	2910	8,468,100
12	3700	13,690,000
13	3110	9,672,100
14	3790	14,364,100
15	3300	10,890,000
16	3290	10,824,100
17	3105	9,641,025
18	2800	7,840,00
19	3305	10,923,025
20	3495	12,215,025
$\Sigma X_i = 66,005$		$\Sigma(X_i^2) = 219,264,225$

*Although the squared values have been written herein for illustration, this is unnecessary if calculator permits each squared number summed in memory as it is computed.

$n = 20$ (number of test results)

$$\frac{(\Sigma X_i)^2}{n} = \frac{66,005^2}{20} = 217,833,000$$

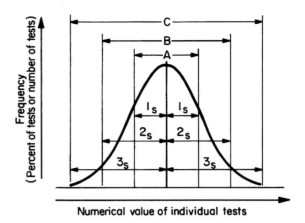

Area	Multiple of Std. Deviation (s)	Tabular Value	Actual Area as Decimal Part of Total Area Under Curve
A	1	.3413	2(.3413) = .6826
B	2	.4773	2(.4773) = .9546
C	3	.49865	2(.49865) = .9973

Fig. A.2.5—Division of standard normal distribution.

$$s = \sqrt{\frac{\Sigma(X_i^2) - [(\Sigma X_i)^2 / n]}{n-1}}$$

$$= \sqrt{\frac{219,264,225 - 217,833,000}{19}}$$

$$s = \sqrt{75,327} = 274 \text{ psi}$$

A3.2—Calculating required average strength f'_{cr}

After the standard deviation has been computed, valuable information is available based on the theoretical normal probability curve. Regardless of the shape of the theoretical curve and the value of s, the area under the curve between $(\bar{X} + s)$ and $(\bar{X} - s)$ will always be 68.2% of the total area under the curve, and the area under the curve between $(\bar{X} + 2s)$ and $(\bar{X} - 2s)$ will be equal to 95.4% of the total. Considering only the half of the curve representing values less than $\bar{X}$, 34.1% of the total area will fall between $\bar{X}$ and $(\bar{X} - s)$, which leaves 15.9% of the area under the curve for values less than $(\bar{X} - s)$, as shown in Fig. A3.1.

These same percentages will apply for the number of tests involved as well as for an area. Therefore, 15.9% of the tests for any normal curve will fall below $(\bar{X} - s)$.

Table A3.1, from ACI 214R, is an adaptation of a table from a math handbook (the normal probability integral) altered to show the percentage of concrete strength tests falling below f'_c as a function of the required average strength f'_{cr}. For example, just as Fig. A3.1 shows that 15.9% of the tests will fall below $(\bar{X} - s)$, Table A3.1 shows that if

$$f'_{cr} = f'_c + s$$

then 15.9% of the tests will fall below f'_c.

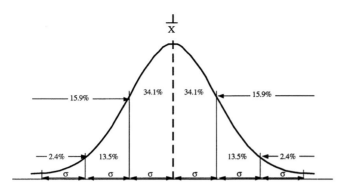

Fig. A3.1—Approximate distribution of area under the normal frequency distribution curve based on deviations from X in multiples of the population standard deviation σ.

Table A3.1—Expected percentages of test lower than f'_c

Required average strength f'_{cr}	Percent of low tests	Required average strength f'_{cr}	Percent of low tests
$f'_c + 0.1s$	46.0	$f'_c + 1.6s$	5.5
$f'_c + 0.2s$	42.1	$f'_c + 1.7s$	4.5
$f'_c + 0.3s$	38.2	$f'_c + 1.8s$	3.6
$f'_c + 0.4s$	34.5	$f'_c + 1.9s$	2.9
$f'_c + 0.5s$	30.9	$f'_c + 2.0s$	2.3
$f'_c + 0.6s$	27.4	$f'_c + 2.1s$	1.8
$f'_c + 0.7s$	24.2	$f'_c + 2.2s$	1.4
$f'_c + 0.8s$	21.2	$f'_c + 2.3s$	1.1
$f'_c + 0.9s$	18.4	$f'_c + 2.4s$	0.8
$f'_c + s$	15.9	$f'_c + 2.5s$	0.6
$f'_c + 1.1s$	13.6	$f'_c + 2.6s$	0.45
$f'_c + 1.2s$	11.5	$f'_c + 2.7s$	0.35
$f'_c + 1.3s$	9.7	$f'_c + 2.8s$	0.25
$f'_c + 1.4s$	8.1	$f'_c + 2.9s$	0.19
$f'_c + 1.5s$	6.7	$f'_c + 3.0s$	0.13

Table A3.1 is useful in establishing the required average strength and determining the probability of low tests occurring when s is known. With s computed from project data and f'_c established, Table A3.1 can be used to compute f'_{cr}. For example, assume that a designer would like to limit the probability of tests falling below 3000 psi to 5%, and the expected standard deviation of the concrete is 560 psi. The concrete should be designed for an average strength of 3925 psi based on the formula from Table A3.1 for 5% low test results

$$f_{cr} = f'_c + 1.65s \text{ (by interpolation)} = 3000 + 1.65(560)$$

$$= 3000 + 925 = 3925 \text{ psi}$$

APPENDIX 4—CONTROL CHARTS ON CONCRETE MATERIALS

Typical control charts for aggregate gradation for individual tests and for range are shown in Fig. A4.1 and A4.2 (Pennsyl-

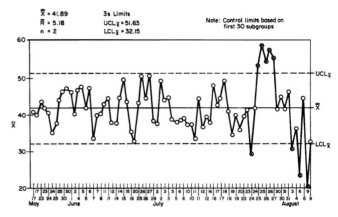

Fig. A4.1—Control chart for average (X), 3/8 in. sieve, percent passing.

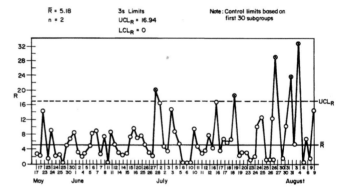

Fig. A4.2—Control chart for range (R), 3/8 in. sieve, percent passing.

vania State University 1974). Plot gradation represents individual percent retained on the particular sieve.

Example calculations and control charts for moving average (five tests) for sand-equivalent test results and for coarse-aggregate material passing the 1 in. sieve are shown in Examples 1 and 2, which are adapted from the California Department of Transportation (1968).

A4.1—Example 1: Calculations to determine moving averages for sand-equivalent test

The procedure and calculations are presented in Table A4.1.

Assume the contract documents require an individual test of not less than 73, and a moving average of not less than 75. The caution zone for this example was arbitrarily set between 75 and 80. The contractor discontinued operations and took steps to correct the deficiency before accepting additional material. Refer to Fig. A4.3 for a plot of the data.

A4.2—Example 2: Calculations to determine moving averages for 1-1/2 x 3/4 in. concrete aggregate (maximum variation of percentage of material passing 1 in. sieve)

The procedure and calculations are presented in Table A4.2.

The contractor's proposed average (job formula) is 26% passing the 1 in. sieve. Assume that the contract documents allow an individual test variation of 14% and a moving average variation of 12% from the average submitted by the

Table A4.1—Example 1: Calculations to determine moving averages for sand-equivalent test

Test no.	Date	Individual test result (minimum 73)	Moving average* Sum† divided by number of results (minimum 75)	Rounded to
1	8/11	79‡	—	—
2	8/14	85	164 ÷ 2 = 82.0	82
3	8/16	84	247 ÷ 3 = 82.7	83
4	8/18	72§	320 ÷ 4 = 80.0	80
§Waived and accepted by engineer.				
5	8/22	80	400 ÷ 5 = 80.0	80
6	8/24	75	396 ÷ 5 = 79.2	80
7	6/25	74	385 ÷ 5 = 77.0	77
8	8/28	68	369 ÷ 5 = 73.8	—‖
‖Materials rejected and results not shown on control charts. Operation discontinued and significant steps taken by contractor to correct deficiency before accepting additional material.				
9	8/29	79	—	—
10	8/31	80	159 ÷ 2 = 79.5	80
11	9/5	81	240 ÷ 3 = 80.0	80
12	9/7	83	323 ÷ 4 = 80.7	81

*Results rounded up to the next whole number using the same number of significant figures as in the individual test result.
†Sum of the five most recent individual test results, including the current test result.
‡Show test result as the first value on the moving-average control chart.

Table A4.2—Example 2: Calculations to determine moving averages for 1-1/2 x 3/4 in. concrete aggregate

Test no.	Date	Individual test result (limits 12 to 40%)	Moving average* Sum† divided by number of results (limits 14 to 38%)	Rounded to
1	6/5	27‡	—	—
2	6/6	24	51 ÷ 2 = 25.5	26
3	6/8	28	79 ÷ 3 = 26.3	26
4	6/12	29	108 ÷ 4 = 27.0	27
—	6/14	35	143 ÷ 5 = 28.6	29
—	6/16	34	150 ÷ 5 = 30.0	30
—	6/20	42§	168 ÷ 5 = 33.6	34
§Waived and accepted by engineer.				
8	6/22	38	178 ÷ 5 = 35.6	36
9	6/26	40	189 ÷ 5 = 37.8	38
10	6/27	42	196 ÷ 5 = 39.2	—‖

‖Materials rejected and results not shown on control charts. The contractor discontinued operations and took steps to correct deficiency before accepting additional material.

*Results rounded up to the next whole number using the same number of significant figures as in the individual test result.
†Sum of the five most recent individual test results, including the current test result.
‡Show test result as the first value on the moving-average control chart.

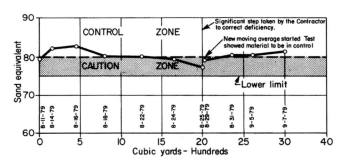

Fig. A4.3—Control chart of moving averages for sand equivalent.

contractor. The caution zone for this example was arbitrarily set between 4% higher than the lower limit and 4% lower than the upper limit. Refer to Fig. A4.4 for a plot of the data.

CHAPTER 3—INSPECTION AND TESTING OF MATERIALS

Materials are inspected and tested to ensure that they meet contract requirements and that they are properly stored, handled, and used in the work. If materials are inspected for acceptance before being shipped to the job site, the field inspector should check their condition once they reach the site, looking for degradation that may have occurred during shipment and storage. The contractor should make available to the inspector all records pertaining to material shipments and tests and certifications.

This chapter discusses sampling and testing procedures for materials commonly used in concrete construction, including cement, aggregates, mixing water, admixtures, reinforcement, curing compounds, and joint sealants. Standard methods of testing concrete are discussed in Chapter 19.

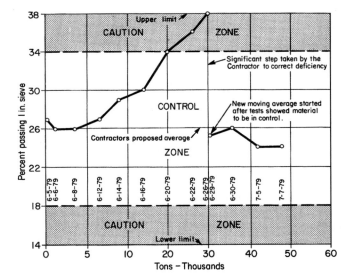

Fig. A4.4—Control chart for moving averages for grading analysis.

3.1—Cement

3.1.1 *Standard types*—Five standard types of portland cement are used in concrete, as specified in ASTM C150:

- *Type I*—A cement for general use when the special properties of other cements are not needed;
- *Type II*—A cement for general use that has moderate sulfate resistance and moderate heat of hydration;
- *Type III*—A cement for use when high early strength is desired;
- *Type IV*—A cement for use when low heat of hydration is necessary, such as for mass concrete; and
- *Type V*—A cement for use when high sulfate resistance is required, such as for structures in contact with soils

or groundwater having high sulfate content, and for concrete in contact with concentrated domestic sewage.

ASTM C150 also specifies three types of air-entraining cement: IA, IIA, and IIIA, which correspond to Types I, II, and III portland cement. These cements contain an air-entraining admixture that is interground with the clinker during manufacture. Air-entraining cement, however, may not be available in all markets, and better control of air content in concrete can usually be achieved by adding the air-entraining admixture at the mixer. For more information on air entrainment and air content, refer to Chapters 5, 6, and 19.

3.1.2 *Blended cements and other hydraulic cements*—Blended cements containing SCMs, as given in ASTM C595, are available in certain localities. These include:
- Type IS: portland blast-furnace slag cement;
- Types IP and P: portland-pozzolan cement;
- Type I(PM): pozzolan-modified portland cement;
- Type S: slag cement; and
- Type I(SM): slag-modified portland cement.

Hydraulic cements set and harden by reacting chemically with water. All portland cements and blended cements are hydraulic cements. ASTM C1157 is a performance specification that lists six types of hydraulic cement:
- Type GU: general use;
- Type HE: high early strength;
- Type MS: moderate sulfate resistance;
- Type HS: high sulfate resistance;
- Type MH: moderate heat of hydration; and
- Type LH: low heat of hydration.

The first comprehensive data on the use of SCMs in concrete in North America were reported in 1937 by Davis et al. SCMs have been widely used in massive structures, such as dams, as well as in structural concrete. They can improve workability, reduce heat of hydration, decrease permeability, improve sulfate resistance, and reduce expansion caused by reactions between certain aggregates and the alkalis in cement. Furthermore, using SCMs can result in greater economy, particularly if the concrete compressive strength is specified at ages exceeding 28 days. ASTM C618 provides requirements for three classes of fly ash and natural pozzolans:
- Class N: Raw or calcined natural pozzolans, such as some diatomaceous earths, opaline cherts, shales, tuffs, volcanic ashes or pumicites, and some clays and shales;
- Class F: Fly ash that is a finely divided residue resulting from the combustion of ground or powdered anthracite or bituminous coal; and
- Class C: Fly ash produced from burning lignite or sub-bituminous coal and generally having sufficient lime content to have some cementitious properties.

Fly ash and natural pozzolans for use in concrete are discussed in more detail in ACI 232.1R and 232.2R.

In the 1980s, an ultrafine pozzolan, silica fume, gained attention as a mineral admixture, leading to the development of a new generation of high-strength, low-permeability concretes, with strengths exceeding 12,000 psi (85 MPa). A by-product of the industrial manufacture of ferrosilicon and metallic silicon, silica fume has extremely fine particles about 1/100 the size of portland cement grains (about the size of cigarette smoke particles). For more information about silica fume, consult ACI 234R. Requirements for silica fume are provided in ASTM C1240.

Requirements for slag cement are provided in ASTM C989, which classifies it as Grade 120, 100, or 80, based on compressive strength results of the slag activity index test with portland cement. Slag cement is a by-product of the steel-making process. Molten slag is drawn off the steel furnace and then rapidly spray-cooled with water, creating a granulated slag material. These granules are then finely ground to a cement-like powder. Unlike other pozzolans, slag cement does not need to be combined with portland cement to develop significant strength and cementing properties. Slag cement is described in more detail in ACI 233R.

3.1.3 *Optional requirements*—ASTM C150 contains additional optional requirements for cement to be applied at the option of the purchaser. They pertain to low-alkali content, false set, 28-day compressive strength, heat of hydration, and sulfate resistance.

3.1.3.1 *Low alkali content*—When the aggregates to be used contain elements known to be destructively reactive with sodium or potassium oxides (the minor alkalies found in cement), using low-alkali cement or an acceptable pozzolan will better ensure the durability and serviceability of the concrete. As defined by ASTM C150, low-alkali cement contains not more than 0.60% of these oxides.

Usually, concrete made with low-alkali cement will be free of objectionable alkali-silica expansion. Petrographic examination of the aggregate is one method of identifying reactive aggregate (ASTM C295). An indication of the potential alkali reactivity of cement-aggregate combinations can be determined by the mortar bar tests (ASTM C227, C1293, or C1260) and, less reliably, by the quick chemical method (ASTM C289).

3.1.3.2 *False set*—ACI 116R defines false set as "the rapid development of rigidity in a freshly mixed portland-cement paste, mortar, or concrete without the evolution of much heat, which rigidity can be dispelled and plasticity regained by further mixing without addition of water." Flash set (or quick set) is defined as "the rapid development of rigidity in a freshly mixed portland-cement paste, mortar, or concrete, usually with the evolution of considerable heat, which rigidity cannot be dispelled nor can the plasticity be regained by further mixing without the addition of water." The presence of false set and flash set can be determined by ASTM C359 (mortar method) or C451 (paste method). These tests, however, do not address all the mixing, placing, temperature, and field conditions that can cause early stiffening.

Although both of these properties are objectionable, flash set is the most detrimental. Once flash set occurs, considerable water must be added to the concrete to regain plasticity, which reduces strength and greatly increases plastic and drying shrinkage cracking. Cement with flash-set properties usually will fail the requirements for time of set in ASTM C150.

False set generally will have no harmful effects on concrete quality, particularly when transit mixing is used. The longer mixing time will restore plasticity without water addition (and often without personnel being aware that false set has occurred). With short mixing times, however, severe false setting may require the addition of slightly more mixing water, which can reduce concrete strength and increase drying shrinkage. With very short mixing times, false set may occur after the concrete has been dumped from the mixer, when it is too late to regain plasticity by additional mixing.

3.1.4 *Sampling and testing procedures*—Cement is usually sampled and tested by the mill, which then issues mill test results and certifications. If the cement remains in bulk storage at the mill for more than 6 months after testing, the material should be retested before shipment. Large users of cement, such as state highway departments, will often sample and test the cement in each silo themselves or hire an independent laboratory to do the testing. Other users typically obtain samples from incoming shipments and have the cement tested by an independent laboratory as necessary.

The plant inspector checks incoming shipments against mill test certificates and owner approvals. Inspectors should also make sure that seals on bulk-shipment cars are unbroken. If the cement is to be sampled at the job site, the inspector should use the following procedures to secure a representative sample of the portion in question in accordance with ASTM C183:

- If the cement is in bags, take a small quantity from one bag for each 100 bags or a fraction thereof. After thoroughly mixing these amounts, choose a sample with a sample splitter or by the quartering method (ASTM C702). Samples for testing should be at least 10 lb (4.5 kg) each (or at least 5 lb [2.3 kg] each if they are to be combined into a 10 lb [4.5 kg] sample);
- If the cement is shipped in bulk, take samples with a slotted sampling tube or by drawing off a considerable quantity of cement at the discharge opening. From the quantity secured, blend a composite sample as described previously;
- Place the cement sample in a clean, dry, waterproof container, and close the cover tightly to exclude air and moisture. Wrapping duct tape around the seam of the cover will improve the seal; and
- Identify the sample clearly, both inside and outside the container. Give the date, name of job, name of inspector, car or lot number, brand of cement, quantity represented by the sample, temperature when sampled, authority or reason for sampling, and tests desired.

3.2—Aggregates

Because nearly 70% of the concrete mass is aggregate, examining and testing aggregates for acceptability involves making necessary tests, seeing that the aggregates are properly stored and handled, and checking batching operations. Minimizing variations in the aggregates as batched is important to the production of quality concrete.

3.2.1 *Specifications*—Standard specifications for aggregates include those for normalweight aggregate (ASTM C33), recommended stock sizes of aggregate for highway construction (ASTM D448), and lightweight aggregate (ASTM C330 and C332). In general, aggregates should be clean, hard, sound, and durable, and particle sizes should be

Fig. 3.1—Lightweight chert can cause popouts.

graded within stated limits. Specifications usually require keeping the chosen grading of fine aggregate reasonably uniform by restricting the range of fineness modulus (refer to the discussion in this chapter under Section 3.2.2.3).

ASTM C33 restricts various deleterious substances to small percentages. The following is a list of these substances and the reasons for restricting them in the final aggregate product:

- *Clay lumps and friable particles*—Unsound particles; may increase water demand if they break down during mixing; affect wear resistance and durability; may cause popouts;
- *Material finer than No. 200 (75 μm) sieve*—Increases demand for mixing water; affects cement paste bond to aggregate; affects the ability to entrain air;
- *Coal and lignite*—May cause detrimental staining of surfaces; cause popouts; interfere with ability of admixtures to entrain air;
- *Soft particles*—Cause popouts; affect durability and wear resistance of concrete; and
- *Lightweight chert (with specific gravity less than 2.40)*—Reduces durability of concrete; causes popouts (Fig. 3.1).

3.2.2 *Sampling procedures*—Sampling methods should be used conforming to the requirements of ASTM D75. Some local variations in the aggregate source should be expected, and samples should be selected so that the effects of these variations are not over- or under-emphasized. In judging test results, the statistical distribution of the amount of undesirable material and the quality-control charts of previous samplings should be considered, as described in Chapter 2. For example, a single clay lump in one sample does not justify rejection of an entire carload of aggregate unless the sample is obviously representative. Aggregate should only be accepted or rejected in accordance with the contract documents for the project.

3.2.2.1 *Where to take samples*—Aggregate samples for testing may be taken from conveyors, bins, cars, barges, or stockpiles, but it is best to select samples that will represent the run of material as batched. Thus, sampling from conveyors or from the discharge openings of bins is preferable (Fig. 3.2).

Fig. 3.2—Where access is convenient and safe, material may be sampled from hopper to bin.

Fig. 3.3—Aggregate stockpiles are difficult to sample properly.

ASTM D75 does not recommend sampling stockpiles because they are difficult to sample properly (Fig. 3.3).

The most representative sample possible is that from a conveyor belt. For fine aggregate, scoopfuls should be taken as the belt goes by to obtain a bucketful from which the test sample can be split or quartered (ASTM C702). For coarse aggregate, samples should be taken from material removed from a short length of belt while the conveyor is stopped. If arrangements cannot be made to stop the conveyor or if no conveyors are used on the job, samples should be taken from the discharge of coarse aggregate from a chute or bin gate. At least several cubic feet of material should be taken, and the test sample quartered from this amount. Such samples are most representative when quartered from material taken from the first, middle, and last of the material to be tested.

If it is necessary to sample from a stockpile, samples should be made up of at least three increments taken from the top third, at the midpoint, and at the bottom third of the column of the pile.

Methods of sampling from belts, bins, and stockpiles are summarized in ASTM D75.

3.2.2.2 *Obtaining representative test samples*—When making up samples taken from surface aggregate in a bin, car, barge, or stockpile, portions should be taken at several separated points. An excess of unrepresentative material should be avoided, such as fines in the center, coarser material toward the edges, and surface materials wetter or drier than average. All material, however, should be represented in the sample. To obtain a sample from under the surface, a board or similar sampling shield should be inserted into the surface of sloping material to prevent it from running down.

When possible, fine-aggregate samples should be taken from damp material to avoid the segregation that occurs in dry sand. Samples should be taken from below the surface, preferably by driving a sampling tube into the sand at several separated points. When sampling for moisture content, nonuniform moisture distribution should be taken into consideration. For example, a mass of wet sand will be wetter toward the bottom a few hours after water has been applied.

If two or more types of sand or coarse aggregate are to be batched separately and blended in the mixer to produce a specified grading, each type should be sampled and tested separately and the results computed for the blended aggregate from the proportions of each type. If necessary, the proportions can be adjusted to obtain the required grading. The computation method for combining two aggregates into a desired grading is identical to that for overall grading of fine and coarse aggregates, illustrated in Table 3.1.

The sample size depends on the type and number of tests to be made and the maximum nominal size of the aggregate. For example, samples of fine aggregate should contain not less than 25 lb (11 kg), and samples of coarse aggregate with a maximum size of 1 in. (25 mm) should contain not less than 110 lb (50 kg). Appendix 5 includes a table showing minimum sample weights for various maximum nominal aggregate sizes.

Some federal specifications for airport pavements require much tighter limits on the amount of deleterious substances in fine and coarse aggregates than those of ASTM C33. These restrictions may range from 0.1% for soft particles and chert to 0.5% for material passing the No. 200 (75 μm) sieve. To provide meaningful results and achieve statistical control of tests (Chapter 2) for compliance with these tight specifi-

Table 3.1—Typical computations of fineness modulus

Percent coarser than sieve number	Sand	No. 4 to 1 in. (4.75 to 25.0 mm) coarse aggregate	Mixture 40% sand and 60% coarse aggregate*
	Cumulative percent by weight retained		
3 in. (75 mm)	—	0	0
1-1/2 in. (37.5 mm)	—	3	2
3/4 in. (19.0 mm)	—	49	29
3/8 in. (9.5 mm)	—	77	46
No. 4 (4.75 mm)	4	96	59
No. 8 (2.36 mm)	15	100	66
No. 16 (1.18 mm)	37	100	75
No. 30 (600 μm)	62	100	85
No. 50 (300 μm)	85	100	94
No. 100 (150 μm)	98	100	99
Total	301	725	555
Fineness modulus	3.01	7.25	5.55

*0.40 × percent of sand plus 0.60 × percent of coarse aggregate.

cations, it is necessary to take much larger sample sizes. To obtain a representative test sample of aggregate from a large sample, a sample splitter or the quartering method should be used (ASTM C702).

Unless the particular test calls for an exact amount of material, the nearest approximate amount resulting from quartering or from the use of the sample splitter should be taken. Adjusting the amount to exactly obtain some arbitrary quantity by adding or removing material may change the average characteristics of the sample, and should not be done unless specifically required by the test method.

If samples are to be shipped to a laboratory for testing, a clean container should be used because even a small amount of some materials (such as sugar or fertilizer) can contaminate the sample. The container should be closed tightly to prevent contamination or loss of fines. As when labeling cement samples for shipment, the sample should clearly be identified inside and outside the container. The date, kind of aggregate, quantity represented by sample, location and other conditions of sampling, authority or reason for test, and kind of test desired should be given.

3.2.2.3 *Tests for grading*—Sieve analysis of aggregates provides information for use in controlling gradation and checking compliance with specified grading requirements.

The amount of the test sample to use for a sieve analysis should be selected in accordance with ASTM C136. To avoid segregation, the sample of fine aggregate should be reduced to the desired size before drying. The sample should be separated into the various sizes with the specified series of sieves, preferably mounted on a mechanical shaker. The sieves should be shaken laterally and vertically, keeping the sample moving continuously over the sieve surfaces. Aggregate fragments on the sieves should not be moved by hand or metal slugs added to the sieves as sieving aids. Sieves, particularly those of small size, should be kept clean and unclogged.

A sieve analysis should be performed only with standard sieves conforming to ASTM E11. With the aid of a conversion chart or graph, it is possible to convert a sieve analysis made using one series of sieves to an analysis using another series of sieves (perhaps having round instead of square openings, or the reverse). The conversion, however, is only approximate. To prevent disputes, the type of sieves specified by the contract documents should be used.

Because aggregates, particularly coarse aggregates, vary considerably within stockpiles and bins, any single test is of limited significance. On a job of moderate size, it may be necessary to perform a sieve analysis once or twice a day and whenever it appears that changes in grading have occurred. If possible, each new test should be averaged with at least two immediately preceding tests on the same material to obtain a more representative analysis of its general run. The average may then be used to adjust the mixture proportioning or to determine if grading requirements have been met.

The fineness modulus is an index number roughly proportional to the average size of particles in a given aggregate; the coarser the aggregate, the higher the fineness modulus. Although the fineness modulus does not indicate grading or distinguish between a single-size aggregate and a

---

PRINCIPAL TESTS ON AGGREGATES

Concrete inspectors are concerned with three types of tests on aggregates:

1. Initial laboratory acceptance tests for suitability as to grading, cleanness (silt and organic impurities), soundness and durability, abrasion resistance, deleterious materials, foreign substances, and mineral composition.

2. Secondary laboratory tests on approved samples to determine physical properties used in mixture proportioning, such as absorption, specific gravity, density, voids, and bulking.

3. Field tests for acceptance or control, such as grading, cleanness, deleterious materials, and moisture content.

The significance of these tests is discussed in ASTM STP 169-C (ASTM International 1994).

---

graded aggregate having the same average size, it does indicate whether one graded aggregate is finer or coarser than another. Thus, it is used for specification and record purposes and is particularly useful for controlling grading and uniformity. One computation method uses the fineness modulus for aggregate proportioning (Portland Cement Association 2002). ASTM specifications require that the fineness modulus of a shipment of fine aggregate not vary more than a certain amount (in some cases, 0.20) either way from the fineness modulus of an acceptably representative preliminary sample. Appendix 5 discusses how to compute the fineness modulus.

3.2.2.4 *Tests for material finer than No. 200 (75 μm) sieve*—The extremely fine mineral material (clay, silt, dust, or loam) occurring in most aggregates causes an increase in the amount of mixing water needed and inhibits the development of entrained air. It also tends to work up to the concrete surface, where it can cause cracking due to shrinkage upon drying. If the fines adhere to the larger aggregate particles, they can interfere with bond between the aggregate particles and cement paste. For these reasons, specifications limit the amount of fine material in aggregates.

The test for percentage of materials finer than the No. 200 (75 μm) sieve is described in Appendix 5 and in ASTM C117. The process begins by washing the fine material through sieves from a weighed sample of oven-dry aggregate. The remaining aggregate is then oven-dried again and weighed to find the amount of fines removed. Refer to ASTM D2419 for a standard field test for fine mineral material in sand.

Often, in the field, a simpler sedimentation test is used to determine the approximate amount of fines in sand (U.S. Bureau of Reclamation 1988) that does not require drying the test sample. For this test, a clear glass jar or bottle (preferably a 32 oz [1 L] graduated prescription bottle) is filled approximately half full with the sand. Clear water is added until the contents reach a level of approximately 3/4 of the volume of the bottle. The container is vigorously shaken, and the contents are allowed to settle for 1 hour. The depth of the layer of fines on top of the sand is measured; roughly 2% by volume (depth) is the equivalent of 1% by mass. If this exceeds a permissible percentage of the depth of sand and fines, the sand is tested by the more accurate method specified.

To determine the amount of clay lumps or other friable particles, use the method described in ASTM C142, which uses material retained on the No. 16 (1.18 mm) sieve (ASTM C117). For fine aggregate, the material is dried, soaked in distilled water for 24 hours, and then the friable particles are broken by squeezing and rolling between the thumb and forefinger. For coarse aggregate, the material is dried and separated into the four sizes, from No. 4 to 1 in. (4.75 to 25 mm). The different sizes are then soaked in distilled water for 24 hours, and the friable particles are broken as for fine aggregate. The residue is separated from each sample by wet sieving over prescribed sieve sizes, and the amount of friable material is computed as a percentage of the sample mass, as shown in Appendix 5.

In coarse aggregate, fine mineral material or crusher dust in objectionable quantities is often clearly visible. ASTM C33 limits materials passing the No. 200 (75 μm) sieve to 1% by weight. This limit increases to 1.5% if the fine material is essentially free of clay or shale. Other criteria also apply, as specified in ASTM C33.

Routine tests are made (daily testing recommended) for material finer than the No. 200 (75 μm) sieve when statistical analysis is desired or when there is reason to believe from the appearance of the aggregate or from sieve analysis that the permissible amount of fine minerals is being exceeded.

3.2.2.5 *Tests for organic impurities in fine aggregate*—A small amount of some types of organic material in fine aggregate can delay or prevent hardening of concrete and drastically reduce concrete strength. Organic matter in sand usually occurs in the form of decayed vegetable matter. In certain locations, tannic acid from pine tree roots leaves deposits on particle surfaces that cannot be removed by washing.

A color-comparison test for the presence and approximate amount of organic material is described in ASTM C40. The use of a slightly damp sample is recommended for the test because an excess of surface moisture dilutes the testing solution, and mechanically dried aggregate may lose some of the organic material in handling or by burnoff from drying at high temperatures. The sample is immersed in a 3% (by weight) solution of sodium hydroxide in a clear glass bottle. After vigorous shaking, the sample is allowed to stand for 24 hours. The color of the supernatant liquid is noted and compared with a standard color solution or standard color plate prepared according to ASTM C40. If the solution is clear, lighter than, or equal to the standard color, the fine aggregate is considered to be free of injurious amounts of organic matter. If the solution is darker than the standard color, organic impurities may be present in quantities sufficient to affect the concrete, and it may be necessary to check the effects of these impurities by comparison strength tests of mortar containing the fine aggregate with and without the organic impurities, in accordance with ASTM C87. Coal or lignite particles in fine aggregate can produce a dark solution in the comparison test, but such particles are permissible in fine aggregate provided that the amounts do not exceed the allowable limits in ASTM C33.

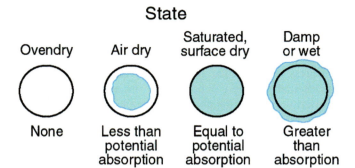

Fig. 3.4—*Range of moisture content in aggregate.*

The frequency of testing for organic matter depends on the condition and homogeneity of the fine aggregate and on the requirements for statistical evaluation (Chapter 2). For washed fine aggregate with a satisfactory record, conducting the color-comparison test as infrequently as once a week may be sufficient. More frequent testing may be necessary if the amount of organic matter is near the permissible limit, if the supply changes, or if the concrete hardens more slowly than normal.

3.2.2.6 *Tests for moisture and absorption*—Tests for moisture content and absorption of aggregates are conducted to:
- Determine the amount of water the aggregates contribute to or absorb from a concrete mixture. An increase of 1% in the moisture content of fine aggregate (for example, from 4 to 5%) can increase concrete slump for normal concrete without water-reducing admixtures and decrease compressive strength; and
- Determine the necessary adjustment in weight or volume to secure uniform quantities of equivalent saturated, surface-dry aggregates in the batches to obtain correct concrete yield. If measurement is by volume, the bulking factor should be known. The bulking factor is the ratio of the surface-dry rodded aggregate density to the difference between the damp loose-aggregate density and the weight of surface moisture unit volume of dry, loose aggregate.

The moisture content of aggregates may be in any of four states, as shown by Fig. 3.4:
- *Oven-dry*—Completely dry and fully absorbent;
- *Air-dry*—Dry at the surface with some interior moisture, but in less than the amount required to saturate the particles (or less than the absorption capacity). Thus, the aggregate is somewhat absorbent;
- *Saturated and surface-dry (SSD)*—Ideal condition in which the aggregate neither contributes water to nor absorbs water from the paste; or
- *Moist*—Contains excess moisture on the surface.

A clear understanding of these conditions is necessary for proper aggregate proportioning and batching.

All computations should be based on aggregate in a SSD condition. Though it is not possible to secure aggregates for construction in this ideal condition, simple arithmetic can be used to convert the measurements on damp or dry aggregate into terms of the equivalent amounts of SSD aggregate. It is only necessary to know the total moisture content and absorption capacity of the aggregate to determine free moisture, as in the following equation

free moisture = total moisture – absorption capacity

If the total moisture is less than the absorption capacity (as in the case of air-dry aggregate), the surface moisture will be negative, and the aggregate will absorb some of the mixing water.

Tests for determining moisture content and absorption of aggregates are described in detail in *Design and Control of Concrete Mixtures* (Portland Cement Association 2002) and in ASTM C70, C127, C128, and C566. The significance of these tests is discussed in ASTM STP 169-C (ASTM International 1994). For laboratory purposes, the tests are made with a precision that will yield results accurate to the nearest 0.1%.

When determining moisture content of fine aggregate in the field, speed is more important than refinement of test results because information on changes in moisture content is needed in time to make appropriate adjustments in batching. Electrical and nuclear equipment for indicating moisture is available that immediately reflects significant changes in moisture content of fine aggregate. This equipment, however, should be periodically calibrated against ASTM tests to confirm accuracy and to indicate necessary equipment adjustments.

3.2.2.6.1 *Absorption*—For practical purposes, the total absorption capacity of a given aggregate does not vary. In hard, dense natural aggregates it usually amounts to 1% or less of the aggregate weight, but it may be higher in some natural aggregates. In manufactured aggregates, such as slag cement, it may be as high as 5%. On small jobs, the absorption value for a given type of aggregate often is assumed, but where careful control is desired, the absorption should be accurately determined.

Coarse-aggregate absorption is determined using procedures in ASTM C127, omitting weighing the sample underwater if specific gravity is not to be measured. A damp sample is brought to the SSD condition, weighed, completely dried by heating, and reweighed. The loss in weight after heating represents the absorption capacity. The surface-dry condition is obtained by wiping the wet particles to remove surface water sheen with an absorbent cloth, as per the procedure.

Fine-aggregate absorption is determined following procedures in ASTM C128, omitting the use of the flask if specific gravity is not measured. A damp sample is brought to the SSD condition, weighed, oven-dried, and reweighed. The surface-dry condition is determined using the cone test described in ASTM C128 and in Appendix 5 of this document. At the point in the series of tests when the pile of fine aggregate begins to slump, the sand is saturated and surface-dry. Another sign that the drying sand has become saturated and surface-dry is when it just ceases to adhere to a clean glass rod or jar. Neither of the tests for the SSD condition is

Fig. 3.5—Drying sand on a hotplate to determine total moisture.

suitable for very coarse fine aggregates or for some very angular fine aggregates.

3.2.2.6.2 *Surface moisture*—To determine surface moisture for either fine or coarse aggregate, a method described in Appendix 5 of this document and in ASTM C566 involves weighing the damp sample, drying it by means of heat, and reweighing it (Fig. 3.5). The surface moisture is calculated by subtracting the established absorption value of the aggregate from total moisture percentage of the damp (stockpile) material. Aggregates that appear to contain no surface moisture should be tested by this method of heating to determine whether surface moisture is either zero or a negative value (indicating that aggregate will absorb mixing water).

A rapid method for determining surface moisture in coarse aggregate involves weighing the damp sample, removing surface moisture by towel-drying to SSD condition, and then reweighing. The moisture loss is divided by the SSD weight to calculate surface moisture.

The surface moisture in fine aggregate can also be determined by the Chapman Flask (ASTM C70). For this test, the specific gravity of SSD aggregate should be known. Several devices using a moisture-reacting powder can be used to provide rapid surface moisture determination of fine aggregate. The accuracy of these tests should be checked daily against the drying methods provided in ASTM C566. Other common methods of determining surface moisture involve weighing the aggregate in air and then under water, or the use of a pycnometer. These methods are described in *Concrete Manual* (U.S. Bureau of Reclamation 1988).

The frequency of testing for moisture content depends on the uniformity of the aggregate supply and the requirements for statistical analysis (Chapter 2). Ordinarily, testing is performed twice a day, and additional tests, particularly of fine aggregates, are made whenever conditions change significantly. To minimize the effect of inaccuracies in sampling and weighing, as large a sample for moisture tests as can be conveniently handled within the time available for drying is used.

3.2.2.7 *Tests for specific gravity*—To convert a given weight of aggregate into terms of solid volume for computing yield, or to convert solid volume to weight for batching purposes, it is necessary to know the specific gravity of the aggregate. Specific gravity is the ratio of the weight of a given solid volume of material to the weight of an equal volume of water. The specific gravity of most aggregates is approximately 2.65, although some suitable aggregates may have a specific gravity of 2.50 or less, and traprock may have a specific gravity of 2.85 or greater. For a given aggregate, the value is generally constant, and can be considered as such for most purposes. As a routine check, specific gravity should be tested about once a month to confirm that no changes have occurred and any time quarry sources change.

To determine the specific gravity of coarse aggregate (ASTM C127), an SSD sample is weighed in air, and then weighed under water by placing the sample in a wire basket suspended from a scale and immersed in water. Next, the sample is oven-dried and reweighed. The sample should weigh, in pounds, at least 10 times the maximum aggregate size in inches; therefore, a 5 lb (2.3 kg) sample would be the minimum required for 1/2 in (12.5 mm) aggregate. The bulk SSD specific gravity should be calculated as shown in Appendix 5 of this document.

To determine the specific gravity of fine aggregate (ASTM C128), a water-filled calibrated flask is weighed; the same flask filled with water and a known weight of SSD fine aggregate is weighed; the aggregate sample is dried with heat; and the dried sample is reweighed. The specific gravity of fine aggregate can also be determined by weighing the sample in air and then in water, as described for coarse aggregate. Other methods of determining specific gravity are described in *Design and Control of Concrete Mixtures* (Portland Cement Association 2002), including a device developed by the U.S. Bureau of Reclamation that can test for moisture, absorption, and specific gravity.

For SSD aggregates, the specific gravity should be based on the surface-dry condition; for materials that will be batched oven-dry, the specific gravity should be based on the oven-dry condition. For aggregates containing other than these two quantities of moisture, the aggregate weight is expressed in terms of either of these given conditions (plus the excess water) and the corresponding dry or SSD specific gravity is used. Many engineers prefer to base recommended mixture proportions on SSD aggregates. ACI 211.1 uses the oven-dry basis for determining the dry-rodded volume of coarse aggregate. The dry volume is converted to weight with the measured, dry-rodded density, and the SSD weight of coarse aggregate is obtained by adding the amount of water absorbed in 24 hours (ASTM C127). Chapter 6 contains additional information on computing aggregate volume.

3.2.2.8 *Tests for voids*—The percentage of voids in aggregate usually is computed from the specific gravity and density, lb/ft^3 (kg/m^3) as follows:

$$\text{percentage of voids} = 100\left[\frac{(\text{specific gravity} \times 62.4) - \text{bulk density}}{\text{specific gravity} \times 62.4}\right] \text{(in.-lb)}$$

$$\text{percentage of voids} = 100\left[\frac{(\text{specific gravity} \times 1000) - \text{bulk density}}{\text{specific gravity} \times 1000}\right] \text{(SI)}$$

When reporting the amount of voids, information about the aggregate's moisture content and compactness should be provided. For dry-rodded aggregate, the method of determining voids is described in ASTM C29/C29M. A previously used method of determining voids by measuring the amount of water required to fill a container of aggregate is subject to error caused by entrapped air.

3.2.2.9 *Tests for bulk density*—The bulk density, or weight per unit volume, of aggregate is used to compute the amount of voids in aggregate, to compute proportions of materials, and to convert bulk-volume quantities to weight, or vice versa.

The bulk density of a given type of aggregate varies with the degree of compaction and with the moisture content. In fine aggregate, the bulking, or fluffing apart, of particles by films of surface moisture may reduce the density as much as 25%.

To provide a uniformly reproducible basis of measurement, ASTM has adopted a standard test method for aggregate bulk density (ASTM C29/C29M). This test consists of compacting the dry aggregate into a cylindrical container of calibrated volume and then weighing the aggregate. In one method of compaction, the container is filled in three layers, each layer rodded 25 times; the excess aggregate is then struck off level with the top of the container. The standard container sizes are:

Maximum size of aggregate	Container		
	Capacity, ft^3	Inside diameter, in.	Inside height, in.
1 in. or less	1/3	8.0	11.5
1-1/2 in.	1/2	10.0	11.0
4 in.	1	14.0	11.2

Note: 1 in. = 25.4 mm and 1 ft^3 = 0.0283 m^3.

For aggregate of maximum size greater than 1-1/2 in. (37.5 mm), the layers of aggregate are compacted by jigging the container instead of by rodding. Although not specifically stated in the standard method, the term "dry" should be interpreted as either SSD or oven-dry, depending on the basis of mixture computation used.

ASTM C29/C29M also provides for measurement of the density of dry loose or damp loose aggregate. This density is determined by filling the standard container heaping full in one layer. The excess aggregate is struck off level with the top of the container without downward pressure, and the aggregate is weighed.

Lightweight aggregate's bulk density should be determined by the shoveling procedure described in ASTM C29/C29M. The aggregate is tested in an oven-dry condition.

Values of aggregate bulk density are generally considered to be accurate within 0.5 to 1%, although it is possible to make bulk density tests with greater accuracy. Aggregate bulk density should be tested for when mixture proportions are made on the bulk density basis and whenever it is apparent that the type or grading of the aggregate has changed.

3.2.2.10 *Bulking factor*—Except when using volumetric batching and mixing equipment conforming to ASTM C685/C685M, batching of aggregates by volume should be avoided because large fluctuations can occur (Portland Cement Association 1994). When volume batching is unavoidable, contract documents should state whether the basis of measurement is damp loose, dry loose, or dry-rodded volume. To compute the number of cubic feet (cubic meter) of damp loose aggregate corresponding to 1 ft^3 (1 m^3) of dry-rodded aggregate—a ratio called the bulking factor—the density should be determined under both conditions and the moisture content of the damp aggregate

$$\text{bulking factor} = \frac{\text{density of surface-dry rodded aggregate}}{\text{density of damp loose aggregate} - \frac{\text{weight of surface moisture in unit volume of damp loose aggregate}}{}}$$

3.2.2.11 *Undesirable substances and properties*—Contract documents may require ASTM tests to check for undesirable substances and properties of aggregates. These tests include C40, C88, C123 (coal and lignite), C131, C142, C227, C289, C295, C535, C586, C666/C666M, and D4791. Substances in aggregate that have a deleterious effect on concrete quality are discussed in detail in ASTM STP 169-C (ASTM International 1994).

Cherts, clayey limestones, and some other aggregates are subject to considerable volume changes during wetting and drying or freezing and thawing. Aggregates susceptible to spalling can be detected by petrographic examination supplemented by visual inspection, sorting, and weighing. Often, these materials also cause high losses in the sodium or magnesium sulfate soundness test and in freezing-and-thawing tests.

Because lightweight chert (specific gravity less than 2.40) contributes to popouts at concrete surfaces, ASTM C33 limits the material in coarse aggregate to 3 to 8%, depending on exposure conditions. Flat or elongated particles also are considered undesirable under some conditions, and can be sorted out of a sample and weighed to determine the percentage. Petrographic examination of aggregates can identify elements that are undesirably reactive with alkalies in cement, such as opal, chalcedony, tridymite, and acid or intermediate volcanic glasses.

Lightweight materials and soft and weak materials usually reduce concrete strength and quality. They can quickly be separated by flotation in a heavy liquid, as described in ASTM C123. Most deleterious substances adversely affecting the chemical activity (setting) of cement are organic and can be detected by the colorimetric test.

3.3—Water
Potable water is the quality criterion typically specified for mixing water. Ordinarily, the presence of harmful impurities such as alkalies, acids, decayed vegetable matter, oil, sewage, or excessive amounts of silt will be known. Water of questionable quality should be tested for its effect on strength and time of set versus water known to be satisfactory using test methods from ASTM C109/C109M and C191. ASTM C94/C94M allows the use of mixer wash water in subsequent batches of concrete if the water complies with optional chemical limits, as confirmed through regular testing of the plant's wash water. This water must also be accurately measured and calculated as part of the allowable mixing water.

Information on water quality requirements are discussed in detail in ASTM STP 169-C (ASTM International 1994).

3.4—Admixtures
The standard specification for chemical admixtures for concrete (water-reducing, retarding, and accelerating) is ASTM C494/C494M. This specification considers seven types of admixtures with various purposes:
- *Type A*: Water-reducing;
- *Type B*: Retarding;
- *Type C*: Accelerating;
- *Type D*: Water-reducing and retarding;
- *Type E*: Water-reducing and accelerating;
- *Type F*: Water-reducing, high-range; and
- *Type G*: Water-reducing, high-range, and retarding.

In addition, ASTM C1017/C1017M covers chemical admixtures for use in producing flowing concrete, and ASTM C260 gives standard specifications for air-entraining admixtures.

These standards provide methods of testing in the laboratory, but the tests are not intended to simulate job conditions. It is important that mixture proportions be prepared with the admixtures and cement to be used on the job. The inspector should verify that admixtures used on the job comply with all contract requirements and that proper storage and dispensing is provided. For example, contract documents should require that calcium chloride (if allowed) be dissolved before being added to the batch to ensure uniform distribution and uniform acceleration of hydration throughout the mixture. Substituting chemical admixture brands, even though complying with ASTM general types, should not be permitted without verification of performance, because some chemical admixtures perform differently in combination and with different cements and/or mineral admixtures.

As a part of the acceptance of admixtures, either a laboratory analysis should be performed or the manufacturer's statements should be relied upon. Admixtures should be inspected to ensure that they:
- Conform to appropriate specifications;
- Are stored without contamination or deterioration;
- Are accurately measured;
- Are introduced into the batch as specified; and
- Perform as expected as far as can be determined from the concrete as mixed and tested.

If possible, make routine quality-control tests of admixtures for specific gravity, pH, and solids (residue by drying) and compare against data supplied by the manufacturer.

3.5—Steel reinforcement
Steel for concrete reinforcement is usually purchased under one of the ASTM specifications listed in Chapter 22. (Refer to Chapter 17 for a discussion of prestressing steel.) Standard practice for the purchase and handling of reinforcement is given in *Manual of Standard Practice* (Concrete Reinforcing

> **PURPOSES OF ADMIXTURES**
>
> ACI 116R defines an admixture as "a material other than water, aggregates, hydraulic cement, and fiber reinforcement, used as an ingredient of concrete or mortar, and added to the batch immediately before or during its mixing." Contract documents may require or permit an admixture to be used in concrete for one or more of the following purposes:
> - Increase workability without increasing water content, or decrease the water content at the same workability;
> - Accelerate the rate of early strength development;
> - Increase strength;
> - Increase surface hardness;
> - Retard or accelerate initial setting;
> - Retard or reduce heat evolution;
> - Modify the rate of or capacity for bleeding;
> - Increase durability or resistance to severe exposure conditions, including application of deicing salts;
> - Control expansion caused by reaction of alkalies with certain aggregate constituents;
> - Decrease capillary flow of water;
> - Decrease permeability to liquids;
> - Produce cellular concrete;
> - Improve penetration and pumpability of grouts and pumpability of concrete;
> - Reduce or prevent settlement, or create slight expansion in concrete or mortar used for filling blockouts under machinery, column and girder spaces, post-tensioning cable ducts, and voids in preplaced aggregate;
> - Increase bond of concrete to steel;
> - Increase bond between old and new concrete;
> - Produce colored concrete or mortar;
> - Produce fungicidal, germicidal, and insecticidal properties in concrete or mortar;
> - Inhibit corrosion of embedded corrodible metal; and
> - Decrease the unit cost of concrete.
>
> Refer to ACI 212.3R and 212.4R for more detailed information.

Steel Institute 1997b). Procedures for inspection of cutting, bending, storing, handling, and placing of reinforcement are described in Chapter 8.

In general, purchase specifications cover the method of manufacture, certain chemical requirements, tests in tension and in bending, surface finish, coating for corrosion protection, marking (for size, grade, and point of origin), and permissible variations in weight. Reinforcement is usually tested for properties at the mill and shipped to the job site in bundles marked with tags. The inspector should verify that each shipment has been tested and certified, that the specified grade has been delivered, and that the reinforcement is not damaged or excessively rusted. If it appears that the reinforcement does not meet the contract requirements, it should not be accepted unless samples sent to a laboratory for testing indicate compliance.

A light film of rust is not objectionable on ordinary reinforcement (in fact, its roughness improves the bond), but a heavy coating with flakes or scales that fly off when the bar is bent or struck with a hammer should be removed. Reinforcement should be clean, and any oil or mortar that has been spilled on it should be cleaned off. Epoxy-coated reinforcement should be produced and installed as provided by ASTM A775/A775M.

Occasionally, Grade 60 or 75 (Grade 420 or 520) high-strength reinforcement will crack or break when bent, especially in cold weather. Any cracked and broken reinforcement should be rejected.

3.6—Curing compounds

Membrane-forming concrete curing compounds are usually used in place of water curing or to provide final curing following a short period of water curing. Clear (with or without a fugitive dye) and pigmented (white or gray) products are available. ASTM C309 provides standard specifications and references for testing methods for curing compounds.

To ensure compliance of curing compounds with contract documents, acceptance should be based on certification by manufacturers or on laboratory testing. Inspection consists of seeing that the compound is:

- Properly labeled and not contaminated, diluted, or altered in any way before application;
- Mixed thoroughly before and during application;
- Applied when concrete surfaces are still damp and full of moisture (formed concrete should be saturated with water before compound application);
- Applied at the specified coverage rate; and
- Kept on the surface for the specified curing time.

See additional discussions on curing in Chapters 10 and 12.

3.7—Joint materials

Joints create openings that should usually be filled or sealed to prevent intrusion of dirt, water, or other unwanted substances. For many years, oil-based mastics or bituminous compounds and metallic materials were the only joint sealants available, and fillers were typically resilient materials such as fiberboard, wood, rubber, or cork. All of these materials are still available today and are used in some instances.

To overcome some of the drawbacks of traditional joint sealants, many new elastomeric materials have been developed in recent years. These materials are flexible rather than rigid at normal service temperatures, and they can be molded in the field or preformed. ACI 504R lists the many types of elastomeric materials and their properties.

The manufacturer will normally test its joint materials to verify performance. The inspector should review the manufacturer's data for conformance to the project specifications or that the material is listed as approved in the project specifications. Once they arrive at the job site, materials should be checked again to see that they are not damaged or contaminated and are properly labeled, stored, prepared, and installed. If laboratory testing is required, representative samples should be prepared to ship to the laboratory.

Instructions for sampling liquid and dry components of a job-mixed mastic joint filler are given in *Concrete Manual* (U.S. Bureau of Reclamation 1988). The liquid component should be thoroughly mixed before taking a sample because the liquid is susceptible to separation. Samples of dry components should be obtained by using a sample splitter or by quartering (ASTM C702). Samples should be shipped in tightly sealed containers.

The successful performance of any joint sealant largely depends on proper installation. Each step in the construction and preparation of the joint to receive the sealant requires careful workmanship and thorough inspection. The contract documents for the work should state the type of sealant, the installation method, and requirements for joint construction and preparation. Each joint should be inspected for cleanness and dryness before placement of backup materials, primers, or sealants. In the absence of specified temperature restrictions, installation of sealants at temperatures above 90 °F (32 °C) and below 40 °F (4 °C) should be avoided.

APPENDIX 5—AGGREGATE SAMPLING AND TESTING

Aggregate quality should be controlled at the point of production using a proper system of quality-control testing. As a check of quality, inspectors can readily perform the following tests in the laboratory.

These are only summaries of sampling and testing procedures; for complete descriptions, refer to Volume 04.02 of *ASTM Standards*.

A5.1—Sieve or screen analysis of fine and coarse aggregate: ASTM C136

1. Sample aggregates in accordance with appropriate method given in ASTM D75;
2. Split sample by quartering or using a sample splitter;
3. Use minimum test-sample weights per ASTM C 136;
4. Dry the sample to a constant weight;
5. Weigh to the nearest 0.005 oz (0.1 g) for fine aggregate and 0.1% for coarse aggregate;
6. Place the sample over nest of sieves or screens arranged with decreasing size of openings from top to bottom;
7. Shake the sieves until not more than 1% of the weight of the residue on any individual sieve will pass the sieve during 1 minute of continuous hand sieving;
8. Weigh the amounts retained on each sieve or screen and record totals on a worksheet similar to that shown in Chapter 20;
9. Calculate the percentage on the basis of the total weight of the sample, including any material finer than the No. 200 (75 μm) sieve determined in accordance with ASTM C117; and
10. Calculate the fineness modulus by adding the cumulative percentages retained on the appropriate set of the following sieves and dividing by 100: No. 100 (150 μm), No. 50 (300 μm), No. 30 (600 μm), No. 16 (1.18 mm), No. 8 (2.36 mm), No. 4 (4.75 mm), 3/8 in. (9.5 mm), 3/4 in. (19.0 mm), 1-1/2 in. (37.5 mm), and larger, increasing in a ratio of 2 to 1.

A5.2—Sampling aggregates: ASTM D75

1. *Sample size*—Minimum composite sample sizes depend on nominal size, as in Table A5.1;
2. *Sampling from conveyor belts*—(a) Secure a minimum of three samples of approximately equal increments selected at random from the entire belt width of a stopped conveyor; (b) place two vertical templates, conforming to the belt, so that the material contained between them will yield the required weight; and (c) scoop all of the aggregate between the templates, and combine with the other samples as required in Step (a);
3. *Sampling from bins or belt discharge*—(a) Secure a minimum of three samples of approximately equal increments selected at random from the unit being sampled, and combine to form the field sample; (b) obtain each sample from the entire cross section of material as it is being discharged; and (c) use a pan supported on rails or a similar device capable of intercepting the entire discharge stream and holding the material without overflowing; and
4. *Sampling from stockpiles*—(a) Avoid sampling from stockpiles if possible; (b) to sample from the face of a stockpile, insert a sample shield of rigid material into the face to prevent material from segregating as it is being sampled; and (c) sample at least three increments from the top third, at the midpoint, and at the bottom third of the volume of the stockpile. Select sample units at random, and combine to form a field sample as required.

Table A5.1

Maximum nominal size of aggregate	Approximated minimum mass of field samples, lb (kg)
No. 8 (2.36 mm)	25 (10)
No. 4 (4.75 mm)	25 (10)
3/8 in. (9.5 mm)	25 (10)
1/2 in. (12.5 mm)	35 (15)
3/4 in. (19.0 mm)	55 (25)
1 in. (25.0 mm)	110 (50)
1-1/2 in. (37.5 mm)	165 (75)
2 in. (50 mm)	220 (100)
2-1/2 in. (63 mm)	275 (125)
3 in. (75 mm)	330 (150)
3-1/2 in. (90 mm)	385 (175)

A5.3—Materials finer than No. 200 (75 μm) sieve: ASTM C117

Prepare a moist sample by mixing the sample and reducing it by splitting or quartering. The minimum test sample after drying to a constant weight is shown in Table A5.2;
2. Place the sample in a container of sufficient size so that it can be covered and agitated vigorously without loss;
3. Agitate the sample with sufficient vigor to completely separate all particles finer than the No. 200 (75 μm) sieve from the coarse particles and to bring the finer particles into suspension;
4. Immediately pour the wash water containing the suspended solids over the nest of two sieves, with No. 16 (1.18 mm) on top and No. 200 (75 μm) on the bottom;
5. Add a second charge of water to the specimen in the container, and repeat agitation and decantation. Repeat this until the wash water is clear;
6. Return all material retained on the nested sieves by flushing to the container holding the rest of the washed sample;
7. Dry the washed aggregate to constant weight at a temperature of 230 ± 9 °F (110 ± 5 °C), and weigh to nearest 0.1% of the weight of the test sample, except if the result is 10% or more; weigh to the nearest whole number; and

Table A5.2—Minimum test sample after drying to constant weight

Nominal aggregate size	Minimum weight
No. 4 (4.75 mm)	0.7 lb (300 g)
3/8 in. (9.5 mm)	2.2 lb (1000 g)
3/4 in. (19.0 mm)	5.5 lb (2500 g)
1-1/2 in. (37.5 mm) or larger	11.0 lb (5000 g)

Table A5.3—Particle sizes and minimum sample weights

Size of particle	Minimum sample weight
No. 4 to 3/8 in. (4.75 to 9.5 mm)	2.2 lb (1000 g)
3/8 to 3/4 in. (9.5 to 19.0 mm)	4.4 lb (2000 g)
3/4 to 1-1/2 in. (19.0 to 37.5 mm)	6.6 lb (3000 g)
Over 1-1/2 in. (37.5 mm)	11.0 lb (5000 g)

Table A5.4

Size of particle making up the sample	Sieve size for removing residue of clay lumps and friable particles
Fine aggregate	No. 20 (850 μm)
No. 4 to 3/8 in. (4.75 to 9.5 mm)	No. 8 (2.36 mm)
3/8 to 3/4 in. (9.5 to 19.0 mm)	No. 4 (4.75 mm)
3/4 in. to 1-1/2 in. (19.0 to 37.5 mm)	No. 4 (4.75 mm)
Over 1-1/2 in. (37.5 mm)	No. 4 (4.75 mm)

8. Calculate A, the percent passing the No. 200 (75 μm) sieve, to the nearest 0.1% as follows

$$A = 100 \times \frac{B - C}{B}$$

where B = original dry weight of sample, lb (g); and C = dry weight of sample after washing, lb (g).

A5.4—Clay lumps and friable particles in aggregates: ASTM C142

The aggregate for this test consists of the material remaining after completion of the test for materials finer than the No. 200 (75 μm) sieve by washing.

1. Dry the aggregate to a constant weight at a temperature of 230 ± 9 °F (110 ± 5 °C);

2. Sieve the dried aggregate to obtain the fine-aggregate portion consisting of particles coarser than the No. 16 (1.18 mm) sieve and smaller than the No. 4 (4.75 mm) sieve. From this, take test samples weighing at least 1 oz (25 g);

3. By sieving, separate the coarse-aggregate portion into test samples of the sizes in Table A5.3;

4. Weigh each test sample and spread it in a thin layer on the bottom of the container, then cover it with water and allow it to soak for 24 ± 4 hours;

5. Break all material that can be broken by squeezing it between thumb and forefinger. Do not use fingernails or press against a hard surface. After breaking all discernible clay lumps and friable particles, separate the detritus from the rest of the sample by wet-sieving over the sizes shown in Table A5.4;

6. Wet-sieve by passing water over the sample through the sieve while agitating until all undersize material has been removed;

7. Remove the retained particles from the sieve and dry at a constant temperature of 230 ± 9 °F (110 ± 5 °C). Allow the particles to cool and weigh to an accuracy of 0.1% of the original test sample; and

8. Calculate P, the percentage of clay lumps and friable particles in fine aggregate or individual sizes of coarse aggregate, as follows

$$P = 100 \times \frac{W - R}{W}$$

where W = weight of test sample, lb (g); and R = weight of particles retained on the designated sieve after test, lb (g); and

9. For coarse aggregate, the percentage of clay lumps and friable particles is the average based on the percentage of clay and friable particles in each size fraction, weighted in accordance with the grading of the original sample before separation for testing. If the aggregate contains less than 5% of any size specified, that size should be considered to contain the same percentage of clay lumps and friable particles as the next larger or next smaller size.

A5.5—Organic impurities in fine aggregate: ASTM C40

1. Obtain a representative test sample of fine aggregate weighing about 1 lb (450 g);

2. Fill the 12 oz (350 mL) graduated glass bottle to approximately the 4.5 oz (130 mL) level with the sample of fine aggregate;

3. Add a 3% NaOH (lye) solution until the volume of the fine-aggregate solution after shaking is approximately 7 oz (200 mL);

4. Stopper the bottle, shake vigorously, and then allow to stand for 24 hours.

5. At the end of 24 hours, hold the test sample and a Gardner color scale or a reference color solution up to a light source and compare the color of the liquid in the bottle to the reference color plates on the Gardner scale. Record the color of the liquid with respect to the reference color plates on the Gardner scale; and

6. If the color of the liquid in the test bottle is darker than that of the No. 3 plate on the Gardner color scale, the fine aggregate being tested may contain deleterious organic compounds, and further tests may be required.

A5.6—Specific gravity and absorption of coarse aggregate: ASTM C127

1. Select the minimum test sample size from the sample by splitting or quartering. Remove all material passing the No. 4 (4.75 mm) sieve;

2. Thoroughly wash the test sample and dry to a constant weight at a temperature of 230 ± 9 °F (110 ± 5 °C). Cool in air at room temperature, and then immerse in water at room temperature for 24 ± 4 hours;

3. Remove the specimen from the water and roll it in a large, absorbent cloth until all visible films of water are removed. Weigh the specimen in this SSD condition to the nearest 0.02 oz (0.5 g) or 0.05% of the sample weight, whichever is greater;

4. After weighing, immediately place the SSD specimen in a sample container (wire basket of No. 6 [3.35 mm] or finer mesh of approximately 1 to 2 gallons [4 to 7 L]) and weigh it in water at 73.4 ± 3 °F (23 ± 2 °C). The container should be suspended in water, shaken to remove entrapped air, and supported on a thin wire;

5. Oven-dry the specimen to a constant weight at a temperature of 230 ± 9 °F (110 ± 5 °C), cool it at room temperature for 1 to 3 hours, and weigh;

6. Calculate the bulk specific gravity on the SSD basis

$$\text{bulk specific gravity (SSD)} = B/(B - C)$$

where B = weight of SSD specimen in air, lb (g); and C = weight of the saturated specimen in water, lb (g); and

7. Calculate the percentage absorption as follows

$$\text{absorption, percent} = 100\,(B - A)/A$$

where A = weight of oven-dry specimen in air, lb (g).

A5.7—Specific gravity and absorption of fine aggregate: ASTM C128

1. Select, by splitting or quartering, approximately 2.2 lb (1 kg) of fine aggregate from the sample;

2. Dry the sample in a suitable pan to a constant weight at a temperature of 230 ± 9 °F (110 ± 5 °C);

3. Allow the sample to cool, cover with water, and let stand for 24 ± 4 hours;

4. Decant excess water, then spread the sample on a flat surface exposed to warm, circulating air to ensure uniform drying. Continue until the specimen is in a free-flowing condition;

5. To check dryness, place a portion of the sample in a conical mold held firmly on a smooth, nonabsorbent surface with the large diameter down. Lightly tamp the aggregate surface 25 times with a metal tamper weighing 12 ± 0.5 oz (340 ± 15 g) and having a flat, circular tamping face approximately 1 in. (25 mm) in diameter, then lift the mold vertically. If surface moisture is still present, the fine aggregate will retain the molded shape. Continue drying with stirring, and test at frequent intervals until the tamped fine aggregate slumps slightly upon removal of the mold. This indicates that it has reached the SSD condition;

6. Immediately introduce 18 ± 0.4 oz (500 ± 10 g) of the SSD fine aggregate into the pycnometer (flask) and fill with water to approximately 90% of capacity. Roll, invert, and agitate the pycnometer to eliminate all air bubbles, and then add water to bring the water level to the calibrated capacity. Keep the temperature at 73.4 ± 3 °F (23 ± 2 °C). Determine the total weight of pycnometer, specimen, and water to nearest 0.005 oz (0.1 g);

Table A5.5—Nominal aggregate size and minimum weight of test sample

Nominal size of aggregate	Minimum weight of test sample
No. 4 (4.75 mm)	1.1 lb (0.5 kg)
3/8 in. (9.5 mm)	3.3 lb (1.5 kg)
1/2 in. (12.5 mm)	4.4 lb (2 kg)
3/4 in. (19.0 mm)	6.6 (3 kg)
1 in. (25.0 mm)	8.8 lb (4 kg)
1-1/2 in. (37.5 mm)	13 lb (6 kg)
2 in. (50 mm)	18 lb (8 kg)
2-1/2 in. (63 mm)	22 lb (10 kg)
3 in. (75 mm)	29 lb (13 kg)
3-1/2 in. (90 mm)	35 lb (16 kg)
4 in. (100 mm)	55 lb (25 kg)
6 in. (150 mm)	110 lb (50 kg)

7. Remove the fine aggregate from the pycnometer, dry to a constant weight at 230 ± 9 °F (110 ± 5 °C), cool in air at room temperature for 1/2 to 1-1/2 hours, and weigh;

8. Weigh the pycnometer filled to its calibration capacity with water at 73.4 ± 3 °F (23 ± 2 °C);

9. Calculate the specific gravity (SSD) of the aggregate as follows

$$\text{bulk specific gravity SSD} = \frac{S}{B + S - C}$$

$$\text{bulk specific gravity dry} = \frac{A}{B + S - C}$$

where A = weight of oven dried specimen in air, lb (g); B = weight of pycnometer filled with water, lb (g); C = weight of pycnometer with specimen and water, lb (g); and S = weight of the SSD specimen, lb (g); and

10. Determine the moisture absorption of the fine aggregate. Immediately weigh the SSD fine aggregate and dry to a constant weight at a temperature of 230 ± 9 °F (110 ± 5 °C). Allow to cool, and weigh to the nearest 0.005 oz (0.1 g). Calculate the percentage of absorption as follows

$$\text{absorption, percent} = 100(S - A)/A$$

where A = weight of oven-dry specimen in air, and S = weight of SSD specimen.

A5.8—Total moisture content of aggregate by drying: ASTM C566

Select a representative sample of the aggregate by splitting or quartering of not less than the weights in Table A5.5;

2. Weigh the moist sample to the nearest 0.1%;

3. Thoroughly dry the sample to a constant weight. If a ventilated oven is used, the temperature should be 230 ± 9 °F (110 ± 5 °C). If a hot plate or heat lamp is used, stir the sample frequently to avoid localized overheating;

4. Allow the dried sample to cool, and weigh to the nearest 0.1%;

5. Calculate P, the total moisture content (percent), of the sample as follows

$$P = 100\,(W - D)/D$$

where W = weight of original wet sample, lb (g); and D = weight of dried sample, lb (g); and

6. The surface moisture (free water) is the difference between the total moisture content and the known absorption of the aggregate.

CHAPTER 4—HANDLING AND STORAGE OF MATERIALS

The quality of concrete depends on the quality of the ingredients used, particularly the cement and aggregates. Problems such as high or low yield, low strength, and deterioration often can be traced to poor handling and storage of these materials.

To help ensure consistent, uniform production of high-quality concrete, inspectors need to confirm that the cement, aggregates, and other materials used are stored and handled properly. In addition to the recommendations given herein, comprehensive guidelines for material transport and handling are provided in ACI 304R.

4.1—Cement

4.1.1 *Storage and hauling of bulk cement*—Bulk cement should be stored in weatherproof, properly ventilated bins to prevent moisture accumulation (Fig. 4.1). The bins should have:
- Smooth interior walls shaped to allow removal of all cement;
- Inverted truncated cones at the bottom;
- Air-diffuser flow pads to facilitate loosening of cement that has settled; and
- A gate and conveyer system for batching, with the conveyor isolated from the elements.

On large jobs, the silos should be periodically emptied and inspected for a buildup of cement, which should be removed before refilling. This inspection should be made whenever a new supplier's cement or a new type of cement is put into an existing silo.

Avoid exposing the cement to air, because moisture in the air causes partial hydration. If cement becomes slightly lumpy during storage, its use may be permitted if most lumps are soft enough to be crushed between the thumb and fingers. If harder lumps exist, they should be removed by screening (provided that the contract documents do not prohibit this practice). If an excessive number of lumps are encountered, the cement should be tested to verify suitability before use. Cement properties to be verified should be confirmed with the engineer, although tests for compressive strength and time-of-set are usually conducted in these circumstances.

Bulk cement is typically transported in tanker trucks, rail cars, or barges. The equipment used to haul cement should be leakproof and free of contaminants. The chutes and boots used for loading transport vehicles should be regularly inspected. To minimize cement loss, the boot should extend into the truck or compartment. Clogged bin vents should be avoided.

4.1.2 *Storage of bagged cement*—Bagged cement may be used on small projects. To protect the bags from both ground

Fig. 4.1—Storage of bulk cement in weatherproof and properly ventilated bins is essential to prevent moisture accumulation.

Fig. 4.2—Keep aggregate piles clean and separate to minimize segregation.

moisture and the elements, it is best to store them in an enclosed building and on pallets above the floor. If the bags need to be stored outside, watertight coverings should be used.

When removing bagged cement from storage, the oldest cement should be used first. If the cement has developed warehouse set, the bags should be rolled out on a hard surface to restore flowability. Cement with lumps that do not break down easily should be discarded.

Additional recommendations are given in ACI 304R.

4.2—Aggregates

To achieve uniform concrete production, it is important to keep the grading and moisture content of aggregates as consistent as possible and to protect them from contamination. Figures 4.2 to 4.4 illustrate methods for aggregate handling and storage. Additional recommendations are given in ACI 304R.

4.2.1 *Storage in stockpiles*—If aggregates are stored in piles on the ground, the following measures should be taken to prevent contamination and segregation:
- Pave the area, lay planks, or leave the ground covered by a bottom layer of undisturbed aggregate several inches deep;
- Do not allow a crane bucket containing other aggregates or materials to swing over the aggregate pile;

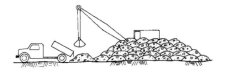

PREFERABLE

CRANE OR OTHER MEANS OF PLACING MATERIAL IN PILE IN UNITS NOT LARGER THAN A TRUCK-LOAD WHICH REMAIN WHERE PLACED AND DO NOT RUN DOWN SLOPE.

OBJECTIONABLE

METHODS WHICH PERMIT THE AGGREGATE TO ROLL DOWN THE SLOPE AS IT IS ADDED TO THE PILE OR PERMIT HAULING EQUIPMENT TO OPERATE OVER THE SAME LEVEL REPEATEDLY.

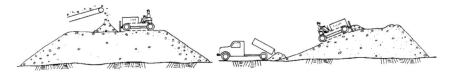

LIMITED ACCEPTABILITY--GENERALLY OBJECTIONABLE

PILE BUILT RADIALLY IN HORIZONTAL LAYERS BY BULLDOZER WORKING FROM MATERIALS AS DROPPED FROM CONVEYOR BELT. A ROCK LADDER MAY BE NEEDED IN SETUP.

BULLDOZER STACKING PROGRESSIVE LAYERS ON SLOPE NOT FLATTER THAN 3:1. UNLESS MATERIALS STRONGLY RESIST BREAKAGE, THESE METHODS ARE ALSO OBJECTIONABLE.

CORRECT

CHIMNEY SURROUNDING MATERIAL FALLING FROM END OF CONVEYOR BELT TO PREVENT WIND FROM SEPARATING FINE AND COARSE MATERIALS. OPENINGS PROVIDED AS REQUIRED TO DISCHARGE MATERIALS AT VARIOUS ELEVATIONS ON THE PILE.

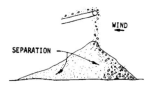

INCORRECT

FREE FALL OF MATERIAL FROM HIGH END OF STACKER PERMITTING WIND TO SEPARATE FINE FROM COARSE MATERIAL.

UNFINISHED OR FINE AGGREGATE STORAGE (DRY MATERIALS)

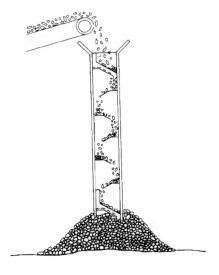

WHEN STOCKPILING LARGE-SIZED AGGREGATES FROM ELEVATED CONVEYORS, BREAKAGE IS MINIMIZED BY USE OF A ROCK LADDER.

FINISHED AGGREGATE STORAGE

NOTE: IF EXCESSIVE FINES CANNOT BE AVOIDED IN COARSE AGGREGATE FRACTIONS BY STOCKPILING METHODS USED, FINISH SCREENING PRIOR TO TRANSFER TO BATCH PLANT WILL BE REQUIRED.

Fig. 4.3—Correct and incorrect methods of handling and storing aggregates.

Fig. 4.4—Concrete mixing plant and aggregate storage. Bins contain different size aggregates.

- Build up piles of coarse aggregate in layers so that aggregate does not run down the slopes at the edges of the pile, causing segregation;
- Avoid excessive handling, which can cause segregation and degradation;
- Allow ample space between piles or use partitions to separate adjacent piles;
- Do not mix or store aggregates from different sources in the same pile;
- Use each pile of aggregate until it is gone. Separate batching produces more consistent, reliable results than attempts to batch materials blended with clam buckets or bulldozers;
- Do not allow aggregate to freefall from drop heights so great that breakage can occur or the wind can separate fine and coarse materials. If necessary, use baffles or rock ladders to break the fall and prevent excessive segregation and breakage; and
- Reduce segregation of fine aggregate dry enough to be free flowing by dampening.

4.2.1.1 *Sand and lightweight aggregates*—The requirements for stockpiling of lightweight aggregate are the same as for normalweight aggregate, except that more care should be exercised to prevent crushing of the aggregate. Because lightweight aggregate can be highly absorbent, it may be necessary to continually soak stockpiles for several days before batching to prevent slump loss of concrete.

Washed sand should be allowed to drain as long as necessary to reach a uniform moisture content in accordance with ACI 304R. Because of sequences in stockpiling operations, large sand stockpiles may have widely varying moisture content, leading to difficulty in maintaining uniform concrete slump and water-cement ratio.

4.2.2 *Storage in bins*—Aggregates of different sizes should be stored in separate bins. When each bin is filled, the aggregate should be dropped vertically into the middle to avoid segregation. If segregation occurs, a baffle, splitter, or rock ladder should be used and the bin kept as full as possible at all times to minimize the tendency for the aggregate to segregate. Bin bottoms should be shaped to facilitate uniform discharge.

4.2.3 *Finish screening*—Finish screening of aggregate is a process that removes excess fines and adjusts proportions of aggregate sizes to conform to the specified grading. Spray bars are sometimes used in the finish-screening process to enhance removal of fines. When finish screens are used in batch plants, many of the concerns with accurate primary screening, handling, and stockpiling of aggregate can be eliminated.

With finish screening, roughly proportionate amounts of each size of aggregate should be fed to the screens rather than a single size at a time.

4.2.4 *Transporting*—Vehicles used to transport aggregate should be clean and free of all contaminants, and the vehicles should be inspected for tightness. Aggregate compartments should be inspected to verify no leakage or holes.

4.3—Supplementary cementitious materials

SCMs should be handled and stored in essentially the same manner as cement. Although many SCMs require tighter storage facilities to prevent leakage, most are generally not as susceptible to deterioration from moisture as is portland cement.

4.4—Admixtures

In practice, most chemical admixtures are delivered in liquid form. They should be protected from freezing. If frozen, proper reblending or remixing of the admixture should be done before the admixture is used, and the manufacturer's recommendations should be followed.

Liquid admixtures should be stored in watertight drums or tanks and protected from freezing. Detailed recommendations for storage and handling are given in ACI 212.3R.

CHAPTER 5—FUNDAMENTALS OF CONCRETE

Quality concrete is a mixture of cement (or cementitious materials), water, aggregates, and, in some cases, chemical admixtures that are designed to perform under expected service conditions (Fig. 5.1). Most quality concrete possesses the same principal requirements.

For freshly mixed concrete, those requirements are:
- *Consistency*—The ability to flow;
- *Uniformity*—An homogenous mixture, with evenly dispersed constituents;
- *Workability*—Ease of placing, consolidating, and finishing; and
- *Finishability*—Ease of performing finishing operations to achieve the specified surface characteristics.

For hardened concrete, they are:
- *Strength*—Resists strain or rupture induced by external forces (compressive, flexural, tensile, torsion, and shear);
- *Durability*—Resists weathering, chemical attack, abrasion, and other service conditions;
- *Appearance*—Meets the desired aesthetic characteristics; and
- *Economy*—Performs as intended within a given budget.

By understanding the nature and basic characteristics of concrete, inspectors are better able to see that these requirements are met. This chapter discusses some of the fundamentals of concrete that affect inspection work. For more details on

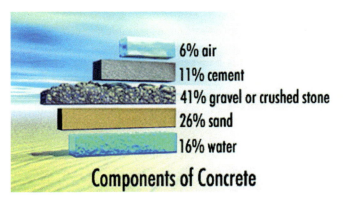

Fig. 5.1—Volumes of components of typical concrete mixture containing an air-entraining agent.

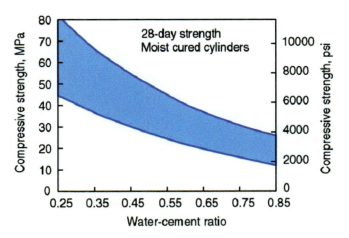

Fig. 5.2—Range of typical strength to water-cement ratio (w/c) relationships of portland-cement concrete on over 100 different concrete mixtures cast between 1985 and 1999. Concrete of lower w/c is stronger and nearly watertight (Portland Cement Association 2002).

principles and practices that affect concrete quality, refer to the latest editions of ACI 224.1R, ASTM STP 169-C (ASTM International 1994), *Concrete Construction Handbook* (Dobrowolski 1998), *Design and Control of Concrete Mixtures* (Portland Cement Association 2002), *Manual of Standard Practice* (Concrete Reinforcing Steel Institute 1997b) and *Standard Practice for Concrete for Civil Work Structures* (U.S. Army Corps of Engineers 1994).

5.1—Nature of concrete

5.1.1 *Components*—Concrete has two major components: paste (or binder) and aggregate, with the individual particles of aggregate typically being embedded in and separated by the paste. The paste is a mixture of cement, air, and water, and the aggregate consists of both fine and coarse materials. The total volume of the mixture is equal to the volume of the cement, water, and aggregate plus the volume of air. In many cases, the concrete will also contain SCMs and chemical admixtures.

Air is introduced into concrete during the mixing process. All concrete, no matter how thoroughly consolidated, contains some entrapped air in the form of scattered voids the size of large grains of sand (40 mil [1 mm] or greater). Entrapped air varies with the size of the coarse aggregate. The amount of entrapped air is generally 1.5 to 2% by volume of concrete with 3/4 in. (19.0 mm) aggregate. The volume of entrapped air will increase as the coarse aggregate size decreases, and it will decrease as the size of the coarse aggregate increases.

In addition to entrapped air, concrete may contain entrained air introduced intentionally by use of air-entraining admixtures. The volume of entrained air will remain constant within a range of 1 to 3% regardless of the size of the coarse aggregate. During mixing of the concrete, air bubbles are stabilized and form many spherical voids within the cement paste approximately the size of larger cement grains and finer sands (less than 40 mil [1 mm]). Entrained air is essential to the freezing-and-thawing resistance of concrete. (Refer to the discussion in this chapter on freezing-and-thawing effects.) In addition to improving freezing-and-thawing resistance, entrained air increases the workability and cohesiveness of concrete and retards segregation and bleeding.

5.1.2 *Water-cementitious material ratio*—The quality of the cement paste contributes most to the compressive strength of concrete. The strength and density of the paste depend primarily on the water-cementitious material ratio (w/cm). In determining this ratio, cementitious material is calculated as the combined mass of the cement, fly ash, ground slag, silica fume, and other pozzolans used in the mixture. Higher water contents dilute the paste and weaken the binder (Fig. 5.2). There are several advantages to maintaining a low w/cm in concrete, including increased compressive, flexural, and tensile strengths; lower permeability; increased resistance to weathering; less volume change; and reduced shrinkage cracking. Refer to Chapter 6 for a discussion on proportioning of mixtures for a specified w/cm.

5.2—Fresh concrete

5.2.1 *Workability*—Fresh concrete needs to be workable during placing and finishing operations. It is important that concrete remains plastic (fluid) throughout the placement without causing segregation or excessive slump loss. Workability depends on the characteristics and quantity of cementitious materials used, w/cm, gradation, shape and surface texture of aggregates, amount of entrained air, and use of chemical admixtures and fibers. The level of workability needed is affected by field conditions, including temperature, haul time, method of placement (conveyor, bucket, or pump), form configuration, and degree of consolidation.

The consistency of concrete is closely related to workability, and is usually measured by the slump test (refer to the discussion in Chapter 7). A high-slump concrete has a wetter consistency than a low-slump concrete.

If the consistency of the concrete is too low, the mixture may appear dry and harsh and be difficult to place and consolidate, resulting in the development of honeycomb (unconsolidated) concrete conditions after hardening. Conversely, a high consistency may lead to segregation, lack of uniformity, lower strengths, and increased permeability of the concrete (unless the high consistency is due to proper use of a high-range water-reducing agent).

Fig. 5.3—Trapped bleed water can cause surface delamination of concrete.

In general, maximum quality for a given concrete mixture will be achieved by placing concrete at the lowest consistency needed to ensure complete consolidation and filling of the form.

5.2.1.1 *Bleeding and settlement*—In newly placed concrete, the solids (cementitious and aggregate) slowly settle, bringing a layer of water, called bleedwater, to the surface. As a result of settlement, the solids at the bottom of the concrete member become more closely packed (Fig. 5.3), and the space occupied by the hardened concrete is slightly less than when it was freshly placed.

Materials selection will affect the rate and capacity of bleeding in concrete. Chemical admixtures, such as water reducers, may increase or decrease bleeding, depending on the type used. The presence of entrained air provides a more cohesive mixture and alters the settlement process, usually reducing the rate and amount of bleeding. The use of supplementary cementing materials with particle sizes finer than portland cement, such as silica fume, will significantly reduce bleeding.

Bleeding is a normal characteristic of fresh concrete, and will not harm concrete that is properly placed and finished. During hot, dry weather, bleeding may help prevent excessive surface drying and plastic shrinkage cracking. A high amount of bleedwater, however, can delay finishing operations because floating and troweling should never be performed while bleedwater is on the surface. Under severe drying conditions, flatwork surfaces may sufficiently stiffen and subsequently be finished before all bleedwater rises to the surface. This bleedwater will be trapped beneath the hard-troweled finish, resulting in blisters or thin cavities and subsequent scaling of the surface. The use of air-entrained concrete in steel-troweled flatwork also increases the potential for late rising bleedwater being trapped beneath the finished surface, and should be avoided.

The same problem can occur in cold weather when concrete is placed on a cold subgrade or metal deck while the finished surface is exposed to much warmer air temperatures from space heaters in enclosed areas. When placing flatwork under conditions of rapid evaporation and drying, workers can minimize these effects by using windbreaks, shade covers, fogging, and evaporation retardants. Finishing should also be delayed as long as possible.

5.2.2 *Consolidation*—Consolidation benefits concrete by reducing the voids between particles, removing unwanted entrapped air, improving bond with reinforcement and other embedded items, and reducing the amount of paste necessary to provide a workable mixture. To ensure uniformity in concrete, consolidation is performed through tamping, rodding, or mechanical vibration. Most consolidation is done using power screeds or mechanical vibrators (external or internal). Equipment should be selected based on concrete mixture characteristics, type of construction, placement method, member size, and reinforcement and formwork conditions. It is important to make sure consolidation is performed correctly. Vibrators should not be held in one place for too long or used to move plastic concrete. Improper consolidation can lead to segregation and loss of entrained air in concrete. Refer to Chapter 9 for more information on proper vibration techniques.

5.2.3 *Hydration, setting, and hardening*—Hydration is a chemical reaction between cement and water that forms new compounds in concrete. Portland cement consists primarily of four compounds: tricalcium silicate (C_3S), dicalcium silicate (C_2S), tricalcium aluminate (C_3A), and tetracalcium aluminoferrite (C_4AF). C_3S and C_2S are responsible for early strength gain and late strength development in concrete, respectively. C_3A contributes to the heat of hydration, and C_4AF primarily governs the concrete's color. All types of portland cement contain the same four compounds, but in varying proportions. During hydration, the calcium silicates, which make up approximately 75% of the weight of portland cement, react with water, forming two new compounds: calcium hydroxide (CH) and calcium silicate hydrate (C-S-H). The principal reaction product of the hydration process is C-S-H, a gel-like substance responsible for the setting, hardening, and strength properties of the concrete. Hydration of cement continues at a diminishing rate as long as sufficient moisture and adequate temperatures are available. If the paste is not kept moist by curing, cement hydration ceases when the evaporable water escapes from the paste.

The rate of hydration and time of set is also influenced by the *w/cm*, cement fineness, use of SCMs and chemical admixtures, and temperature of materials.

5.2.4 *Heat of hydration*—Hydration is accompanied by liberation of heat. Some of the heat generated escapes through the concrete surface, but part of the heat is retained and raises the concrete temperature. Excessive temperature rise can reduce ultimate concrete strength and produce undesirable stresses that cause cracking as the temperature drops. This is particularly true where temperature differentials exist between parts of the concrete mass and where members are partially restrained.

In mass concrete, the heat escapes slowly, and the temperature rise will be somewhat similar to that shown in Fig. 5.4. Temperature rises in mass concrete can be controlled by cooling materials and handling equipment, reducing cement content, and using supplementary cementing materials and chemical retarding admixtures. Other measures include

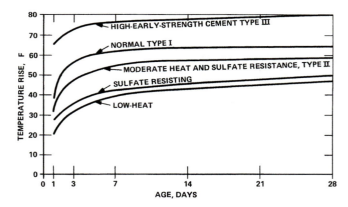

Fig. 5.4—Temperature rise of concrete for various types of cement when no heat is lost (376 lb cement per yd³).

scheduling placements during cooler periods of the day, placing in lifts, and using embedded pipes with circulating cool water. Refer to Chapter 16 for information on proportioning mass concrete mixtures.

5.3—Hardened concrete

5.3.1 *Curing*—Curing is essential to the strength development and durability of concrete. Continued cement hydration is dependent on the availability of moisture in concrete, a favorable ambient temperature, and sufficient space for hydration products to form. If the concrete's relative humidity drops to approximately 80% or its temperature falls below freezing, hydration and strength gain will cease. Figure 5.5 shows the relationship between strength gain and moist curing. Adding moisture to dry concrete will resume cement hydration to some degree and continue strength gain. To maximize concrete quality, however, curing should be continuous for the period specified.

Curing is performed by either sealing in the original water in the concrete mixture or by supplying additional water. Chapter 10 provides more detailed information on curing and the protection of concrete during normal, hot, and cold weather conditions.

5.3.2 *Strength*—Concrete is very strong in compression, but relatively weak in tension. Concrete will crack when loads or stresses due to restraint exceed its tensile strength. Structural reinforcement is used in concrete to transfer the tensile and shear stresses from the concrete to the steel, allowing the concrete to withstand higher levels of stress.

Concrete is often specified to achieve a 28-day compressive strength f_c' expressed in pounds per square inch (psi) or megapascals (MPa). The amount of strength gain depends on cementitious properties (C_3S and C_2S content), w/cm, hydration, and age of the concrete. Under standard laboratory conditions, approximately 50 to 70% of the ultimate compressive strength of concrete (made with Type I cement) will be reached during the first week, and approximately 95% during the first month.

5.3.3 *Durability*—Although strength is an important characteristic of concrete, other qualities can be even more important for concrete exposed to harsh environments. In general, the principal causes of deterioration in concrete are

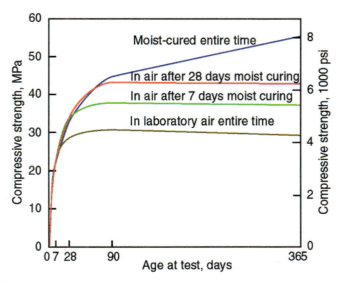

Fig. 5.5—Concrete strength increases with age as long as moisture and favorable temperature are present for hydration of cement (Portland Cement Association 2002).

corrosion of reinforcing steel, exposure to freezing-and-thawing cycles, alkali-silica reaction (ASR), and sulfate attack. All of these problems start with exposure to moisture. Thus, increasing watertightness and reducing the permeability of concrete are the keys to durability. Permeability, or the amount of moisture migration (gas, vapor, and liquid forms) in concrete, is a function of the w/cm, density of the paste, and aggregate gradation.

5.3.4 *Chemical attack*—Concrete can deteriorate when exposed to aggressive groundwater, acidic rain or condensation, seawater or salt spray, raw sewage, and acids or caustics. The effects vary, but usually include staining, erosion, deterioration of hydration products, cracking, and spalling. Chemicals can also aggressively attack reinforcing steel and other embedded metallic elements.

5.3.4.1 *Sulfates*—Sulfate attack is one of the most common forms of chemical attack. Sulfates of sodium, potassium, and magnesium, present in alkaline soils and water (rain, groundwater, or seawater), can cause expansion of concrete by reacting with the hydration products in the cement paste. Sodium or potassium sulfates attack calcium hydroxide and form gypsum, which then reacts with calcium aluminate hydrate to form ettringite. Gypsum, however, has a limited solubility, and the rate of attack is relatively slow. Magnesium sulfates can cause more severe deterioration. In addition to reacting with calcium hydroxide and hydrated calcium aluminates, magnesium sulfate will also attack and decompose calcium silicates. Other forms of sulfate attack include salt crystallization and thaumasite. Sulfate salt crystallization causes severe pressure in concrete pores. Thaumasite is very similar to ettringite. The formation of thaumasite eventually results in a softening within the cement matrix, causing disintegration of the concrete. Thaumasite forms in concrete as a result of a reaction between calcium carbonate and calcium silicates with sulfates (van Aardt and Visser 1975). The resistance of concrete to sulfate attack can be improved by using a sulfate-resisting portland cement with a low C_3A content, such as a Type II cement with moderate resistance, or a Type V cement with high resistance. Resistance to all forms of chemical attack can be improved by reducing the *w/cm* (0.40 or less).

5.3.4.2 *Corrosion*—Ideally, concrete will protect embedded reinforcement because of its highly alkaline nature. In a high-pH environment, a dense layer of iron oxide will form on the reinforcement, protecting it from corrosion. If chloride ions are present, however, they can penetrate this protective film. Chloride attack may occur from internal sources, such as contaminated mixing water, aggregates, or the use of a calcium chloride accelerator, or from external sources, such as exposure to seawater or deicing salts. If the protective film is disrupted, the steel will corrode if given sufficient moisture and oxygen. The corrosion product formed is rust, which expands to take up approximately four times its original volume. This increase in volume causes internal stresses to build within the concrete, resulting in cracking and spalling.

To protect from chloride ingress and corrosion, concrete should be kept watertight and have low permeability. Resistance to corrosion can be increased by using a low *w/cm* (0.40 or less), using SCMs and corrosion-inhibiting admixtures, providing adequate curing, and increasing the concrete cover (the distance between the exposed concrete surface and the reinforcing steel). Other methods of protection include surface treatments, coating the reinforcement, corrosion-resistant or stainless steel reinforcement, sacrificial anodes, and cathodic protection.

5.3.5 *Freezing-and-thawing effects*—When concrete is subjected to repeated freezing-and-thawing cycles without protection, damage may result in the form of cracking, scaling, or crumbling. Water expands approximately 9% upon freezing. When concrete is saturated above its critical level (91.7%) and the temperature drops below freezing, the expansion of water produces osmotic and hydraulic pressures that exceed the tensile strength of the paste or aggregates. Failure in the paste occurs when the pressure exceeds the tensile capacity of the capillaries and pores in the paste, causing the cavities to expand and rupture. If porous, low-density aggregates are used in the concrete, water can accumulate in the stone, and resulting pressures from freezing and expansion will burst the aggregate and surrounding paste. If the deleterious aggregate is located near the surface of the concrete, popouts can result.

The use of air-entraining admixtures greatly improves the resistance of concrete to damage from freezing-and-thawing cycles. Entrained air protects concrete by providing pressure-relief chambers comprised of many small, closely spaced air voids. For adequate durability, the following air-system characteristics are recommended:
- Sufficient volume (Table 5.1);
- A specific surface area of 600 in.^{2}/in.^{3} (24 mm^2/mm^3) or greater; and
- A calculated spacing factor between air voids of less than 8 mil (0.2 mm).

Table 5.1—Recommended average air content, %

Exposure	Nominal maximum aggregate size					
	3/8 in. (9.5 mm)	1/2 in. (12.5 mm)	3/4 in. (19 mm)	1 in. (25 mm)	1.5 in. (37.5 mm)	2 in. (50 mm)
Mild	4.5	4.0	3.5	3.0	2.5	2.0
Moderate	6.0	5.5	5.0	4.5	4.5	4.0
Severe	7.5	7.0	6.0	6.0	5.5	5.0

Current field practices measure only the volume of air in fresh concrete, and do not include air-void size or spacing. Spacing and air-void size can be determined in fresh concrete by an air-void analyzer test, and in hardened concrete from petrographic analysis (ASTM C457). The recommended air contents for frost-resistant concrete contained in ACI 201.2R provide about 9% of air in the mortar fraction for severe exposure and about 7% of air for moderate exposure.

The amount of air required in concrete for each level of exposure and nominal maximum aggregate size is shown below. ACI 301 and 318 allow a 1% reduction from Table 5.1 values for concrete with strength exceeding 5000 psi (35 MPa). Refer to Chapters 6 and 7 for a discussion of other aspects of air entrainment.

5.3.6 *Alkali-aggregate reactivity*—Expansive reactions can occur in concrete between highly alkaline environments and reactive mineral aggregates. The two most common are alkali-carbonate reaction (ACR) and alkali-silica reaction (ASR).

ACR is related to a de-dolomitisation, or the breakdown of dolostones and dolomitic limestones, and the associated expansion of the coarse-aggregate particles.

ACR is a serious, but rare, variety of alkali-aggregate reactivity. ASR is associated with the dissolution of silica (SiO_2) in the aggregate, which reacts with the alkali hydroxides in the pore solution released during cement hydration. The reaction product of ASR is the formation of an alkali-silica gel in the aggregate and surrounding cement paste. This gel has the capacity to swell from absorbing additional moisture, causing cracking, spalling, and displacement in hardened concrete. For ASR to occur, there must be reactive silica in the aggregate, a sufficient amount of alkalis in the concrete pore solution, and sufficient moisture. To control ASR, nonreactive aggregates should be used when possible, and the total alkali content of the concrete should be limited. Use of SCMs (pozzolans) in quantities proven to control ASR can allow the use of locally available aggregates deemed to be potentially reactive.

5.3.7 *Volume changes*—Concrete volume changes over time due to changes in moisture, temperature levels, and stress. The volume of concrete, when initially hardened, is the largest it will ever be during the life of the concrete. Following initial installation and curing, concrete that is allowed to dry will generally lose volume until an equilibrium state of moisture is reached. Most of the volume change that develops in concrete is due to drying shrinkage of the paste, which occurs as excess water not needed for hydration evaporates. To a lesser extent, concrete will also lose volume due to autogeneous shrinkage of the paste as cement hydration progresses. Many factors influence the degree of drying

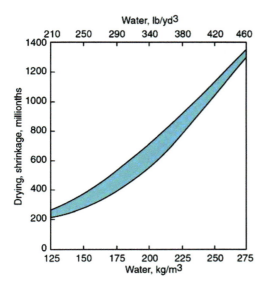

Fig. 5.6—Relationships between total water content and drying shrinkage of concrete. A large number of mixtures with various proportions are represented within the shaded area of the curves. Drying shrinkage increases with increasing water contents (Portland Cement Association 2002).

shrinkage that develops, including w/cm, type of cementitious products employed, choice of aggregates, method of curing, temperature, and time. Generally, keeping the water content as low as possible is the largest factor in limiting the shrinkage that will occur for a given concrete mixture. Typically, concrete mixtures will shrink approximately 2/3 in. per 100 ft (60 mm per 100 m), or approximately 0.06% (ACI 224.1R). Figure 5.6 plots correlations between total water content, cement content, and projected shrinkage of hardened concrete for different mixtures made with the same materials. Aggregate quality, source, or both can have major influence on a given mixture's resistance to shrinkage of the cement paste.

When concrete is restrained by a subgrade, reinforcement, or connections with other members, it may crack as a result of drying shrinkage stress or the combined effects of thermal and drying stresses. If the surface of the hardened concrete dries more rapidly than the interior, differential stresses develop that may cause a network of shallow plastic shrinkage cracks to form on the surface. Unequal drying of opposite surfaces of slabs can cause curling, most noticeably at edges and corners.

When concrete is subjected to external loads, it will deform elastically. If the load is sustained over time, the stress will cause creep (permanent deformation) to develop. The amount of creep that develops depends on the ratio of sustained stress to concrete strength, the humidity of the environment, the dimensions of the element, and the composition of the concrete.

Volume changes in concrete due to thermal stresses can also have a great impact on a structure's behavior. Concrete expands when heated and contracts when cooled. Restraint of this thermal contraction can cause cracking. A temperature change of 100 °F (56 °C) will induce a volume change of approximately 0.06% when calculated from a general coefficient of thermal expansion of 0.0006 per 100 °F (56 °C) (similar to the change noted previously for concrete as it dries from the saturated to dry state).

Because volume change, stress, and cracking are inevitable, a number of practices and techniques aimed at minimizing and controlling concrete cracking have evolved. In reinforced concrete construction, the reinforcement employed for load resistance will generally control the development of shrinkage or thermal-induced cracks. In unreinforced concrete or concrete with light reinforcement, such as slabs, joints of proper spacing and depth should be formed or saw-cut as soon as possible after the concrete sets to prevent excessive amounts of random cracks from developing. The use of light reinforcement (mesh) helps to keep cracks that do form tightly closed. In members requiring no structural reinforcement, light temperature steel is often employed simply to control thermal cracking by producing many small, insignificant cracks instead of a few large, objectionable cracks.

CHAPTER 6—PROPORTIONING AND CONTROL OF CONCRETE MIXTURES

The required properties of concrete are governed by its use, and these properties should be reflected in the contract documents for the job. To ensure that the desired properties are achieved, it is necessary for inspectors to see that the appropriate mixture proportions are used (Fig. 6.1).

This chapter presents the factors to consider and general principles to follow when proportioning concrete mixtures. Detailed procedures for mixture proportioning are covered in ACI 211.1, 211.2, and 211.3R. Proportioning of concrete mixtures containing one or more admixtures is discussed in ACI 212.4R. Chapter 16 of this manual details proportioning of special types of concrete, including lightweight, high-density, mass, and shrinkage-compensating.

6.1—Factors to consider

The major factors to consider when proportioning concrete mixtures include:
- Durability and strength (compressive or flexural);
- Placeability;
- Appearance; and
- Setting and strength-gain characteristics.

Durability and strength, as required in the contract documents, dictate the w/cm for given materials. Durability requirements might include specific requirements for high strength, high density, abrasion resistance, or low permeability.

Placeability (the workability and consistency of the concrete) is primarily dictated by the proposed concrete placing methods (buckets, chutes, buggies, pumps, or conveyors), the placing conditions (form depth and dimensions, amount of reinforcement, and accessibility), and the consolidation method (spading or internal or external vibration).

Appearance is an important consideration for architectural concrete. Characteristics to examine include the color and type of the coarse aggregate, color and type of the cement, and proportion of fine-to-coarse aggregate (particularly in the case of exposed-aggregate concrete).

Setting time and strength gain can be controlled by using various combinations of portland cement, blended cements,

Fig. 6.1—A concrete inspector evaluates mixture proportions to ensure that they meet design requirements.

chemical admixtures, and pozzolans. Many of the properties of Types III, IV, and V portland cements can be duplicated by the addition of appropriate SCMs, and admixtures to Types I and II portland cements. Various types of chemical admixtures have been developed to control a wide range of concrete properties, including air content, water reduction, set acceleration and retardation, corrosion resistance, and paste rheology.

6.2—Methods of specifying concrete proportions

Generally, contract documents require that the concrete develops a certain strength at a given age within a limiting range of consistency (slump). A maximum *w/cm* or minimum cement content is often specified as an additional precaution to ensure the necessary durability, impermeability, or workability.

6.2.1 *Strength specifications*—In contract documents, the specified compressive strength usually is designated as a minimum strength. Even when the required concrete mixture proportions are used, however, an occasional strength test result falling below this minimum can occur. This may lead to confusion and arguments that result in unnecessary modifications to the mixture proportions or requirements for additional testing.

When evaluating the results of concrete strength tests, consider that concrete production is susceptible to variations in ingredients, batching, and testing. Codes for structural design permit occasional strength test results as much as 500 psi (35 MPa) below the specified (design) strength (up to 10% below for concrete designs greater than 5000 psi [35 MPa]). Chapter 2 of this manual and ACI 318 and 301 discuss the issue of occasional low strength tests.

Because of variability in concrete, as well as variability in test specimens and test results, it is necessary to proportion concrete for an average strength far enough above the specified strength f_c' that all but a small proportion of the concrete will equal or exceed f_c' (ACI 214R). The amount of overdesign needed for proposed mixtures should be statistically based on the previous record of concrete production or the minimum required by code. Some contract documents require an overdesign of up to 1500 psi (10 MPa) when data from previous records indicate extreme variability in production or when no previous data are available. In the absence of such contract requirements, the mixture proportions should be prepared as discussed in Chapter 2. After collecting a history of field strengths for the established mixture proportions and attaining proper control of field operations, it may be possible to adjust mixture proportions.

For pavement concrete, the same approach is used except that the flexural strength, rather than compressive strength, of standard test beams may be required.

6.2.2 *Prescriptive specifications*—A specification may prescribe the proportions of materials by one or a combination of the following four methods:

1. A fixed or a minimum amount of cement (and admixture, if specified) per volume of concrete;
2. A fixed or a maximum *w/cm*;
3. Fixed proportions of cementitious materials, fine aggregate, coarse aggregate, and admixtures; and
4. Limits on the proportion of fine-to-total aggregate (for example, 35 to 50%).

Methods 1 and 2 are often combined, and are more typical than Methods 3 and 4. The slump is also typically specified. For air-entrained concrete, it is common to specify the range of total air in the concrete mixture. If Methods 1, 2, and 4 are combined, specified values, including strength, should be consistent.

If the specified slump range cannot be obtained with the minimum cementitious materials content and maximum *w/cm* specified, making the following adjustments can remedy the situation:

- Increase the cementitious materials content;
- Change the aggregate types (from angular to rounded, which will reduce the water requirement);
- Change the relative proportions of fine-to-coarse aggregate; and
- Use a water-reducing admixture.

Note that if building codes or project specifications require compliance with ACI 318, then concrete proportions should also satisfy all ACI 318 requirements, such as those for durability given in Chapter 4 of ACI 318.

6.3—Proportioning for specified strength or *w/cm*

When proportioning to determine the quantity of cement per unit volume of concrete that will produce hardened concrete of the specified strength and durability (assuming proper curing), the quantity of cement needed depends on:

- Cement type and quality;
- Quantity and quality of SCMs;
- Maximum *w/cm* or water-cement ratio (*w/c*);
- Consistency of mixture;
- Use of admixtures, alone or in combination;
- Maximum aggregate size and grading;
- Other characteristics of the aggregate, such as particle shape and surface texture; and

- Rate of strength gain desired.

6.3.1 *Cement types*—Contract documents typically state the type of cement to be used. For ordinary purposes, a normal portland cement (ASTM C150, Type I or ASTM C1157, Type GU) or a blended cement (ASTM C595 or C1157) may be required. Special situations may call for the use of moderate-heat (Type II), high-early-strength (Type III), or high-sulfate-resisting (Type V) cements. When the choice of cement is other than Type I or III, considerations other than strength often dictate the type to use. Nevertheless, a minimum concrete strength at a specified age typically is required. (Refer to the discussion on cement in Chapter 3.)

6.4—Concrete with supplementary cementitious materials

To properly assess the advantages and disadvantages of using SCMs (refer to discussion on SCMs in Chapter 3), laboratory studies of proposed concrete mixtures using actual job materials should be conducted. SCMs generally reduce the rate of early strength development, so concrete strengths at early ages may be less than those of comparable mixtures containing only portland cement at the same water content. Strengths at later ages, however, will be nearly equal or higher. Class N pozzolans may increase mixing water requirements, and thus, at least theoretically, drying shrinkage, but this is not evident in field structures. Class F (fly ash) pozzolan, on the other hand, usually reduces mixing water requirements.

SCMs have proven to be effective in reducing ASR and sulfate attack in concrete. The actual effects of SCMs intended for use can be checked against control mixtures made using 100% portland cement. Detailed information on the use of SCMs in concrete is given in *Concrete Manual* (U.S. Bureau of Reclamation 1988), *Standard Practice for Concrete for Civil Work Structures* (U.S. Army Corps of Engineers 1994), and *Fly Ash, Silica Fume, Slag, and Natural Pozzolans in Concrete* (Malhotra 1992).

6.4.1 *Mixture proportioning and control*—Because SCMs are often used to produce concrete with unique or enhanced properties, it is desirable to test resulting mixtures for these properties before using them in structures. The recommendations of ACI 211.1 for trial mixture proportioning should be followed using the established specific gravity of each material in calculations.

Generally, the dry mass of portland cement, pozzolans, and other cementitious materials are combined to determine the total cementitious content of a mixture and, subsequently, the *w/cm*. Because portland cement and fly ash may be similar in color, batching and storage equipment should be carefully identified to differentiate between these materials. Silica fume is sometimes combined with water in a slurry solution to facilitate handling and batching during mixing operations. The amount of water added to the batch as part of the slurry solution should be determined and properly accounted for in calculating the *w/cm*.

6.4.2 *Water-cementitious material ratio*—If aggregate of approximately the same grading and maximum size and from the same source is used—and other materials remain the same—the potential compressive strength of concrete is nearly constant when the *w/cm* is held constant. If, however, different aggregates and different gradings (particularly the maximum aggregate size) are used, the necessary *w/cm* for a desired slump and strength will vary somewhat, and it may change substantially when admixtures are used. Nevertheless, average relationships of *w/cm* and strength usually are sufficient for proportioning a trial or starting mixture.

The necessary *w/cm* should be estimated from a published table of *w/cm* and corresponding strength values. Table 6.1 (adapted from ACI 211.1) gives approximate strengths for different *w/cm* for air-entrained and non-air-entrained concrete. Strengths obtained from these *w/cm* will generally be conservative. If there is a history of usage of given materials, a review of available test information can provide useful data, particularly as to the water requirement and strength level of the concrete.

6.4.3 *Aggregate selection*—After determining the *w/cm*, the next step is to find the aggregate proportions that will provide a workable mixture with a minimum amount of paste.

The nominal maximum aggregate size is usually limited by placing conditions, dimensions of structural members, and spacing of reinforcing steel. Well-graded and well-shaped aggregate of large maximum size has less total volume of voids than smaller aggregate. Therefore, concretes made with larger aggregates that are properly graded and shaped require less mortar and, thus, less water, per unit volume (Fig. 6.2). Generally, for average strength requirements, the largest maximum aggregate size that is most economical and consistent with the dimensions of the structural elements and placing conditions should be used. When high-strength concrete (8000 psi [55 MPa] or greater) is desired, best results are usually obtained with smaller maximum aggregate sizes of 3/4 or 1/2 in. (19.0 or 12.5 mm).

When conducting mixture proportioning studies and concrete tests to evaluate materials, the inspector should assess the properties of mixtures having various blends of fine and coarse aggregate. The most favorable combination of fine and coarse aggregates should be selected that provides the required concrete properties such as workability and strength. To enhance concrete finishability and pumpability, the amount of pea gravel or flat and elongated small coarse aggregate should be minimized. Also, the percentage

Table 6.1—Relationships between *w/cm* and compressive strength of concrete

Compressive strength at 28 days,[*] psi (MPa)	*w/cm*, by weight	
	Non-air-entrained concrete	Air-entrained concrete
6000 (40)	0.41	—
5000 (35)	0.48	0.40
4000 (28)	0.57	0.48
3000 (21)	0.68	0.59
2000 (14)	0.82	0.74

[*]Values are estimated average strength for concrete containing not more than percentage of air shown in Table 6.3.4(a) of ACI 211.1. For constant *w/cm*, the strength of concrete decreases as air content increases.
Note: Relationship assumes maximum aggregate size of approximately 3/4 to 1 in. (19.0 to 25.0 mm); for a given source, strength produced for given *w/cm* will increase (as will cement content) as maximum aggregate size decreases. Refer to Sections 3.5 and 6.3.2 of ACI 211.1.

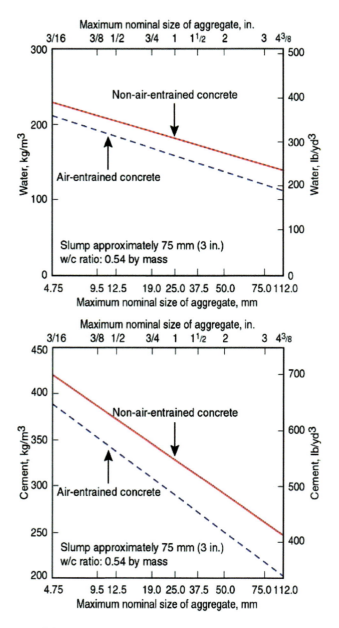

Fig. 6.2—Cement and water contents in relation to maximum size of aggregate for air-entrained and non-air-entrained concrete. Less cement and water are required in mixtures having large, coarse aggregate (Portland Cement Association 2002).

of sand should be kept as low as is practical while still providing the needed fine sizes for good workability and minimal bleeding. Low sand content usually minimizes the water requirement of the concrete and gives the most economical proportions, provided that gap-grading does not occur due to lack of intermediate sizes. Strictly minimizing the proportion of fine aggregate, however, is not always advisable, particularly if using more fine aggregate makes the concrete noticeably easier to place, consolidate, and finish. Low sand contents are desirable in low-slump concrete to maximize strength, but higher sand contents are needed to minimize segregation in higher-slump concretes.

For each percentage point increase in the sand-aggregate ratio, the water requirement increases by approximately 1%.

Other factors being the same, aggregates composed of angular particles require more paste than equal proportions of smooth, rounded particles. When concrete strength is the criterion, however, rounded aggregates are not always preferable because crushed aggregates normally give higher strengths at a given *w/cm*. When durability requirements do not govern the *w/cm*, it is sometimes permissible, on the basis of test data, to use a higher *w/cm* with crushed material than would be required for rounded material. Thus, with a different aggregate, a different *w/cm* may sometimes be used to achieve the same strength. For pavement concrete designed on the basis of the modulus of rupture (flexural strength), angular aggregates usually are preferable.

Whenever possible, the concrete mixture should be based on trial mixtures using job materials. All mixtures should be workable. The minimum properties generally determined from tests of trial batches are strength (flexural or compressive), slump, percent air, and density. When these tests cannot be made, use recommendations based on experience, such as those of ACI 211.1 and 211.3R as well as *Design and Control of Concrete Mixtures* (Portland Cement Association 2002). Although these recommendations do not take into account differences in strength that may arise from differences in aggregate or cement characteristics, they are conservative enough to provide sound results. The initial mixture proportions should be adjusted as necessary based on test results and observations during production.

6.4.4 *Air entrainment*—For moderate-strength concrete of a given *w/cm*, air entrainment will reduce strength about 2 to 6% for each percentage point of added air (Klieger 1952). Less mixing water, however, will be required to provide the same slump because the small air bubbles help to lubricate the mixture. Thus, if slump and cement content remain constant, a lower *w/cm* results, which will partly compensate for the strength loss.

Coarse-aggregate proportions are usually the same for both air-entrained and non-air-entrained mixtures, but less fine aggregate is needed because the air increases the volume of mortar and improves workability. With some lean (low-cement-content) mixtures, air entrainment may actually increase the compressive strength by reducing the *w/cm*.

6.4.5 *Quantity of paste*—For given materials, the optimum mixture proportions will use the least amount of total water per unit volume of concrete to obtain the required slump and workability. With a fixed *w/cm*, using mixtures having the least paste reduces material costs. Typically, the cementitious material in the paste is the most costly ingredient of the concrete, so using more paste than required unnecessarily adds to the cost of the concrete. When using expensive aggregates, however, the cost of the paste should be balanced against the cost of the aggregate.

Minimizing the paste is also desirable because water in the paste is the primary cause of shrinkage as the concrete hardens and dries. The more water (and the more paste), the greater the drying shrinkage. Also, cement produces heat as it hydrates. Therefore, high cement contents can produce an undesirable temperature rise and crack-producing temperature differentials (Fig. 6.3).

The quantity of paste required in a unit volume of concrete depends on:
- The *w/cm* of the paste;
- Consistency of the fresh concrete;
- Grading of the aggregate (including chemical content in some cases);
- Shape and surface texture of the aggregate particles;
- Amount of entrained air;
- Chemical and mineral admixtures;
- Nominal maximum aggregate size;
- Proportion of fine aggregate to total aggregate;
- Characteristics of the cement; and
- Amount, type, and quality of other cementitious materials.

The nominal maximum size of well-graded aggregate is the principal factor determining paste requirements. Typical relationships are shown in Table 6.2.

In mass concrete, nominal maximum aggregate sizes up to 6 in. (150 mm) can be used advantageously, but sizes larger than 6 in. (150 mm) may make mixing and handling more difficult.

6.4.6 *Proportion of fine-to-coarse aggregate*—ASTM C33 allows fairly wide limits in the grading of any particular coarse or fine aggregate. The principal requirement of combined gradings is that the optimum proportion of each aggregate be established for all aggregates to be combined. This optimum is generally the percentage that gives the required workability with the least amount of water per unit volume of concrete at the selected *w/cm*. Because the curves are relatively flat, and a small percentage deviation from optimum will not result in significant variations, it is sometimes advisable to use 1 or 2% more fine aggregate than the optimum to ensure adequate workability.

6.4.6.1 *Grading of fine aggregate*—For desirable finishability and workability of fresh concrete, fine aggregate needs to contain an adequate percentage of fines. The required percentage depends on the quantity and composition of the paste. For high-cement-content concretes, coarsely graded sands may be satisfactory because the cement helps provide the needed fines. With low-cement-content concretes, however, the fine sand particles are necessary for workable mixtures. In air-entrained concrete, deficiencies in sand grading affect workability less than they do in non-air-entrained concrete. Efficiency in achieving desired levels of air entrainment, however, is often affected by the grading of the sand. Other cementitious materials also affect the paste characteristics that contribute to workability.

6.5—Proportioning for resistance to severe exposure conditions

When concrete needs to resist freezing, it is important to include air entrainment and exclude unsound aggregates. To judge the degree of aggregate soundness, the service record of the aggregate should be reviewed if possible. For critical construction, having the proposed aggregates examined by a qualified petrographer should be considered (per ASTM C295) because the pore structure characteristics of aggregate

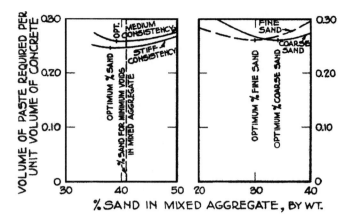

Fig. 6.3—The optimum percentage of fine aggregate, or that which requires the least paste, is somewhat smaller for stiffer consistencies and finer sands. The optimum percentage of fine aggregate also is less for richer mixtures and for larger maximum aggregate sizes. (These diagrams are for illustration only, to show trends. Each diagram is based on a different series of tests.)

Table 6.2—Typical paste requirements for various nominal maximum aggregate sizes

Nominal maximum aggregate size, in. (mm)	Paste fraction in unit volume of concrete	Cement,* lb/yd^3 (kg/m^3)
3/8 (9.5)	0.40	750 (445)
3/4 (19.0)	0.30	565 (335)
1-1/2 (37.5)	0.26	490 (291)
3 (75)	0.22	415 (246)
6 (150)	0.21	395 (234)

*w/cm = 0.58 by weight.

can affect the freezing-and-thawing resistance of concrete. The durability performance of different aggregates in concrete can be compared by ASTM C666/C666M or C682 test methods.

6.5.1 *Paste quality*—The quality of the paste is critical when concrete is subject to severe exposure conditions. For best results, potable mixing water, the appropriate cementitious materials, and the recommended *w/cm* should be used. Tables 6.3 and 6.4 give maximum recommended values of *w/cm* for special exposure conditions. For more information on durability, refer to ACI 201.2R and Chapter 4 of ACI 318.

Where prolonged exposure to water is expected, a low-water-content paste should be provided to reduce permeability, absorption, and the effect of leaching. If one side of the concrete member is in contact with moist earth and the rate of moisture transmission from the moist-earth side to the exposed side of the member is a concern, the paste should be dense enough to reduce this transmission rate and, hence, the average degree of saturation of the concrete member. The use of SCMs can contribute to producing a less permeable paste. They work by chemically combining with the free lime present in the concrete, thus preventing leaching of the lime as well as providing cementing properties.

6.5.2 *Required air entrainment*—Air entrainment of concrete is necessary to provide resistance to the effects of

Table 6.3—Requirements for special exposure conditions

Exposure condition	Maximum w/cm,* by weight, normalweight-aggregate concrete	Minimum f_c', normalweight- and lightweight-aggregate concrete, psi (MPa)
Concrete intended to have low permeability when exposed to water	0.50	4000 (28)
Concrete exposed to freezing and thawing in a moist condition or to deicing chemicals	0.45	4500 (31)
For corrosion protection of reinforcement in concrete exposed to chlorides from deicing chemicals, salt, saltwater, brackish water, seawater, or spray from these sources	0.40	5000 (35)

*When both Tables 6.3 and 6.4 are considered, the lowest applicable maximum w/cm and highest applicable minimum f_c' shall be used.
Note: This table adopted from ACI 318, Table 4.2.2.

Table 6.4—Requirements for concrete exposed to sulfate-containing solutions

Sulfate exposure	Water-soluble sulfate (SO_4) in soil, percent by weight	Sulfate (SO_4) in water, ppm	Cement type	Maximum w/cm,* by weight, normalweight-aggregate concrete	Minimum f_c', normalweight- and lightweight-aggregate concrete, psi (MPa)
Negligible	$0.00 \leq SO_4 < 0.10$	$0 \leq SO_4 < 150$	—	—	—
Moderate†	$0.10 \leq SO_4 < 0.20$	$150 \leq SO_4 < 1500$	II, IP(MS), IS(MS), P(MS), I(PM)(MS), I(SM)(MS)	0.50	4000 (28)
Severe	$0.20 \leq SO_4 < 2.00$	$1500 \leq SO_4 < 10,000$	V	0.45	4500 (31)
Very severe	$SO_4 > 2.00$	$SO_4 > 10,000$	V plus pozzolan‡	0.45	4500 (31)

*When both Tables 6.3 and 6.4 are considered, the lowest applicable maximum w/cm shall be used.
†Seawater.
‡Pozzolan that has been determined by test or service record to improve sulfate resistance when used in concrete containing Type V cement.
Note: This table adopted from ACI 318, Table 4.3.1

Fig. 6.4—When concrete contains insufficient entrained air, freezing and thawing can cause scaling of the top surface followed by continued deterioration of concrete.

freezing-and-thawing cycles (Fig. 6.4). Though other factors are important to concrete quality and durability, such as aggregate quality, cement content, w/cm, type and amount of admixtures, concrete consistency and density, and curing, none provide as much resistance to freezing effects as the proper amounts and distribution of entrained air. Chapter 5 provides recommended ranges of air for concrete (containing various maximum sizes of coarse aggregate) subject to severe exposure. The effectiveness of voids produced by an air-entraining admixture is due to their relative smallness and spacing in the paste for a given total amount of air. Air-entraining admixtures are designed to produce a large number of small (nearly microscopic) voids in the paste at a calculated spacing factor of 8 mil (0.2 mm) or less. This close spacing allows relief of the pressure that develops when moisture in the concrete expands during freezing.

6.5.3 *Aggregate proportions*—To minimize the loss of compressive strength of the concrete caused by the addition of entrained air, the percentage of the fine aggregate in the mixture should be adjusted so that, with the proper amount of entrained air, the paste content is at a minimum for the required consistency. First, the mixture should be proportioned as would be done without an air-entraining admixture. Then, allowing for the effect of adding an air-entraining admixture, the proportion of fine aggregate and water should be reduced to produce concrete having the same coarse-aggregate content and consistency as it would have without entrained air.

With a mixture proportioned to include the proper amount of entrained air, the ability of concrete to resist the effects of freezing depends mainly on the quality of the paste and the porosity and pore characteristics of the aggregate particles, particularly the coarse aggregate. Refer to Chapter 5 for further information on the effects of freezing and thawing.

6.5.4 *Proportioning by absolute volume*—Proportioning concrete by absolute volume assumes that the volume of compacted fresh concrete is equal to the sum of the absolute volume of all ingredients plus the volume of air voids within the concrete.

Absolute volume is sometimes called solid volume, particle volume, or displacement volume. If a container is filled exactly to its top with solid material, such as gravel, sand, or cement, the volume of the container represents the bulk volume of the material it contains. The particles piled together in the container do not fit each other exactly, so there are tiny voids between them. Thus, the absolute volume of all of the particles plus the total volume of the spaces between particles equals the bulk volume.

The actual amount of solid material in a given bulk volume of aggregate varies with its grading and degree of consolidation. The absolute volume of solid material in a given weight of aggregate, cement, or SCM depends on its specific gravity. For water, absolute volume is the same as bulk volume.

When aggregate, cement, and water are mixed to produce a batch of fresh concrete, the cement-water-sand mortar fills the spaces between the coarse particles. Thus, if the concrete is compacted to remove most entrapped air, the total volume of the concrete is the sum of the absolute volumes of the ingredients, including the entrapped air. If entrained air is included in the mixture, the volume of the total air is included in the sum of absolute volumes.

Because it is not practical to batch aggregate, cement, or SCMs by absolute volume directly, the desired absolute volume should be converted into terms of weight for purposes of batching. Conversely, proportions by weight should be converted into terms of absolute volume for computing yield (typically cubic yards or cubic feet of fresh concrete).

6.5.5 *Computing absolute volume and percentage of solids*—The absolute volume of a quantity of a material can be computed from its mass and specific gravity. For aggregate, the same absolute volume will be calculated whether using bulk dry specific gravity or SSD specific gravity, provided that dry specific gravities are used with dry aggregates and that SSD specific gravities are used with SSD aggregates. The following examples use SSD specific gravity. Absolute volume (SSD) is generally based on the density of water at 60 °F (16 °C), 62.4 lb/ft^3 (1000 kg/m^3)

$$\text{absolute volume, ft}^3 = \frac{\text{weight, lb}}{\text{specific gravity} \times 62.4} \quad \text{(in-lb)}$$

$$\text{absolute volume, m}^3 = \frac{\text{weight, kg}}{\text{specific gravity} \times 1000} \quad \text{(SI)}$$

Thus, 100 lb of cement having a specific gravity of 3.15 (an average value) contains

$$\frac{100}{3.15 \times 62.4} = 0.509 \text{ ft}^3 \text{ of solid material}$$

If specific gravity and density in pounds per cubic foot of an aggregate in a given condition of compaction are known, the percentages of solids and voids in the aggregate can be computed as

$$\text{percent solids} = \frac{\text{surface dry unit weight, lb/ft}^3}{\text{specific gravity} \times 62.4} \times 100 \quad \text{(in-lb)}$$

$$\text{percent solids} = \frac{\text{surface dry unit weight, kg/ft}^3}{\text{specific gravity} \times 1000} \times 100 \quad \text{(SI)}$$

Then

$$\text{percentage of voids} = 100 - \text{percentage of solids}$$

Therefore, a SSD rodded aggregate weighing 110 lb/ft^3 and having a SSD specific gravity of 2.65 contains

$$100 \times \frac{110}{2.65 \times 62.4} = 66.5\% \text{ solids}$$

$$100 - 66.5 = 33.5\% \text{ voids}$$

6.5.6 *Example of proportioning by absolute volume*—Concrete mixtures are commonly proportioned by absolute volume. Weight procedures described in ACI 211.1 also are frequently used. The following example calculations summarize procedures for proportioning by absolute volume.

Specification requirements:
- f'_c = 3000 psi;
- Slump = 3 to 4 in.;
- Entrained air content = 5%;
- Exposure conditions = maximum; and
- w/cm = 0.50 (Table 6.3).

Material data:
- Cement: Type I, ASTM C150, with a specific gravity of 3.15;
- Fine aggregate: Natural, specific gravity (SSD) = 2.62; fineness modulus = 2.60;
- Coarse aggregate: Crushed granite, No. 57, per ASTM C33; specific gravity (SSD) = 2.67; density, SSD rodded = 96.6 lb/ft^3; and
- Admixture: Air-entraining admixture.

Trial mixture data (basic procedures from ACI 211.1):

1. Estimate water requirement from past experience with materials being used, or use approximate value given in Table 6.3.3 of ACI 211.1. Water-content estimate for air-entrained concrete with 1 in. nominal maximum aggregate size and 3 to 4 in. slump is 295 lb/yd^3 of concrete;

2. The minimum cement content is set because a *w/cm* of 0.50 by weight is specified (cement content = 295/0.50 = 590 lb/yd^3). (Note: From past experience or average *w/cm*-to-compressive strength relationship, the maximum *w/cm* specified should indicate that the specified compressive strength might be obtained. Several trial mixtures should be proportioned, varying the *w/cm* and, thus, cement content, up to the maximum.);

3. Aggregate quantities are determined by several methods. Concrete with satisfactory workability is produced when a given volume of coarse aggregate, on a dry-rodded (or SSD-rodded) basis, is used for a unit volume of concrete. Table 6.3.6 of ACI 211.1 gives the approximate dry-rodded volume of coarse aggregate per cubic yard of concrete based on maximum aggregate size and fineness modulus of fine aggregate. For the example material, the plotted volume ratio is 0.69 (volume ratio = 1 – void ratio). Therefore, the weight of coarse aggregate equals 0.69 × 96.6 × 27 = 1800 lb/yd^3 of concrete. SSD-rodded density can be substituted for dry-rodded density when determining the coarse aggregate volume ratio if aggregate absorption values are low;

4. Fine-aggregate content is then determined by difference, using absolute-volume computations in Table 6.5.

CONCRETE MIX PROPORTIONING WORK SHEET

ANALYSIS OF AGGREGATES

COARSE AGGREGATE					FINE AGGREGATE				
SUPPLIER Crushed Granite No.57					SUPPLIER Natural				
SIEVE NO.	WT. RET.	% RET.	% PASS	ASTM C33 SPECS.	SIEVE NO.	WT. RET.	% RET.	% PASS	ASTM C33 SPECS.
2½"					⅜"**	0.0	0	100	100
2"					NO. 4**	10.4	2	98	95-100
1½"*	0.0	0	100	100	8**	46.9	9	91	80-100
1"	1.1	2	98	95-100	16**	130.4	25	75	50.85
¾"*	13.4	29	71	-	30**	265.9	51	49	25.60
½"	23.4	50	50	25-60	50**	396.3	76	24	10.30
⅜"*	38.1	82	18	—	100**	505.8	97	3	2-10
NO. 4*	42.9	92	8	0.10	200	512.5	98.3	1.7	3.0 MAX
NO. 8*	46.1	99	1	0.5	PAN	521.4	—	—	
PAN	466	-	-		F. M.		2.60		2,3-3.1
F. M.		7.02			SPECIFIC GRAVITY		2.62		
SPECIFIC GRAVITY		2.67			UNIT WEIGHT, % GE		—		
UNIT WEIGHT, % GE		966							

CLIENT: ______

PROJECT: ______

JOB NUMBER ______

LABORATORY NUMBER: ______

DATE MIX PREPARED: ______

TO DETERMINE E. M. VALUE

*ADD 400 TO % RETAINED ON SCREENS. DIVIDE SUM BY 100.

**ADD % RETAINED ON INDICATED SCREENS, DIVIDE SUM BY 100.

MIX DATA	
MINIMUM CEMENT FACTOR, SKS/YD.	
MAXIMUM WATER-CEMENT RATIO,	0.50
SPECIFIED AIR CONTENT, % BY VOL.	4-6

MIX COMPUTATION DATA

MATERIAL	PER YARD	SOLID VOLUME WEIGHT	VOLUME IN MIX
CEMENT:	590 lbs	÷ 3.15×62.4 =	300 ft³
		+	=
WATER	295 LBS	+ 62.4	= 4.73 ft³
AIR	5%	× 27	= 1.35 ft³
TOTAL			9.08 ft³

FINE AGGREGATE	COARSE AGGREGATE
27.00 - 10.80 - 9.08 = 7.12 ft³	0.69 × 96.6 × 27 = 1800 LBS.
7.12 × 2.62 × 62.4 = 1164 LBS	10.80 × 2.67 × 62.4 = 1800 LBS.

TRIAL BATCH WEIGHTS

CEMENT ______

FINE AGG. ______

COARSE AGG. ______

WATER ADMIXTURE ______

TYPE: ______

ACTUAL SLUMP, INCHES ______ ACTUAL AIR CONTENT, % ______

WATER ADDED, LBS. ______ WATER RET., LBS. ______

TEST CYLS. REQUIRED: ______ WET WEIGHT, PCF ______

CEMENT, TYPE & SUPPLIER ______

Fig. 6.5—Sample worksheet for concrete mixture proportioning.

Table 6.5—Absolute-volume computations

Material	Computation	ft³ per yd³ of concrete
Cement, absolute volume	590/(3.15 × 62.4)	3.00
Water	295/62.4	4.73
Total air content	5% of 27	1.35
Coarse aggregate	1800/(2.67 × 62.4)	10.80
Total	—	19.88
Fine aggregate, absolute volume	27 – 19.88	7.12

Therefore, the SSD weight of fine aggregate equals 7.12 × 2.62 × 62.4, or 1164 lb/yd³ of concrete; and

5. Check performance in terms of water requirement and workability. Adjust proportions and make additional trial batches, as necessary, to achieve desired yield and workability while adhering to specified requirements for the mixture. Figure 6.5 shows a worksheet that can be used to summarize the computations and estimates made.

6.6—Control of concrete proportions

Contract documents often specify a definite or maximum *w/cm* value and the types and amounts of admixtures to be used. Compliance to a maximum *w/cm* will often result in concrete strengths well above the specified minimum strength of concrete given in the specification. Table 6.1 shows conservative strength values that can be obtained for various *w/cm*. To allow for field fluctuations and to avoid exceeding a maximum *w/cm* specification, trial mixture proportions should be prepared at the highest permissible slump and temperature. The resulting concrete should provide a strength that exceeds f_c' by the amount specified in ACI 318 or 301.

6.6.1 *Laboratory batch quantities*—The amount of water to be added to the batch is estimated based on SSD aggregate. The value should be corrected for the free surface moisture contained on the aggregate. The water content of admixture solutions should be considered as part of the mixing water when dosage rates exceed 10 oz per 100 lb (625 g per 100 kg) of cementitious material. Table 6.6 provides an example of a laboratory computation of batch quantities of an air-entrained concrete mixture, with field corrections.

If the cement content is specified and the *w/cm* is known, the quantities of aggregate per cubic yard (cubic meter) of concrete should be computed: total absolute volume of aggregate is equal to 27 ft³ (1 yd³ [1 m³]) less the volume of water, less the absolute volume of cement, less the volume of entrained and entrapped air (known or estimated). The desired sand-aggregate ratio or coarse-aggregate content (as shown in the previous example of proportioning by absolute volume) should be used to determine the fine and coarse aggregate quantities.

6.6.2 *Field batch quantities*—Batch quantities furnished to field jobs generally give the weight of each ingredient. These quantities may be in proportions relative to a unit proportion of cement, but more often, the amount of each ingredient will be stated in actual mass per cubic yard of concrete, assuming that the aggregates are SSD. If proportions are provided in terms other than mass, they should be converted to weight quantities before batch weights are computed.

Because conditions in the field vary from those in the laboratory, the laboratory may need to adjust proportions in

Table 6.6—Sample computation of batch quantities for air-entrained concrete and correction for surface moisture of aggregate*

Property	Cement (3.15)†	Water (1.00)†	Fine aggregate (2.62)†	Coarse aggregate (2.67)†	Air content	Total	Computed density, lb/ft³
Material mass in pounds per batch (SSD aggregate)	590	295	1164	1800	(5.0%) volume	3849	—
Absolute volume, ft³	3.00	4.73	7.12	10.80	1.35	27.00	142.6
Total moisture in aggregates, percent by mass (determined by drying with heat)	—	—	1.9	1.0	—	—	—
Absorption capacity of aggregates, percent by mass (determined by test)	—	—	0.2	0.5	—	—	—
Surface moisture in aggregates, percent by mass (total moisture less absorbed moisture)	—	—	1.7	0.5	—	—	—
Surface moisture in aggregates, lb (percent times aggregate weight divided by 100)	—	—	20	9	—	29	—
Adjusted material mass per batch, lb	590	266	1184	1809	—	3849	142.6

*$w/c = 0.50$.
†Specific gravity, SSD.
Note: The use of admixtures other than air-entraining is not considered in this example. SSD condition is the percentage of moisture at which the aggregate with neither draw mixing water from the paste nor supply additional mixing water to the paste. If surface moisture of aggregates is determined directly by the test, lines for absorption and total moisture are omitted. 1 lb/yd³ = 0.593 kg/m³; 1 yd³ = 0.765 m³; 1 lb = 0.00445 kN.

the field to achieve the desired workability, strength, entrained air content, or cement content. After any necessary adjustments have been made to accommodate the concrete proportions to field conditions, the mass of aggregate and water should be adjusted to take into account changes in aggregate moisture content from the SSD condition. Table 6.6 illustrates corrections for surface moisture present in aggregates. Because aggregate moisture contents change, even throughout the day, periodic checks of aggregate moisture and subsequent batch adjustments are required.

6.6.3 *Field control of selected proportions*—Select concrete proportions to provide the necessary workability, strength, and durability for the particular application. Workability, according to ACI 116R, is the ease and homogeneity with which concrete can be mixed, placed, compacted, and finished. Workability also encompasses consistency, which is the relative ability of freshly mixed concrete to flow. Consistency is usually measured in terms of slump: the higher the slump, the wetter the mixture consistency, and the more easily the concrete will flow during placement.

Required workability is dictated by placing requirements. Because variations in materials, weather, and other conditions on the job site are unavoidable, consistency will change even if a fixed amount of each constituent material is strictly maintained (Fig. 6.6). Nevertheless, the more uniform the grading and moisture content of the aggregates, the less adjustment required.

Concrete supplied to the job should be as uniform in consistency as possible. If necessary, adjustments should be made in the amounts of water added at the mixer in accordance with procedures called for in the contract documents. When projects require inspectors to inspect batching and mixing procedures, they are responsible for verifying that proper adjustments to aggregate and mixing water proportions are made. These adjustments should be based on tests for changes in moisture content and, if appropriate, aggregate grading so that concrete nearly uniform in consistency will be obtained. The concrete should also vary as little as practical

Fig. 6.6—When concrete dries too fast or the consistency is too stiff, the screeding operation cannot close the surface.

in w/cm, cement content, strength, or basic proportions in accordance with contract requirements and good practice.

In air-entrained concrete, if the air content is not kept within tolerances, excessive variations in yield, workability, slump, water and cement contents, w/cm, strength, and durability will result. Tests should be conducted frequently to verify proper air content (refer to ACI 311.5 [reprinted in Appendix B of this document] for recommended testing frequencies). Adjustments in the dosage of air-entraining admixtures added at the mixer may be required for major changes in aggregate grading and for higher concrete temperatures and slump. Increases in the amount of fines in the fine aggregate, including the use of SCMs, can require an increased dosage of air-entraining admixture to maintain a specified air content. Chemical admixtures (such as water reducers and retarders) generally entrain some air; thus, if they are used in conjunction with an air-entraining admixture, less than the normal dosage of the admixture may be required. The required dosage of air-entraining admixture to maintain a specified air content increases with increasing temperatures, and vice versa. It will also increase with the use of low-slump concrete (2 in. [50 mm] or less), high cement contents, and high-early-strength concrete.

Air-entraining portland cements (Types IA, IIA, and IIIA, per ASTM C150) are produced with an interground air-entraining admixture. Although convenient for some purposes, their use makes it difficult to accurately adjust air content to compensate for changes in aggregate grading, amount of admixture, or temperature. If less air content is desired, the adjustment is likely to be complex, requiring less cement, a different cement, or the addition of non-air-entraining cement. If greater air content is needed, an air-entraining admixture can be added, but air-entraining cements are difficult to adjust by such additions. Also, the air-entraining capability of these cements decreases with age, making control of air content with these cements even more difficult. For these reasons, air-entraining cements are rarely used.

The moisture content of the aggregate as batched should be routinely monitored, particularly the fine aggregate, using moisture meters. The moisture content should be checked several times daily and whenever there is an indication that it has changed. Also, the calibration of moisture meters should be checked daily and any necessary adjustments made. Prescribed aggregate proportions or aggregate quantities for a certain cement content remain uniform only when the aggregate and water batch weights are adjusted to take into account the amount of water in the aggregate. For example, to compensate for the weight of surface and free water contained in each aggregate, increase the aggregate batch weight above the specified surface-dry weight while decreasing the weight of the mixing water by the same amount. To maintain a nearly constant *w/cm* when the water content of aggregate changes, it is necessary to adjust the weights of the aggregates and the weight of the mixing water.

6.7—Computations for yield

Yield is defined in ASTM C138/C138M as the volume of concrete produced from a mixture of known quantities of the component materials. ACI 116R defines yield as the volume of freshly mixed concrete produced from a known quantity of ingredients, or the total weight of ingredients divided by the density of the freshly mixed concrete. Yield computations are used to determine actual cement content or to check batch-count volume against observed volume in place. If the quantity of total mixing water is obtained, the *w/cm* can be determined for verification.

6.7.1 *Density method*—Yield, as determined by the density method, is described in ASTM C138/C138M. In this standard, yield is calculated using the density of concrete determined by weighing a sample of the fresh concrete in a container of known volume. It automatically takes into account entrapped or entrained air and is independent of the specific gravity of the ingredients.

To compute yield by the density method, consolidate concrete in the test container either by rodding or vibrating, depending on the slump, as described in ASTM C138/C138M (unless a specific method is stated in the contract documents). Calculate the yield, in cubic feet of concrete per batch, as follows

$$\frac{\text{total weight of materials, lb, in batch}}{\text{density of concrete, lb/ft}^3}$$

The quantity of cement (or other ingredients) per cubic yard of concrete is then the batch weight of cement multiplied by $27/(\text{yield in ft}^3)$.

Assume, for example, the weights of materials in a typical batch are:

Cement	2950 lb
SSD sand	5820 lb
SSD coarse aggregate	9000 lb
Water	1475 lb
TOTAL	19,245 lb

The density of the concrete was measured as 142.6 lb/ft^3. Therefore

$$\text{yield} = \frac{19,245}{142.6} = 135.0 \text{ ft}^3 \text{ of concrete}$$

Using the yield calculated and batch weight of the cement, its quantity per cubic yard can be determined

$$\text{cement content} = 2950 \times 27/135.0 = 590 \text{ lb/yd}^3 \text{ of concrete}$$

A simplified method of making these computations in the field is to base the cement content and yield per batch on the actual weight of a sample of the freshly mixed concrete as delivered and batch weights of materials used. For any size mixer batch, the yield, in cubic feet, is

$$\frac{\text{total weight of all materials in batch, lb}}{\text{density of concrete, lb/ft}^3}$$

and the cement content, in pounds per cubic yard, is

$$\frac{27 \times \text{weight of cement per batch} \times \text{density of concrete}}{\text{total weight of batch}}$$

or

$$\frac{27 \times \text{weight of cement per batch}}{\text{yield, ft}^3 \text{ per batch}}$$

It is not necessary to know the moisture content of the aggregate because the moist weight of the aggregates and actual weight of added water can be used to calculate the total batch weight.

For greater accuracy, determine the density from the average of at least three measurements, each taken from a sample of sufficient size to be tested in a 1/2 ft^3 container. For a mixture containing large aggregate (such as mass concrete), determine density in larger containers, perhaps 1 ft^3 or larger. In all cases, consolidation of the sample should represent procedures used on the job. Care should be taken not to overvibrate the concrete in the sample container. Take each sample in the manner described in ASTM C172, and test in accordance with ASTM C138/C138M or C567 (lightweight concrete), as appropriate.

Whatever methods are used to mix, deliver, and place the concrete, the volume of concrete presumably placed in the structure will be more than that calculated from the yield tests and computations and from the computed theoretical volume within the forms. Some increase in required volume or loss in delivered volume is inevitable due to foundation overexcavation, spreading of forms, loss of entrained air, wastage and spillage, or concrete lost in washout after each truckload. Except for the latter, these losses are not the responsibility of the concrete supplier because they are out of the supplier's control. With experience, it is possible to estimate the extra concrete needed to compensate for possible losses and for required volume increases on a particular project. When conformance with ASTM C94/C94M is a contract requirement, it mandates that testing for acceptance be performed by ACI-certified (or equivalent) personnel. Similarly, strength testing in the lab requires ACI-certified personnel (or equivalent). Other materials testing activities will also probably require conformance with applicable ASTM standards such as ASTM C1077. The inspector should be aware of any such requirements and may need to verify compliance.

CHAPTER 7—BATCHING AND MIXING

Batching and mixing operations should produce uniform concrete containing the required proportions of materials. To ensure such uniformity, inspectors should see that:

- All ingredients are kept homogeneous before and during batching;
- The equipment used will accurately batch the required amounts of material and allow the amounts to be easily adjusted when required;
- The required proportions of materials are maintained from batch to batch;
- All materials are introduced into the mixer in proper sequence;
- All ingredients are thoroughly combined during mixing, and the cement paste completely coats all aggregate particles; and
- The concrete discharged from the mixer is uniform and homogeneous within each batch and from batch to batch.

This chapter provides an overview of the equipment and procedures used for batching and mixing of central- and transit-mixed concrete. Batching and mixing procedures are covered in detail in ACI 304R and ASTM C94/C94M. When conformance with ASTM C94/C94M is a contract requirement, it requires that testing for conformance shall be done by ACI-certified (or equivalent) personnel. Similarly, strength testing for conformance requires certified (or equivalent) personnel. Other materials testing activities may require conformance with applicable ASTM standards, such as ASTM C1077. The inspector should be aware of any such requirements and may be called upon to verify compliance.

Fig. 7.1—Control console of automatic batch plant.

7.1—Batching operations

Batching can be done manually, semi-automatically, or automatically. Automatic batching is preferred because it lessens the variability of the concrete production process. In manual batching, all weighing and batching of the concrete ingredients are done by hand or by mechanized weigh batchers that require operators to observe scales or water meters to control weighing and cutoff of ingredients. Manual plants are acceptable for small jobs that do not require fast batching rates. Rapid batching can exceed a manual plant's capacity, resulting in weighing inaccuracies.

In a semi-automatic batching system, manually operated push buttons or switches are used to open aggregate bin gates for charging batchers. When the designated weight of material has been delivered, the gates close automatically. Interlocks prevent batcher charging and discharging from occurring simultaneously.

In an automatic batch plant, a single starter switch electrically activates automatic batching of all materials (Fig. 7.1 to 7.3). Interlocks interrupt the batching cycle when the scale has not returned to zero balance or when preset weighing tolerances are exceeded.

7.1.1 *Measurement tolerances*—The tolerances of batch weight measurements of ingredients for ready mixed concrete are provided by ASTM C94/C94M or by contract documents. The allowable tolerance for weighing cement depends on the amount to be weighed and the applicable specifications. If the quantity exceeds 30% of the full capacity of the scale, ASTM C94/C94M specifies a batching tolerance of 1% of the required weight at that weighing point. If smaller weights are to be batched, the tolerance is not less than the required weight nor more than 4% in excess of the required weight.

If aggregates are weighed in individual hoppers, the allowable batching tolerance is ±2% of the required weight. In a cumulative aggregate weigh hopper, the cumulative weight after each successive weighing should be within 1% of the cumulative required weight at that weighing point, provided that the cumulative weight exceeds 30% of the scale capacity. If the cumulative weight of aggregate is less than 30% of capacity, the allowable tolerances are ±0.3% of

Fig. 7.2—Permanent batch plant.

Fig. 7.3—Portable batch plant.

the scale capacity or ±3% of the required cumulative weight, whichever is less.

ASTM C94/C94M requires that water added to the batch be measured and batched to an accuracy of 1% of required total mixing water. The total water, which includes aggregate surface moisture, ice if used, water in admixtures, and any wash water, should be measured to an accuracy of ±3%. For admixtures, the batching tolerances are ±3% of the required amount.

7.1.2 *Weighing equipment*—Weigh hoppers should be constructed so that materials discharge easily and completely by gravity, with no materials sticking to the hopper.

For all semi-automatic and automatic plants, interlocking devices should ensure that:
- The charging device can open or start only when the scale indicates zero load and the weigh hopper discharging gate is closed; and
- The discharging gate can open only when the desired weight is in the weigh hopper and the charging device is closed.

In cumulative automatic batchers, interlocks are used to ensure that the scale returns to zero before batching starts and that each material is within tolerance before the next can be weighed. The batch plant operator should never circumvent interlocks without notifying the inspector and obtaining permission to proceed.

Weigh hoppers for aggregates should be built so that the contained material can be inspected easily and the aggregate can be sampled. If aggregates cannot be sampled from the hoppers, take a sample from the belt of the conveyer system using the aggregate sampling procedures described in Chapter 3.

Figure 7.4 shows desirable and undesirable arrangements of batching hoppers. The inspector should check that all working parts, particularly the knife edges, are in good, clean condition, free from friction, readily accessible for inspection and cleaning, and protected from falling or adhering material and other contamination. The inspector should also verify that locking devices protect all nuts that could work loose in operation and that the weigh hopper and gates are tightened against leakage.

Provisions for adjusting the amount of materials in a given batch, including the removal of excess material, should be made. The weighing mechanism and dials indicating when the correct amount of material is in the hopper should be clearly visible to the batch plant operator and the inspector. Further details on weigh hoppers are provided in the National Ready Mixed Concrete Association's (NRMCA) "Plant Certification Checklist" (2002).

Scales for batching concrete ingredients may be beam, springless dial, electric, hydraulic, or load cells, but they should conform to applicable sections of the current edition of NRMCA's "Plant Certification Checklist." Beam scales should be equipped with a balance indicator sensitive enough to show movement when a weight equal to 0.1% of the nominal capacity of the scale is placed in the batch hopper. Pointer travel should be a minimum of 5% of the net-rated capacity of the largest weigh beam for underweight and 4% for overweight. Scale accuracy limitations of NRMCA meet the requirements of ASTM C94/C94M.

7.1.2.1 *Check tests*—Each plant should have an adequate amount of test weights (usually at least 10 standard 50 lb [23 kg] weights) meeting NRMCA's "Plant Certification Checklist" requirements for calibrating and testing of weighing equipment. Using these test weights, the scales should be checked up to the full amount of the batches. First, the scale should be balanced at zero load; when the scale has been checked to the limit of the weights, the weights should be removed and enough material should be placed in the weigh hopper to produce the same scale setting. Then the weights should be reapplied to check the scale at higher loads. The scale reading should be recorded for each weight increment, and the scale adjusted to read correctly. Scale adjustment is best done by a scale technician.

At least twice during each shift, manually operated scales should be balanced at zero load. Automatic batchers with zero interlocks should be checked for proper cutoff. The scale and weigh hopper should be inspected frequently for

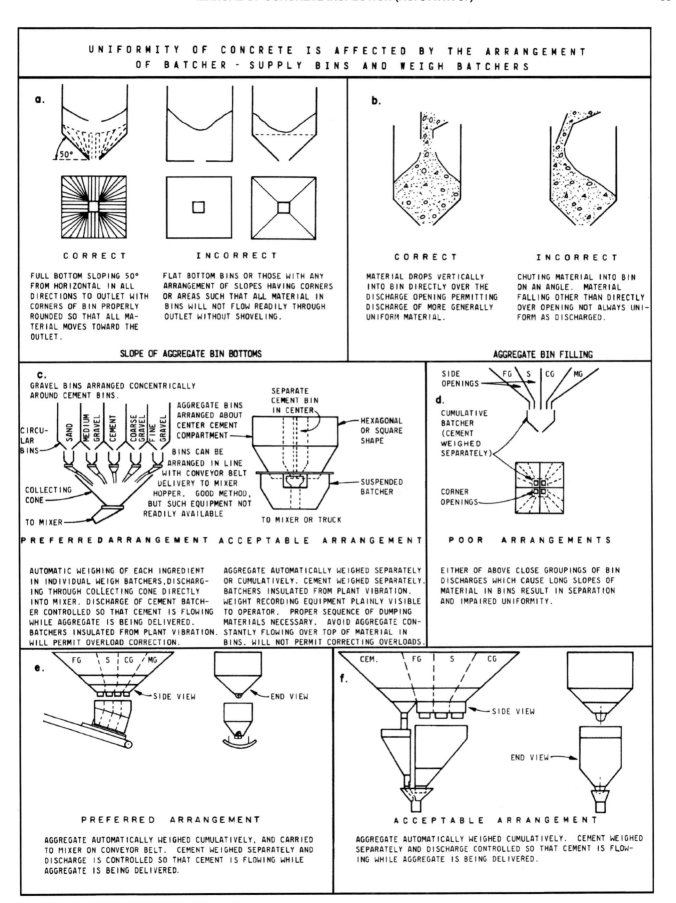

Fig. 7.4—Correct and incorrect methods of batching (ACI 304R).

Fig. 7.5—A concrete inspector checks a batch plant weigh scale.

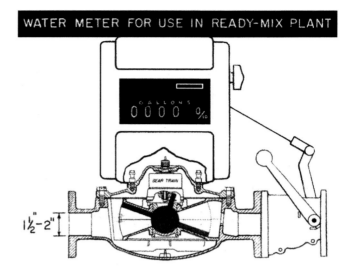

Fig. 7.6—Meter shows quantity of water delivered to mixer. Some meters automatically shut off flow when the desired quantity has been delivered.

signs of sluggishness, inaccuracy, damage, or sticking materials that do not discharge (Fig. 7.5).

In plants equipped with automatic feeding and cutoff and graphic or digital recorders, a check test involves applying known loads in increments with the aid of test weights (after first setting the scale to zero) and comparing the actual load with the corresponding readings of the beam or dial of the recorder. The scale mechanism should then be adjusted to comply with the actual weighing within specified tolerances (usually the smallest division of the scale). Test the cutoff mechanism during regular batching operations by bringing the cutoff setting on a given scale up to the normal setting in several increments for a number of successive batches and comparing the dial readings at cutoffs with the cutoff settings. In some plants, it will be necessary to adjust both the main and "dribble" feed. Adjust the recorder to be within the allowable tolerance (usually the smallest division of the scale), and adjust the cutoff mechanism to conform to allowable weighing tolerances.

7.1.3 *Batching equipment*—ACI 304R recommends that the batch plant conform to the size of the project and that storage bins are of adequate size to accommodate the plant's production capacity. Desirable and undesirable arrangements of batching equipment for large installations are shown in Fig. 7.4. The inspector should verify that:
- Compartments in bins adequately separate the various concrete materials. Shape and arrangement of aggregate bins should prevent aggregate segregation, overflow, and breakage (Fig. 7.5);
- Aggregate bins have adequate separate compartments for fine aggregate and for each required size of coarse aggregate. Each compartment should discharge material into the weigh hopper efficiently and freely, with minimum segregation;
- Each cement and SCM bin has a dust seal between the bin and the weigh hopper. The dust seal should be installed so that it does not affect weighing accuracy; and
- Weigh hoppers have smooth-operating clamshell or undercut radial-type bin gates. Power-operated gates should be used to charge semi-automatic and fully automatic batchers and have a dribble control to obtain the desired weighing accuracy.

7.1.4 *Measuring water*—Batch plants usually meter (Fig. 7.6) or weigh water. In older plants, water may be measured by volume in a calibrated tank. The tank or meter should be calibrated by measuring or weighing sample batches of water drawn out for various settings of the device. Modern meters operate well over a wide range of pressures, but are inaccurate at very low flow rates.

To ensure that no unmeasured water goes into the mixer, check for any leakage into the pipe leading to the mixer, either from the measuring device or from any connections or valves. In addition, a valve arrangement that allows unmeasured water to flow into the mixer from a tank that is being charged or discharged should not be permitted.

For proper control of mixing water, it is necessary to account for the free water from the aggregate. Moisture meters or probes installed in the aggregate hopper often are used to monitor the water content of fine aggregate. To calibrate a moisture meter, readings obtained from the meter should be compared with readings obtained by weighing a representative aggregate sample before and after drying in an oven or over a hot plate (ASTM C566). The meter can then be adjusted as necessary.

7.1.5 *Measuring admixtures*—Liquid admixtures may be dispensed into the mixer by weight or by volume. Some manufacturers of fluid admixtures supply dispensers that inject the proper dosage into the mixing water or into the fine aggregate. In any case, dispensing equipment should conform to ASTM C94/C94M and provide for visual confirmation of the correct volume for each batch. The equipment also should discharge admixtures slowly so there is no possibility of an inadvertent double dosage (ACI 212.4R).

Liquid admixtures (and powdered admixtures dissolved in water before use) should be added into the stream of mixing water being batched into the mixture. If a concentrated fluid admixture is used, preparing it as a dilute solution before batching results in better accuracy.

Powdered admixtures used without first being dissolved in water should be weighed. These admixtures, however, should only be used when absolutely necessary because they are difficult to batch and properly blend into concrete mixtures. Most contract documents prohibit batching powdered admixtures by bulk volume because large fluctuations can occur. Powdered admixtures to be used in small quantities should be packaged in advance.

If two or more admixtures are used in the concrete, they should be added separately to avoid intermixing until they combine with the batch water that is already in contact with the cement. Taking this precaution will prevent possible chemical interaction of the admixtures, which could cause partial solidification of the admixtures, diminish the efficiency of either admixture, or adversely affect the concrete. If multiple admixtures are required, consult with the admixture supplier for charging sequence recommendations.

7.1.5.1 *High-range water-reducing admixtures*—High-range water-reducing admixtures (HRWRAs) (Types F and G) can be charged into the batch at the plant or at the job site immediately before final mixing and discharge. In the field, before discharge of these admixtures, it is important to confirm that the water content of the concrete mixture is within design parameters. A slump test is often used for verification; however, slump is a relative indicator of the w/c and not an actual measurement. HRWRAs increase concrete slump dramatically and, therefore, slump is no longer a reliable relative indicator of the w/c. The inspector should verify that the amount of water has not exceeded the mixture proportion as verified on the truck ticket and any additions.

Before adding HRWRA at the job site, it should be verified that the slump of the concrete mixture is within established parameters. Some suppliers equip their mixer trucks with admixture tanks; others add the HRWRA manually. In either case, the quantity and brand and type of admixture being introduced should meet the requirements of the approved mixture proportions and the batch should be thoroughly mixed in accordance with the admixture manufacturer's recommendations. Many HRWRA manufacturers express their mixing requirements in terms of a minimum mixing duration (often 5 minutes). The number of revolutions of the drum should be monitored to ensure that the total number of revolutions (before and after charging the mixture with HRWRA) does not exceed the 300 maximum drum revolutions specified in ASTM C94/C94M. If the travel time from the plant to the site is such that thorough mixing of a job-site-added HRWRA cannot be accomplished without exceeding this parameter, uniformity tests should be performed at the maximum anticipated mixing duration. The architect/engineer should then decide whether the specification can be exceeded. If a central mixer is used and the HRWRA is to be added at the site, the transport vehicle should be qualified as a mixer, not merely as an agitator as defined in the National Ready-Mixed Concrete Association (NRMCA) Manual.

ACI 301 specifies a slump of 2 to 4 in. (50 to 100 mm) before HRWRA is added and a maximum of 8 in. (200 mm) at the point of delivery after the admixture is added. Project specifications, however, may allow slumps higher than 8 in. (200 mm), although concretes having slumps greater than 9 in. (230 mm) may not be suitably cohesive for measurement using ASTM C143/C143M.

7.1.5.2 *Accelerators*—Some contract documents provide for the use of accelerators to increase the rate of hydration and accelerate the early-age strength gain of concrete. They accelerate the setting and hardening of concrete, thus expediting the start of finishing operations and reducing the time that protection is needed.

Calcium chloride is often used as an accelerator, and many proprietary admixtures, primarily water-reducing admixtures, used as accelerators contain calcium chloride as the active ingredient. Calcium chloride and other chloride-containing accelerators can have severe side effects on concrete, including corrosion of metals, increased shrinkage, increased susceptibility to sulfate attack, increased alkali-aggregate reaction (AAR), and mottling (marking with spots or streaks of different colors). In warm weather, the use of accelerators can result in rapid setting of the concrete, making finishing difficult or impossible.

All calcium chloride used should meet the requirements of ASTM D98. If calcium chloride or other admixtures containing chloride ions are used in reinforced concrete or in concrete containing steel embedments, the total chloride ion content of the concrete should meet the requirements of ACI 318. Accelerators containing little or no chlorides are available to minimize the introduction of additional chloride ions into the concrete. Accelerators should always be thoroughly dissolved in water before being introduced into the mixture. Dry calcium chloride should never be added directly to a mixture.

Calcium chloride should not be used as an admixture in concrete that will be exposed to severe sulfate-containing solutions, as stated in ACI 318, or in prestressed concrete.

7.2—Mixing operations

Mixing can be performed in central- or site-mix plants, truck mixers, pavement mixers, or portable mixers at the placing site. Some projects use a combination of mixing methods. Whatever method is used, it is essential for materials to be uniformly distributed throughout the mixture and for all aggregate surfaces to be well-coated with cement paste. To accomplish this within a reasonable time, the mixer should be:

- Clean and in good condition;
- Properly designed, particularly as to type and number of blades;
- Charged correctly and not overloaded; and
- Operated at the optimum speed recommended by the manufacturer.

The valves controlling the mixing water should not allow leakage into the mixer.

7.2.1 *Central or site mixing*—Before concrete mixing begins, the inspector should examine the mixer to ensure that the mixing blades and interior of the drum are clean, the blades are not worn more than 10% as measured as a percent

Fig. 7.7—Plant mixer rating plate shows minimum capacity of mixed concrete and rotating speed of drum.

of the original radial height, and the batch timer and counter are working properly. The mixer drum should be watertight. A reference plate attached to the mixer should state the maximum capacity of the mixer and the mixing speed (Fig. 7.7). The NRMCA *Quality Control Manual*, Section 3, "Plant Certification Checklist," provides guidance on quality control of production facilities and delivery vehicles.

7.2.1.1 *Charging the mixer*—It is best to feed water into the mixer over the full period of charging the mixer with dry material, beginning just before and ending just after this loading operation. All dry materials should be fed at the same time and as rapidly as possible without any loss of materials, either as spillage or dust, during charging.

When aggregates are batched by weight, the batch weights should be adjusted periodically during batching to compensate for variations in aggregate moisture content (ACI 311.5 recommends the minimum testing frequencies). Contract documents or supplementary instructions should make clear what aggregate conditions the mixture proportions are based on, and whether "dry" aggregate means oven-dry, air-dry, or saturated surface-dry (SSD). The selected basis should be used throughout batch computations.

At the beginning of a run, the amount of moisture in the aggregate may be higher than average because the aggregate is usually drawn from the bottom of the bin, where water in the aggregates may collect. During cold weather, aggregates stored in heated bins for extended periods, such as overnight, are sometimes dried to less than an SSD condition. To correct for these variations, make adjustments until conditions stabilize, preferably by removing and then refilling bins with saturated, aggregate well-drained of known moisture content.

7.2.1.2 *Water temperature*—Water may be any temperature that consistently produces concrete of the required temperature without causing setting problems or formation of cement balls. Other causes of cement balls are the introduction of cement ahead of coarse aggregate, worn mixer blades, hot aggregate or cement, and delayed mixing in truck mixers.

7.2.1.3 *Duration of mixing*—The necessary mixing time varies with the size and type of mixer. In the absence of contract document requirements for mixing time, the requirements for stationary mixers given in ASTM C94/C94M should be used, which call for a minimum mixing time of 1 minute for batches of 1 yd^3 (0.76 m^3) or less and an additional 15 seconds for each additional cubic yard (cubic meter) or fraction thereof. It should be determined if shorter mixing times are possible or longer mixing times are required based on results of performance tests, as described in ASTM C94/C94M. Shorter mixing times may be acceptable provided that the mixing time is sufficient to produce uniform concrete. Longer mixing times may be needed to bring results of performance tests up to standard, particularly if poor or dirty equipment is used (although use of such equipment should not be permitted). In large central mixers, failure to obtain well-mixed concrete is usually the result of an inefficient charging procedure, sequence, or worn mixing blades.

A maximum permissible time of actual mixing should be established (ACI 304R). If the batch is to be delayed beyond this time, the mixer should be operated only at intervals. When delays occasionally occur, continuing to mix the batch for several minutes can be beneficial because it improves uniformity and strength. Excessive mixing, however, is harmful because the grinding action can affect the soft aggregates and alter the setting and hydration performance of the cement. Discharge of the concrete should be completed within 1-1/2 hours or before the drum has revolved 300 revolutions, whichever comes first, after introduction of the mixing water to the cement and aggregates or the introduction of the cement to the aggregates. These limitations can be waived by the purchaser if, after the time and revolution limitations have been reached, that it can be placed without the addition of water to the batch.

The effect of mixing time on air content requires particular attention. Generally, the total air content increases by approximately 1% when the mixing time increases from 1 to 5 minutes. It then remains constant for the next 5 minutes of mixing; after 10 minutes, however, further mixing causes gradual air loss. Chapter 5 discusses the differences between entrained and entrapped air. It is the entrained air that should be retained in concrete.

To ensure that concrete is mixed for the proper length of time:

- Controls should be provided so that the batch cannot be discharged until the required mixing time has elapsed. At least three-quarters of the required mixing time should take place after the last of the mixing water has been added;
- The mixing time should be checked against the batch time on the delivery ticket for the first truckload, when sampling test specimens, and whenever the mixture appears segregated;
- The mixer should be efficiently time charged and discharged so that high rates of output can be attained without slighting mixing time. The entire batch should be discharged before the mixer is recharged (except for multiple-drum paving mixers); and

- The mixer should be operated at the speed recommended by the manufacturer.

7.2.1.4 *Mixer uniformity*—If the last portion of a batch being discharged from the mixer contains an excess of coarse aggregate, correct the condition by adjusting the mixer, the charging sequence, or the size of the batch. Segregated batches that vary from the uniformity requirements of ASTM C94/C94M, Table A1.1, should not be used. To check coarse aggregate uniformity, samples taken from different portions of the batch as discharged should be washed out, as described in Chapter 19. The amount of coarse aggregate in one part should not differ greatly from that in another. Segregated concrete batches should be rejected and not incorporated into the structure.

The uniformity of a concrete mixture is a direct function of the condition of the mixing equipment. Without proper maintenance, mixers will produce nonuniform concrete. Degradation of mixer uniformity can be rapid, even occurring within a few loads if proper care is not taken. ASTM C94/C94M lists six standard tests for uniformity that are to be performed on two concrete samples representing the first and last portions of the batch being tested. The tests and the permissible maximum differences between the samples are shown in Table 7.1. Test results conforming to the limits of five of the six tests indicate uniform concrete.

Mixer uniformity tests vary. The U.S. Bureau of Reclamation (1988) requires two samples from the first and last portions of the batch, similar to the requirements of ASTM C94/C94M. The samples are tested for variance in quantity of coarse aggregate and density of air-free mortar, as described in Chapter 19. The U.S. Army Corps of Engineers requires three samples taken from the first, middle, and last portions of the batch. The samples are tested for density of air-free mortar, quantity of coarse aggregate, water content, and cement content.

7.2.1.5 *Transporting equipment*—Central-mixed concrete may be transported in truck mixers or agitators or in suitable nonagitating containers. Nonagitating equipment should have smooth, watertight, metal bodies with gates for control of discharge. Covers should be provided to protect concrete from the weather. Uniformity requirements for nonagitated concrete are the same as those discussed previously.

7.2.2 *Ready mixed concrete*—ACI 116R defines ready mixed concrete as central-mixed, truck-mixed, or shrink-mixed concrete manufactured for delivery to a purchaser in a plastic and unhardened state. (Shrink-mixed concrete is partially mixed in a stationary mixer and then delivered to a separate mixer for final mixing.) ASTM C94/C94M is the standard specification that applies to ready mixed concrete. The previous discussion in this chapter on central and site mixing applies to ready-mix operations that use a central mixing plant. This section covers concrete that is truck-mixed or shrink-mixed and delivered to the job in truck mixers or agitators. ASTM C 94/C 94M mixer uniformity tests also apply to truck mixers.

7.2.2.1 *Methods of ordering*—ASTM C94/C94M provides for three optional ordering methods. The purchaser specifies the method to be used, along with the maximum

Table 7.1—Permissible differences between tests

Test		Maximum permissible difference
Density of concrete (calculated air free)		1 lb/ft^3 (16 kg/m^3)
Air content (volume percent of concrete)		1%
Slump	4 in. (100 mm) or less	1 in. (25 mm)
	4 to 6 in. (100 to 150 mm)	1.5 in. (40 mm)
Coarse aggregate content (weight percent)*		6.0%
Density of air-free mortar		1.6%
Compressive strength (7 days)		7.5%

*Ratio of weight of aggregate retained on and washed over a No. 4 (4.75 mm) sieve to the total weight of the concrete sample.
Note: Adopted from ASTM C94/C94M, Table A1.1.

aggregate size, type of aggregate (normal or lightweight), slump, and air content. For each ordering method, the procedures for determining proportions that will provide the required concrete quality will differ. Refer to ASTM C94/C94M for a detailed description of each option.

Whatever option is used, each set of mixture proportions should be clearly labeled to facilitate identification of each mixture delivered to the project. This designation should appear on the delivery ticket. ASTM C 94/C 94M specifies 10 mandatory and eight additional (if required by contract documents) items of information to be provided on the delivery ticket.

All automated and many semi-automated batch plants are equipped with recorders. These devices monitor the automated batching process and produce a complete record of:
- The quantity of all ingredients in the batch (other than those added separately from the automated or semi-automated process, such as flaked ice);
- The bin, silo, admixture dispenser, and water source used;
- The amount of aggregate moisture measured in the bin via a direct reading or input based on laboratory measurement; and
- The time of day when batching was initiated and completed.

A recorder clearly distinguishes when a batch computer is overridden and quantities are manually charged. It also maintains an electronic copy of all batch ingredients that can be downloaded to a computer disk for archiving. In addition, the recorder usually can be coupled to a printer to produce a hard copy of the batch ingredients. The inspector can request that a copy of the recorder printout accompany each concrete delivery so that an immediate verification of ingredients, tolerances, design *w/cm*, and batch times can be made. The correlation between the recording and the mixture proportions should be checked.

7.2.2.2 *Requirements for truck mixers and agitators*—ASTM C94/C94M requires that each truck mixer and agitator have an attached metal plate (Fig. 7.8) giving the gross volume of the drum, the capacity of mixed concrete, and the minimum and maximum rotating speeds of the drum, blades, or paddles. When concrete is transit- or shrink-mixed, the volume should not exceed 63% of the total volume of the drum or the container. If the concrete is central-mixed, the volume of concrete in the truck mixer or agitator should

Fig. 7.8—A truck mixer must not mix and transport a batch larger than the capacity shown on its rating plate. If it operates as an agitator only, it can transport a larger amount.

not exceed 80% of the total volume. Each truck mixer or agitator should be equipped with counters or other means of verifying the number of revolutions before discharge.

All truck mixers should be capable of combining the ingredients into a thoroughly mixed and uniform mass within the specified time or number of revolutions. Agitators should be capable of maintaining the mixed concrete in a uniform mass and discharging concrete with a satisfactory degree of uniformity. To check uniformity, slump tests should be performed from samples taken after discharge of approximately 15 and 85% of the load. If these values differ more than the permissible amount required in the ASTM C94/C94M uniformity test described in Chapter 19 (Table 19.1), the full series of uniformity tests should be performed and adjustments made to correct any nonuniform conditions such as repair/replacement of mixer blades, use of a longer mixing time, a smaller load, or a more efficient charging sequence.

Mixers and agitators should be examined frequently to detect changes in condition due to worn blades or accumulations of hardened concrete, and to ensure there is no leakage from the water tank into the truck mixer.

7.2.3 *Volumetric batching and mixing*—Some central- and site-mix plants use volumetric batching equipment combined with continuous mixing equipment. Volumetric batching and mixing can also be provided by trucks that carry aggregate, cement, and other ingredients in separate compartments and mix fresh concrete at the point of placing. ASTM C685/C685M provides specifications for volumetric batching and mixing. Further guidance is given in ACI 304.6R.

7.2.3.1 *Measuring materials*—Cementitious material, fine and coarse aggregates, water, and admixtures should be accurately measured materials fed into the equipment in a uniform flow. When proportioning by volume, counters, calibrated gate openings, or flow meters should be used to control and determine the quantities of ingredients discharged. The proportioning and indicating devices should be checked individually by following the equipment manufacturer's recommendations. The manufacturer's recommendations also should be followed for operating the equipment and for calibrating and using the various gauges, revolution counters, speed indicators, and other control devices.

7.2.3.2 *Mixing mechanism*—For continuous mixing, an auger or other type of mixer suitable for mixing concrete should be used to meet the consistency and uniformity requirements of ASTM C685/C685M.

7.3—Plant inspection

At central- or site-mixing plants, inspectors' duties include:
- Verifying proper storage conditions of all materials;
- Verifying the condition of all mixers and plant equipment;
- Monitoring the plant's quality-control activities to the requirements of the ready-mix supplier's quality plan;
- Verifying the type and amount of all materials specified;
- Verifying proper adjustments to mixture proportions for moisture in aggregates;
- Verifying the performance of required material tests;
- Observing batching and mixing operations;
- Verifying the accuracy of all weighing and measurement systems;
- Testing concrete at the plant for slump, air content, temperature, and density, if required; and
- Casting of strength specimens and transfer of specimens to testing laboratory at proper time, if required.

7.3.1 *Control of water content*—Because control of water content is important to ensuring the desired concrete quality, it should be continually watched to see that proper charging of water takes place and that proper adjustments are made for the moisture content of the aggregates. If the plant has moisture meters for measuring moisture in the aggregates, they should be checked for proper calibration and correct use, and their readings compared with other physical indicators of moisture content.

Concrete consistency can vary when waterline pressure fluctuates. When pressure is low, the measuring tank does not fill completely before discharging. This variation can be prevented with the use of interlocks and other methods, as discussed previously.

The inspector should verify that the mixer is empty after cleaning and before batching. Any cleaning water not properly drained can enter the following batch. When specifications allow the use of wash water in subsequent batches, it should be verified that this water is properly accounted for in calculating total water.

It should be arranged with the producer to have truck-mounted water tanks filled before each delivery. Although ASTM C94/C94M requires that the tank level be recorded before the truck leaves the plant, the possibility of additional water being introduced to the load during transit is minimized if the tanks are always filled before truck dispatch. The water level remaining in the tank can easily be verified upon concrete delivery. During freezing weather, the waterlines may need to discharge continuously or be fully drained to avoid freeze-up.

To minimize concrete spillage on public streets, most suppliers instruct their drivers to wash the hoppers and fins of trucks before they leave the batching yard. Some suppliers erect wash racks specifically for this purpose; others use the water stored in the truck tank. Some water will always enter the batch when the truck is cleaned, but the amount can be reduced considerably when the operator exercises care. Inspectors should monitor this activity and insist that the amount of water entering the batch be minimized. Any water that does enter the batch should be estimated and accounted for.

7.3.2 *Control of air content*—Correct air content is best obtained with the use of a carefully batched air-entraining admixture. As discussed in Chapter 6, the use of air-entraining cements often leads to erratic air contents that are difficult to correct.

When air entrainment is required, maximum and minimum percentages are based on the maximum coarse aggregate size. Appropriate amounts are shown in Chapter 5, and specific recommendations can be found in ACI 211.1, 211.2, and 301. The amount of entrained air per volume of concrete required is usually less for aggregate of larger maximum size because the paste content is less.

Contract documents may require certain air-content limits. If improved workability and cohesion of the mixture are desired rather than freezing-and-thawing resistance, less air may be needed. When entrained air is not needed for durability, it may also be unnecessary for workability, particularly when mixture proportions are favorable and when pozzolans or water-reducing admixtures that entrain a slight amount of air are used. For certain applications, contract documents may not allow the use of entrained air, usually when maximum density or extra-high strength is required or when shake hardeners are to be applied to surfaces.

Regardless of the total amount of air required, it is important that the concrete contain a quantity of air uniformly close to the desired amount batch after batch. Too much air can reduce strength without a compensating improvement in durability, and too little may fail to provide the desired workability and durability. If the sand and water content are at a minimum for a certain amount of air entrainment, a drop in air content can cause a serious loss of workability. Too much entrained air can cause excessive stickiness during slab finishing operations.

One or a combination of factors can cause on-the-job variations in air content from a given dosage of air-entraining admixture:
- A change in the concentration of the air-entraining admixture;
- Changes in the brand or type of cement, pozzolan, or other admixtures;
- Variations in the temperature of the mixture, slump, or length of mixing. A given dosage of admixture will produce less air when the temperature of the mixture rises, mixing is excessively prolonged in truck mixers, slump is lower, or there is an increase in cement or pozzolan fineness; and
- Changes in the fineness modulus of the fine aggregate.

Changes in air content that will be caused by changing conditions should be anticipated. To verify that correct amounts of air are being obtained, routine tests should be performed once or twice per shift, or more often if there is reason to suspect a change. When frequent and quick indications of approximate air content are desired, a small pocket-type air indicator can be used, but it should not be used as a basis for accepting or rejecting batches. In case of doubt, the concrete should be tested by one of three approved methods. These methods and the advantages of each are discussed in Chapter 19. Several meters are available for direct measurement of air content of a representative sample of concrete in a properly filled container. Volumetric and pressure-meter methods are described in Chapter 19 and in ASTM C173/C173M and C231.

The significant amount of air in concrete is the entrained air that remains after the concrete has been placed and consolidated. Loss of air that occurs as concrete is handled, transported, and vibrated after placement may not be reflected by air-content tests of samples taken at the mixer (except for the extent to which consolidation of the test sample in the air-meter container represents consolidation in the forms). Fortunately, the first air lost as a result of these manipulations consists of the larger bubbles of entrapped air that do not contribute to freezing-and-thawing durability. Long agitation or mixing in truck mixers and excessive vibration, however, can seriously reduce the amount of air, especially when the initial amount is less than that recommended in Table 5.1.

The concrete placement method can also affect air content. When pumped, most concrete mixtures lose some entrained air into solution in the mixing water due to the pressure of pumping. Some air-entraining admixtures, however, seem to increase the quantity of entrained air as the mixture passes through the pump chamber.

When concrete is pumped through a typical boom truck, the boom's position can influence the total air content. If the boom is nearly vertical (as is the case when the point of placement is close to the pump), the freefall of concrete through the pump line creates a vacuum, causing an evacuation of entrained air from the concrete. Regardless of the sampling location defined in the contract documents, inspectors should perform check tests of concrete after it has been pumped in this manner to confirm that the air content at the point of placement into the forms meets specified limits.

Occasionally, and certainly when there is a possibility that significant air loss has occurred, the amount of air in the concrete should be tested after it has been vibrated in place. It is particularly important that the surface and upper portion of pavement and bridge deck slabs and the exposed surfaces of hydraulic structures contain the specified amount of entrained air. If more than one-fourth of the amount of air is lost, the practices causing excessive loss should be corrected or compensating entrained air added. Care should be taken when adjusting air. If placing conditions change, the concrete in place may end up with a high air content and lower strength.

7.3.3 *Control of temperature*—As noted in Chapter 9, high temperatures within plastic concrete can cause excessive evaporation and difficulties in placing and finishing. High concrete temperatures combined with atmospheric factors, such as high winds and low humidity, can also lead to excessive or rapid drying and plastic shrinkage. In hot weather, or

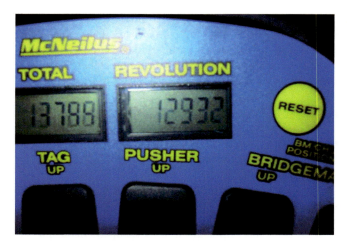

Fig. 7.9—A mixer revolution counter helps maintain uniform concrete consistency. Seventy to 100 revolutions are permitted at designated mixing speed for ready mixed concrete.

under conditions contributing to rapid stiffening of the concrete, the purchaser may specify that the discharge of the concrete be completed in less than 90 minutes. Typically, most contract documents limit the temperature of concrete as placed. Thus, it is essential to measure the temperature of the mixture before placement. Several kinds of thermometers are suitable for this purpose. The concrete temperature should be recorded each time a slump test or air-content test is conducted or when compression test cylinders are made.

If plastic concrete temperatures are expected to be less than 50 °F (10 °C), and if low temperature is causing difficulty, warm water may be used for mixing water. Cool temperatures well above freezing, however, can be advantageous to the ultimate quality of concrete by reducing cracking. (Refer to Chapter 16 for effects of high temperatures on heavy concrete sections, such as mass concrete.)

7.4—Placing inspection

When monitoring the mixing of ready mixed concrete delivered to the job in a truck mixer, the revolution counter of the truck should be checked (Fig. 7.9) to confirm that the number of revolutions at mixing speed is within the prescribed limits (usually 70 to 100). Any additional revolutions should be at agitating speed. The consistency of the delivered concrete should be determined, and the required test specimens taken.

7.4.1 *Control of slump loss*—Slump loss during the time interval between mixing and placing is sometimes a serious problem with ready mixed concrete because it can result in the need to add extra water with a subsequent increase in total water content. Slump loss greater than 1 in. (25 mm) is objectionable because it usually creates a demand for a higher initial water content to provide greater initial slump to compensate for the slump loss that will occur before placing. Slump loss increases as the time between the start of mixing and the placing of concrete increases. It is also aggravated by higher temperatures, absorptive aggregates, severe false setting of cement, richer mixtures, and improper use of accelerators.

To reduce slump loss, the following measures may be taken:
- If concrete temperatures are high, the temperature of the concrete and surrounding materials should be lowered by using chilled water or adding ice to the mixing water; spraying aggregate, forms, and subgrade with water; shading and dampening ready-mixed mixer drums; and shading materials. Concrete will also stay cooler if placed at night or during early-morning hours. In hot weather, lower early-morning air temperatures, lower wind velocity, and higher humidity may not only reduce slump loss, but may also help reduce plastic shrinkage;
- If a long haul is at fault, arrangements should be made to add water and mix concrete after trucks arrive at the forms, or at least delay mixing until a few minutes before truck arrival;
- If there is too much of a delay between discharge of the first and last part of each batch, the batch size should be reduced;
- The cement should be as cool as possible and free of any tendency toward false set (see discussion of false set in Chapter 3);
- The *w/cm* should be carefully checked and the cement requirements increased as necessary to correspond with increases in unit water content; and
- The use of retarders should be considered. Slump loss, however, may sometimes be greater with retarding admixtures.

7.4.2 *Control of consistency*—One of the most important duties of the inspector is to observe the consistency of concrete at the mixer, in conveying devices, and especially at the forms. As stated in Chapter 6, it is desirable to keep the consistency of concrete constant for a given kind of work to simplify conveying, placing, and finishing.

Radio or telephone communication between personnel at the discharge point of ready mixed concrete and personnel at the central mixing or batch plant allows faster changes in proportions that may be required to obtain proper workability or consistency, especially if truck mixers are used. A consistency or slump meter can be of great help to the mixer operator or ready-mix truck driver in attaining the desired slump. Meters that record the power usage of the mixer (wattage and oil pressure) can also be useful. A relationship between the power used and the desired slump of the concrete can then be established so that adjustments to the mixing water can be made readily if the power readings are too high.

The general tendency of some contractors is to make the consistency of concrete as wet as possible because a wet consistency will reduce the labor of placing (but not necessarily the overall labor requirement). The use of a wetter consistency, however, results either in lower-strength concrete or in a greater cement requirement to keep the *w/cm* at the required level. Furthermore, concretes with higher water contents experience greater shrinkage and are more likely to segregate. Therefore, the mixture should be only as wet as absolutely necessary for proper placement. Whenever a high

Fig. 7.10—The slump test is made on a smooth, non-absorbent surface such as a sheet of metal. Protect the surface from jarring by nearby equipment. Hold down the slump cone with both feet.

slump is desirable, using an admixture to make the consistency more fluid without diluting the *w/cm* should be considered.

Consistency is usually regulated by varying the amount of water added at the mixer based on observed or tested consistency of previous batches. If aggregates are uniform in moisture content and grading, there should be little need to vary the amount of added water. Nevertheless, the water-measuring device should not be locked at a fixed quantity because unavoidable variations in aggregate moisture content would then result in variable consistency. Most modern plants are equipped with moisture meters and, in the case of central mixing drums, slump indicators that assist the operator in making decisions about adjustments to mixing water from load to load. The inspector should verify that the approved maximum *w/cm* value is not exceeded whenever mixing water adjustments are made.

Consistency should be tested periodically for record purposes and to determine compliance with contract documents. Consistency, however, should not be judged by tests alone; observe the workability of the concrete and suitability of its consistency by how the concrete responds during placement and vibration in the forms. A test should be made and recorded when wetter concrete is subject to rejection.

The slump test, made in accordance with ASTM C143/C143M, is typically used to test consistency. The procedures for this test are described in Chapter 19 and Fig. 7.10. The result of a single slump test should not be the only basis for rejection, because the test itself is subject to considerable variation unless all requirements of ASTM C143/C143M are strictly followed. For example, the indicated slump may be too high if the base beneath the slump cone is subjected to jarring or too stiff if the base is rough or dry. Contract documents usually set permissible slump tolerances, but some documents may simply state that it shall not exceed a fixed value. In the absence of specific contract provisions, ASTM C94/C94M can be used.

When a truck mixer is approved for mixing and delivery of concrete, water should not be added after the initial introduction of mixing water for the batch unless concrete arrives at the job site with a lower slump than specified. If water is added to bring the slump within specified limits, the design *w/cm* should not be exceeded (otherwise, the batch should be rejected), and the water addition should be recorded and noted on the delivery ticket. The added water should be injected into the mixer under the pressure and flow needed to meet ASTM C94/C94M uniformity requirements. The drum or blades should be turned an additional 30 revolutions or more at mixing speed until uniformity is within limits. Do not add water to the batch at any later time.

If the time between the mixing of concrete and its final placement in the forms is too long, the mixture is likely to have stiffened so much that it cannot be consolidated satisfactorily. The degree of stiffening beyond which the concrete cannot be consolidated will depend on the nature of the placement and the vibration method used. More than a slight amount of stiffening calls for correction if the time between mixing and placement cannot be reduced.

Other factors leading to premature stiffening include:
- Excessive evaporation of mixing water by sun or wind;
- Unanticipated high absorption of mixing water by aggregates;
- High temperatures of one or more of the ingredients;
- Improper use of an accelerator; and
- Use of cement or cement-admixture combinations with premature stiffening characteristics.

Faulty performance of vibrators sometimes gives a false indication of stiffening. In hot climates, retarders are routinely added to concrete to prevent premature stiffening, but they can increase the rate of slump loss.

7.4.3 *Measuring concrete quantity*—Concrete may be measured by volume in the receiving hopper or forms, by weight, or by adding the absolute volumes of cement, water, air, and aggregates (Chapter 6). According to ASTM C94/C94M, the basis of sale for ready mixed concrete is the volume yield of the batch determined by dividing the total weight of the materials batched by the density of freshly mixed, unhardened concrete. Although such measurements are of interest primarily to the producer and contractor, inspectors can use them as a check on batch quantities, and thus, on cement content.

CHAPTER 8—INSPECTION BEFORE CONCRETING

Common imperfections and unsatisfactory results in concrete construction are often due to inadequate preparation for the work. Close inspection at various stages before concrete placement begins is critical to ensuring that the desired results are achieved.

The first step is to conduct a thorough preliminary study of the project, becoming as familiar as possible with site conditions, the contract documents, and relevant requirements of any referenced or related specifications and building codes. Before concrete is placed in a given section of the work, it is necessary to inspect the excavation, forms, shores, reshores, reinforcement, and embedded items to verify that they meet specified requirements. It also is important to make sure that preparations have been made to form construction joints and to cure and protect the concrete.

8.1—Preliminary study

The following actions should be taken to become familiar with conditions at the site and the requirements of the contract documents:
- Confirm that the contract documents correlate with each other and with any special instructions;
- Examine any shop drawing details and erection or placement drawings, and verify that they conform to contract document requirements;
- Check reinforcement details and other details for potential constructibility problems; and
- Confirm that drawings bear appropriate approval stamps and are the latest revisions.

The general organization of the work and the facilities and equipment being used should be observed. Particular attention should be given to subgrade compaction equipment and procedures; concrete batching, mixing, transporting, and placing facilities; construction-joint planning; and equipment for vibrating concrete.

To help prevent problems during concrete placement, all parties involved in the placing operations should meet to establish clear lines of communication. Methods to be used for curing, form stripping, shoring, reshoring, and testing should be reviewed, along with safety regulations.

8.2—Stages of preparatory work

Inspection before concrete placement should continue throughout the following three stages of the work:
- *Preliminary*—A preliminary inspection should be made when excavation has been completed and forms have been built. (Note: Before placement of forms, it should be confirmed that the subgrade has been approved.) If form dimensions and stability are satisfactory, the contractor can then clean the foundation, coat the forms, and install any reinforcement and fixtures;
- *Semifinal*—When everything is in place for concreting, a detailed inspection should be made of foundations, forms, reinforcement, and all equipment or parts to be embedded in the concrete. If the installations are satisfactory, the work is ready for final inspection; and
- *Final*—The final inspection should be made immediately before concrete is placed. Check that forms and fixtures have not been displaced and that surfaces are clean and, if specified, wetted. All pertinent items on the checklist (described at the end of this chapter) should be properly signed for.

Assuming that requirements for materials, mixture proportions, and working conditions (weather, time, lighting, equipment, access for concrete delivery, and curing protection) have been met, the contractor can then proceed to place the concrete.

8.3—Excavations and foundations

Excavated surfaces upon or against which concrete is to be placed (Fig. 8.1) should conform to the specified location, dimensions, shape, compaction, and moisture requirements. It should be verified that:
- The footing/foundation area has been checked and approved for bearing design by a geotechnical engineer;
- The foundation material has been moistened to provide a nonabsorbent surface for concrete placement. All standing puddles of water should be removed before concrete placement, as this water will bleed into the concrete and reduce the *w/cm* of the concrete;
- Provisions have been made for drainage where necessary;
- The slope of excavations is stable or shored to ensure worker safety during concrete placement and vibration. (Note: OSHA requires daily inspections of excavations and adjacent areas by a competent person.); and
- For drilled pier foundations, refer to the requirements of ACI 336.1.

When structural concrete must be placed directly against the earth, the condition of the subgrade is vital. Unless the

Fig. 8.1—This excavated trench will serve as the form for concrete grade beams to be placed without wood forms. The condition of the subgrade is vital. Unacceptable conditions, such as those depicted, require correction prior to concrete placement.

structure or slab is being placed on controlled fill or is supported by another mechanism, such as piles, the usual requirement is for the earth to be undisturbed. Clay soil subgrades should be protected from water penetration before placement of concrete. Water-softened or disturbed soils will need to be undercut, replaced, and compacted with structural fill material as defined in the project specifications or the geotechnical report before concrete placement.

Depending on local climate conditions, the expected duration of the preparation work, and the nature of the clay soils as described in the project geotechnical report, it may be necessary to place a mud mat or seal slab on the subgrade. Typically installed at a nominal thickness of 2 to 4 in. (50 to 100 mm), this base layer of concrete should be placed as soon as possible after the bearing surface is exposed so it will seal the subgrade from moisture gain or loss and preserve the subgrade's integrity. Mud mats may also be required with expansive rock, such as shale, or with massive foundations that must support a large amount of reinforcement. Placing bar supports on a mud mat provides much greater stability than placing them directly on the subgrade. Although mud mat concrete need only develop strength equal or superior to the strength of the soil, a 1500 to 2000 psi (10 to 14 MPa) concrete is commonly used.

8.3.1 *Building slabs-on-ground*—The inspector should ensure that the subgrade is compacted and tested as required by contract documents. The type of subgrade material dictates the type of compaction equipment to be used. Cohesive materials (clays) are best compacted by rollers or tamping equipment. Cohesionless materials (sand and other granular materials) are best compacted by vibrating compaction equipment. Particular attention should be paid to compaction along edges of foundation walls. Backfill in trenches and ruts should also be thoroughly compacted and tested for density. Soft spots and all portions of the subgrade that might later be subject to settlement or swelling should be eliminated, including fissures, inclined layers, clay layers, and water-bearing sand layers.

For more information on slabs for buildings, see Chapter 12. Pavement slabs are discussed in Chapter 13.

8.3.2 *Building foundations*—For major foundations, approval of the bearing area by the resident engineer, architect/ engineer, or geotechnical engineer usually is required before concrete may be placed. In rock excavations, the surface of the rock should be sound, completely exposed, perpendicular to the direction of load, and of a capacity required by the design. Some designs may require keying the footings into the rock. If unsuitable soils (either natural soil or fill) are encountered at a planned bearing elevation, the resident engineer, architect/engineer, or geotechnical engineer should be contacted for resolution. A copy of the geotechnical report should be maintained at the project site for reference by all parties. The project geotechnical report should be consulted for further information.

Surfaces against which concrete will be placed should be clean and moist, but not soft. Preferably, water or an air-water jet should be used to clean rock surfaces of foundations, followed by an air jet to remove excess water. No pools of water should remain. If new concrete will be placed on or against previously cast concrete, the surface of the older concrete should be free of oil, grease, foreign matter, and laitance. The contract documents may require the use of wet sandblasting or high-pressure water blasting. They may also call for roughening. If a thoroughly clean surface, comparable with a fresh break, is obtained, roughening is not necessary to obtain bond. A clean joint surface approaching dryness without free water is best for bond strength (U. S. Bureau of Reclamation 1988). Free surface water will increase the w/cm and, therefore, weaken the mortar or cement paste in new concrete adjacent to old concrete.

8.3.3 *Underwater placements*—Placement of concrete under water is a difficult operation to accomplish and to inspect. Therefore, it should not be done unless permitted by the contract documents or the architect/engineer. For further details on underwater concreting, refer to Chapter 15.

8.3.4 *Pile foundations*—If piles will support the concrete, their installation should be inspected by a specialist. The concrete inspector should check that the number and location of piles are correct and that none of the piles deviates more than the specified tolerance from the designed alignment.

8.4—Forms for buildings

Achieving sound concrete with visually pleasing exposed surfaces requires that forms be tight and properly aligned, coated for release, clean, and well-maintained. Before concreting begins, the forms, shores, and bracing that will support the concrete should be inspected so that errors can be corrected without delay. The contractor should be advised whenever the security or rigidity of the forms is in doubt. If the contractor does not correct the situation, the contractor, the owner, and the engineer should be given written documentation of the problems and a description of possible remedial actions.

Also, the inspector should check that any foreign material inside the forms (such as sawdust, dried mortar, snow, and ice) has been removed immediately before concrete placement. Because foreign material is likely to accumulate in corners

Fig. 8.2—True, tight, well-balanced forms will resist bulging as concrete is placed and vibrated.

and other places difficult to reach, it may be necessary to vacuum or blow out the debris. In deep, narrow forms, the contractor should provide holes for cleaning and inspection at the most effective locations, which are usually at the bottom and at joint levels of the forms. These holes should be closed before concreting starts or just before the concrete reaches their level.

8.4.1 *Form tightness and alignment*—Forms should be checked as soon as possible after erection for any lack of tightness. Smooth form surfaces with tight joints and no holes will prevent mortar from escaping during concrete vibration.

The forms should result in hardened concrete that meets the required dimensions, alignment, and surface finish. Although the project engineer will set the governing points of line and grade, additional measurements may be needed from and between these points. A job-built template can serve as a convenient, accurate way of checking dimensions and alignment. An accurate straightedge of proper length should also be at hand. Irregularities usually can be detected by careful sighting, but plumb lines and stretched lines or wires may be necessary in some locations.

Joints in forms should be secure, even, and tight. To avoid unsightly offsets and mortar leakage at horizontal construction joints, workers should reset the forms to overlap the concrete only 1 in. (25 mm) or so. The use of ample tie bolts close to the joint will ensure that the forms are held tightly to the hardened concrete during placement and vibration of adjacent concrete. If climbing or jump forms are used (Fig. 8.2), the forms should be raised tight against the concrete already cast.

Form ties and spacers should not leave metal near an exposed surface nor bend if workers climb on them. Wire ties should be used only on light work, and holes for the wires should be as small as possible. Where appearance of the structure is important, form lines and form ties should be arranged to make neat patterns. The designer should approve the layout before work begins.

When forms are tight and their surfaces are dense and impervious, air bubbles may form or water may accumulate at the surface, resulting in minor surface defects in the concrete. These defects principally affect the appearance of the concrete, and are not serious unless they occur in architectural exposed concrete.

8.4.2 *Shoring*—The number, type, and location of shores and bracing should be verified. Many contract documents and code jurisdictions require shoring systems to be designed by licensed design professionals. If bottom forms sag even when shored as specified, the contractor should be advised of the situation and corrections made and documented. Adequate bearing area on the ground can prevent settlement of shores. During concreting operations, it may be necessary for the contractor to adjust the shoring using screw jacks to maintain proper elevations.

Shores supporting successive stories should be placed directly over those below, and shores should be installed for the number of stories required to carry total loads. Shores for cantilevers often are critical, and should be adequate to support all loads. Construction joints may create temporary cantilevers until the member is completed. Concrete on both sides of such joints should be supported until all of the concrete has developed enough strength to permit the member to carry its own weight. Similarly, inspectors should be alerted to the need for additional shoring under temporary construction loads such as material or equipment storage. See also Chapter 10.

8.4.3 *Preventing bulging and settlement*—After they are filled with concrete, the locations and dimensions of forms may not be the same as when they were erected because the weight of concrete, workers, and placing equipment can cause settling, sagging, or bulging of the forms. Fresh concrete, when vibrated, exerts the maximum pressures shown in Tables 8.1 and 8.2, which are taken from ACI SP-4 (Hurd 2005).

Because it is usually impossible to force a form back into position after it has bulged or slipped while being filled, use of proper bracing and form ties and sufficiently stiff form members is essential. If possible, a carpenter should be assigned to each placement so that adjustments can be made to the forms and shores when necessary. To give early warning of any movement or deflection of the forms, installing "telltale" arrangements (string lines and plumb lines left in place during placing operations) at several locations on the forms should be considered, particularly where settlement or deflection is most likely. Immediate actions can then be taken to stop these movements or deflections, or at least to control them within prescribed limits. During concrete placement, a worker should monitor telltales to check for and stop leaks and to check and tighten forms, accessories, and bracing as required.

Settlement and sagging can be controlled by building camber into the form. One widely used rule is to camber floor and beam forms 1/4 in. per 10 ft (6 mm per 3 m) span. Cambering of the forms is the responsibility of the contractor.

8.4.4 *Coating for release*—Before concrete is placed, check that all contact surfaces and edges of forms are wetted or coated with a nonstaining form oil or other suitable release agent. The coating should not be applied so thickly that staining or softening of the concrete surface will occur. It is important for form coatings to be applied before reinforce-

Table 8.1—Maximum lateral pressure for design of wall forms

Rate of placement R, ft/hour	p, maximum lateral pressure, lb/ft², for temperature indicated					
	90 °F	80 °F	70 °F	60 °F	50 °F	40 °F
1	600 lb/ft² governs					
2						
3					690	825
4			664	750	870	1050
5	650	712	793	900	1050	1275
6	750	825	921	1050	1230	1500
7	850	938	1050	1200	1410	1725
8	881	973	1090	1246	1466	1795
9	912	1008	1130	1293	1522	1865
10	943	1043	1170	1340	1578	1935

Note: Do not use design pressures (in lb/ft²) greater than 150 × height in feet of fresh concrete in forms. Table applies only for normalweight concrete made with Type I cement, no admixtures or pozzolans, slump no more than 4 in., and vibration depth of 4 ft or less. 1 ft = 0.305 m; (°F – 32)/1.8 = 1 °C; 1 lb/ft² = 4.88 kg/m².

Table 8.2—Maximum lateral pressure for design of column forms

Rate of placement R, ft/hour	p, maximum lateral pressure, lb/yd², for temperature indicated					
	90 °F	80 °F	70 °F	60 °F	50 °F	40 °F
1	600 lb/ft² governs					
2						
3					690	825
4			664	750	870	1050
5	650	712	793	900	1050	1275
6	750	825	921	1050	1230	1500
7	850	938	1050	1200	1410	1725
8	950	1050	1178	1350	1590	1950
9	1050	1163	1307	1500	1770	2175
10	1150	1275	1436	1650	1950	2400
11	1250	1388	1564	1800	2130	2625
12	1350	1500	1693	1950	2310	2850
13	1450	1613	1822	2100	2490	3000
14	1550	1725	1950	2250	2670	
16	1750	1950	2207	2550	3000	
18	1950	2175	2464	2850		
20	2150	2400	2721	3000		
22	2350	2625	2979			
24	2550	2850	3000			
26	2750	3000				
28	2950					
30	3000	3000 lb/ft² maximum governs				

Note: Do not use design pressures (in lb/ft²) greater than 150 × height in feet of fresh concrete in forms. Table applies only for normalweight concrete made with Type I cement, no admixtures or pozzolans, slump no more than 4 in., and vibration depth of 4 ft or less. 1 ft = 0.305 m; (°F – 32)/1.8 = 1 °C; 1 lb/ft² = 4.88 kg/m².

ment is placed to avoid coating the reinforcement and preventing bond with the concrete.

8.4.5 *Form reuse and maintenance*—Reusable forms can be made of various materials, including wood, plastic-coated plywood, steel, and glass fiber-reinforced plastic. The latter type is widely used for architecturally exposed concrete because it can be shaped to any desired contour. For more details about forms for architectural concrete, see Chapter 14. Construction of forms for structural and architectural concrete and specially formed surfaces are described in ACI SP-4 (Hurd 1995) and ACI 347.

The inspector should make sure that forms do not contain cracks, nicks, dents, bulges, loose joints, or deformations that prevent proper fit. Forming crews should avoid marring form surfaces throughout construction and other work preliminary to concreting. Before reuse, forms should be cleaned and, if necessary, reconditioned. Typical maintenance requirements include filling open seams, planing warped boards, straightening metal facings, and rematching joints. Metal forms should not be sandblasted or abraded.

8.5—Reinforcement

Reinforcement should be checked as soon as possible for proper strength, grade markings, size, bending, horizontal and vertical spacing, concrete cover, location, adequacy of support, and surface condition. Do not wait until reinforcement has been wired in place (and is more expensive to alter) before checking it. Improper placement or omission of reinforcement can lead to severe cracking, steel corrosion, excessive deflections, and even failure. For detailed information and illustrations, refer to *Placing Reinforcing Bars* and *Manual of Standard Practice* (Concrete Reinforcing Steel Institute 1997a,b).

8.5.1 *Cutting and bending*—It should be verified that all bending details are correct. Pay particular attention to seismic criteria. Hooks of 135 degrees, rather than 90 degrees, are required. Unless closer limits are stated in the contract documents, straight bars have a length tolerance of 1 in. (25 mm). Bent bars usually are measured by using outside-to-outside dimension of the bar, but some organizations use center-to-center dimensions.

If reinforcement will be job-fabricated, the bending pin diameter should not be less than the recommended sizes in *Manual of Standard Practice* (Concrete Reinforcing Steel Institute 1997b) and ACI 301. Different pin diameters are required for different grades of steel (Grades 40, 60, and 75 [280, 420, and 485]). If several bars will be bent alike, check the first one bent (preferably by placing it in the forms) before the others are bent. Bending bars around a pin of too small a diameter may induce cracking of the bar at the bend.

Bending or straightening of the steel in a manner that would weaken the material should not be allowed. Because heating can change the characteristics of the steel, reinforcement should be heated for bending only when approved by the design engineer. If heating is permitted, the steel should not be heated above 1500 °F (816 °C), and should be allowed to cool slowly. If bars being bent by heating are partially embedded in concrete, the heating or bending process should not be allowed to damage the concrete surrounding the bar. For additional requirements, refer to ACI 301. Bending or heating of prestressing steel should never be permitted (refer to Chapter 17).

8.5.2 *Storage and handling*—Except for prestressing tendons, a thin, adherent film of rust or mill scale is not objectionable on reinforcing steel because it increases the bond of the steel to concrete. Storage conditions that might

Fig. 8.3—Reinforcing steel is ready for placement of a slab.

Fig. 8.4—A certified inspector checks the placement of the reinforcement.

cause excessive rusting of the steel, however, should be avoided. Coatings that are objectionable and require removal before concrete placement include paint, oil, grease, dried mud, and weak dried mortar or concrete. (If mortar is difficult to remove with a wire brush, it is probably harmless and need not be removed.)

Special care should be taken in the handling of epoxy-coated steel. The reinforcement should be staged on cribs and lifted with nylon slings to prevent coating damage. In sunny regions, it may be necessary to store epoxy-coated reinforcement in shaded areas because differences in the thermal expansion characteristics of the steel and the coating can cause coating failure. Any breaks in the coating, including cut ends, should be repaired. Many specifications require that epoxy reinforcement coating be opaque when storage periods exceed 60 days. Additional requirements for handling, inspection, and repair of epoxy-coated reinforcing steel are given in *Manual of Standard Practice* (Concrete Reinforcing Steel Institute 1997b).

Post-tensioned tendons should be stored so that sheathing is maintained in a clean and undamaged condition. Tendons should be inspected for any damage such as splits or punctures in the sheathing before concrete placement, and it should be verified that repairs are performed satisfactorily before placement.

8.6—Installation

8.6.1 *Cover depth*—Reinforcement is embedded at a minimum distance (clearance) from the surface of the concrete to prevent buckling under certain conditions of compressive load, rusting when exposed to moisture, or loss of strength when exposed to fire. Therefore, it is essential for reinforcement to be properly spaced, spliced, tied, and embedded to give the required clearance to all concrete surfaces (Fig. 8.3 and 8.4). The CRSI references (1997a,b) and ACI 318 give detailed information on these matters.

For a particular application, cover should be checked at all locations to ensure compliance with drawing requirements. Tolerances given in ACI 117 may be applied to specified cover values unless noted otherwise.

Minimum concrete cover around steel reinforcement as stated in ACI 350 is required for concrete exposed to corrosive liquids or vapors, deicing salts, or a marine environment. Clearance should be checked under sagging horizontal bars midway between supports and at stirrups and column ties that project beyond other reinforcement. On structural slabs, especially bridge decks frequently exposed to deicing salts, adequate clearance between the top of the slab and the top of the reinforcement is critical.

8.6.2 *Splicing, welding, and anchoring*—Unless permitted by the contract documents, splicing of bars or welded-wire fabric should be performed only with the approval of the engineer. The location and lengths of all lap splices are usually shown on project drawings or defined in structural drawing notes. Contact the engineer as necessary for clarification of requirements as necessary. To facilitate concrete placement, bar splices should be staggered whenever possible. The splices in column ties around the four corners of the column should be staggered instead of one above the other. Refer to the CRSI references (1997a,b) for lap lengths, splicing methods, location and orientation of splices, and provisions for lateral support. Be aware that epoxy-coated bars require longer laps than uncoated bars.

Welding of reinforcement will lower the strength of the bar, and is prohibited in many contract documents. If welded splices are permitted, however, the weld should be the required size and length and the bars should not be burned or reduced in cross section. A certified welder should do all welding in strict accordance with American Welding Society (AWS) D1.4. The welder's certification should be current and all preheating requirements stipulated in AWS D1.4 should be carefully followed. A written welding procedure should be required. Tack welding should not be permitted unless allowed by the contract documents, because tack welds weaken bars at the tack location. Tack welds should be treated the same as splice welds in terms of welder certification and preheating requirements. If many welds are made, weld testing is advisable.

Make sure bars are anchored wherever necessary. Anchoring methods include extending the bar beyond the point of no stress, bending it around another bar or steel member, or bending it into a 90-degree or semicircular hook of specified minimum radius.

8.6.3 *Congestion*—Where reinforcement is too congested to permit concrete placement, openings can be provided by temporarily crowding bars to each side and then replacing them to their designated positions, or, if approved by the engineer, leaving them in that position permanently. To permit satisfactory placement of concrete around reinforcing bars, the nominal maximum aggregate size in the concrete mixture, especially where reinforcement is congested, should not exceed 3/4 of the minimum clear spacing between bars.

If there is more than one mat of reinforcement, bars should align vertically above each other in both horizontal directions to minimize interference with concrete placement and consolidation. Symmetrical location of reinforcement will also aid future alterations to the structure, such as core drilling, by minimizing the amount of reinforcing steel that is cut. Provisions for preplanned openings in the upper mat for drop chutes will help to prevent scattering and segregation of the concrete.

Vibrator heads should fit between bars in congested areas. The use of smaller vibrators may require a reduction in the spacing of vibrator insertion locations and an increase in vibration time. Before concrete placement begins, it should be verified that all vibrators are working and can achieve the proper frequency and amplitude. Refer to ACI 309R for details of how to check equipment performance.

8.6.4 *Support*—All reinforcement should be held firmly in place before and during the casting of concrete. Devices that will prevent displacement during construction include concrete blocks, metal or plastic chairs, spacer bars, and wires. Rocks, wood blocks, or other unapproved objects to support the steel should not be used.

The quantity and strength of bar supports and spacers should be sufficient to support both reinforcement and construction loads. To prevent settlement of horizontal reinforcement into base soil or indentation into soffit forming, chair or block supports should be installed every 5 or 6 ft (1.5 to 1.8 m). For specific recommendations, refer to *Manual of Standard Practice* (Concrete Reinforcing Steel Institute 1997b). For pavement slabs or mass concrete, a permissible alternative to supporting reinforcement on chairs is to place relatively stiff concrete up to a given level, then to lay the mesh or bars on the surface before placing the rest of the concrete. When pavement concrete is placed in a single course, wire fabric or bar mats can be laid in proper horizontal alignment on the full depth of struck-off concrete and machine depressed with special equipment to proper elevation (refer to Chapter 13). Workers should never lift or hook reinforcement in a thin slab from the bottom of the slab up to its prescribed level during concreting. This operation is seldom, if ever, done properly, and it results in uncertain location of the steel, with much of it ending ineffectively on the slab bottom. Specifications generally require chairs or some other supports to ensure proper location of reinforcement.

Some spacers expose more metal than necessary at the surface. To prevent surface staining from rust, contract documents may require that no corrodible metal be left in the concrete within a stated distance of the surface. To meet this requirement, stainless steel or plastic bar supports or supports having plastic tips can be used (or concrete bricks, if permitted).

Fig. 8.5—Improperly placed steel can deflect under the weight of workers and fresh concrete.

The inspector should check that bars are tied at sufficient intervals so they will stay in place during concrete placement and consolidation. Tie spacing depends on the position of the mat and the construction traffic the mat will be exposed to. Usual practice is to tie a 20 ft (6 m) length of bar in six to eight spots. An 18-gauge (MW 120) or heavier wire should be used to tie bars. Workers should twist the ends of tie wires so they project away from the concrete surface.

During the concreting operation, reinforcement, especially light temperature steel, can become displaced by heavy loads of fresh concrete (especially in deep forms), by the weight of workers, and by tools used to consolidate the concrete (Fig. 8.5). Constant attention is required to prevent this displacement and to detect and correct any displacement that does occur.

8.7—Embedded fixtures

Before concrete is placed, it should be verified that all anchor bolts, inserts, pipe sleeves, pipes, conduits, wiring, flashings, manhole-cover frames, instruments, and other embedded fixtures and mechanical equipment are firmly fixed in position. Do not allow built-in fixtures to displace reinforcement, except as shown on the contract documents, because that could appreciably reduce the strength of the construction. Moving or relocating reinforcement beyond the specified tolerances to avoid interferences should be approved by the architect/engineer according to ACI 301.

In general, conduits 1 in. (25 mm) or less in diameter do not significantly reduce the compressive strength of concrete. Conduits coming together at a particular location, however, may require extra reinforcement above and below to minimize excessive cracking at that spot. More than one conduit should not be allowed to take space in the cross section of a column or to be massed together at the face of a column unless permitted by the contract documents or specifically approved by the engineer. Because conduits (and pipes) will float on freshly placed concrete as its level rises in the forms, they should be secured against both vertical and lateral displacements.

Embedded metals, other than steel reinforcement, can cause galvanic action and corrosion unless coated to isolate them. In particular, aluminum should not be embedded in

Fig. 8.6—Contraction joints are designed to control the location of cracks by forming planes of weakness.

reinforced concrete unless it is fully coated. If wood inserts must be embedded in the concrete, make sure they are soaked or sealed before concreting; otherwise, the wood tends to swell and could split the concrete.

For detailed requirements regarding embedded fixtures, refer to ACI 318.

8.8—Joints

The various types of joints used in concrete construction serve different functions:

- Isolation joints are designed to allow the structure on each side of the joint to move independently. If they are omitted, compressive or tensile forces can crush, crack, or otherwise damage the concrete. These forces develop by thermal or chemical expansion, shrinkage, applied loads, or differential settlement. Reinforcement should not cross isolation joints;
- Contraction joints are purposely made planes of weakness designed to control the location of cracks that might otherwise occur randomly due to contraction of concrete from drying shrinkage and temperature drop (Fig 8.6). Reinforcement may be continuous or discontinuous in contraction joints, depending on the design of the structure; and
- Construction joints are created at necessary interruptions of the concrete placement (refer to Chapter 9 for a detailed discussion). If the construction joint is also a contraction joint, the reinforcing steel should be treated the same as in a contraction joint.

Recommended practice regarding design, location, and construction of joints is given in ACI 224.3R, 301, 318, and 504R. Because of their function, isolation and contraction joints are working joints. Construction joints are not working joints unless they are also designed and constructed for that purpose. Materials for filling working joints are discussed in Chapter 3, and detailed information on elastomeric joint sealants is provided in ACI 504R.

Smooth dowels may be required across some joints (refer to Chapter 13). If so, they should be aligned carefully and lubricated on one end to allow joint movement, with the lubricated end fitted with an expansion cap. Similarly, sliding joints require a bond breaker to be installed before adjacent concrete is cast.

Before concreting, the specified initial joint opening should be allowed for; be on the lookout for construction conditions that may later interfere with joint movement or with proper water drainage. Filler should not be installed until debris is removed from the joint opening and the joint surfaces are clean.

8.9—Final inspection before placing

When making the final inspection before concrete placement, the inspector should:

- Check all bracing and shoring to ensure that it has not been loosened or misplaced;
- Verify that the subgrade has not been disturbed by preplacement construction activities;
- Inspect all forms for damage and mortar tightness, and verify that tie bolts are tight at construction joints;
- Check reinforcement for completeness and proper placement, and make sure that the specified cover will be obtained;
- Confirm that all embedded inserts are properly sized, located, and mounted and that they are protected from contamination;
- Check forms and construction joints for cleanliness and absence of surface moisture. Forms should be free of foreign material, such as standing water, ice, and dirt;
- See that forms have been properly oiled. If re-oiling is required, reinforcement and other embedded items should be kept clean and completely free of oil;
- Check the weather forecast for the day of the concrete placement. If extremely hot, cold, windy, or low-humidity conditions are forecast, precautions should be taken to protect fresh concrete, and materials and equipment for protection should be readily available at the job site; and
- Verify that all preparations have been completed. It is not good practice for crews to be finishing preparations while concrete placing operations are beginning.

8.10—Checklist

During inspection of preparatory work, a checklist should be used to ensure that no details have been overlooked. Figure 8.7 shows an example of a comprehensive checklist used successfully by the contractor on one project to control release of concrete placements. The form lists, on one sheet, many of the relevant items discussed in this chapter. The columns at the right are for the signatures or initials of project engineers and supervisors to certify the readiness of each item for inspection. Space is also provided for the inspector's signature once approval is given to place the concrete. For additional guidance in developing an inspection checklist, refer to Chapters 20 and 21 of this manual and ACI 304R and 311.4R.

The best location for the checklist is in a convenient and protected spot on the job site where anyone can quickly verify the status of the preparations. After giving authorization

CONCRETE PLACEMENT CHECKOUT SHEET						
Unit No._____	Pour No._____		Elev. From_____	Elev. To_____		DATE_____
Feature_____			Sta. From_____	Sta. To_____		

CHECKOUT ITEM	FOREMAN	CONTR. ENGR	ENGI-NEER	DATE	OK TO PLACE CONCRETE:
Rock Foundations					CONTRACTOR'S ENGINEER_____ Date_____ Time_____
Forming & Blockouts					
OK to close thin walls					INSPECTOR_____ Date_____ Time_____
Line & Grade					
Final Cross Sections					SUP'T_____ Date_____ Time_____
Sandblast					Pour Started Date_____ Time_____
Reinforcing Steel					Pour Completed Date_____ Time_____
Embedded Metal Items					
6-in. Waterstops					Computed_____ Line Cu. Yds._____
9-in. Waterstops					Computed Backfill Cu. Yds._____
PIPING					Total Computed Cu. Yds._____
Sewer					
Drain					Cu. Yds. Grout_____
Air					Cu. Yds. Concrete_____
Water					
Other					Total Cu. Yds. Placed_____
ELECTRICAL					Cu. Yds. Waste at Pour_____
Conduit & Boxes					
Ground Wire					Cu. Yds. Waste at Plant_____
Technical Installations					Total Cu. Yds. Plant_____
Anchor Bolts					
Joint Filler Material					Overbreak, Cu. Yds._____
Concrete Cleanup					

Fig. 8.7—Example of checklist for work preliminary to concrete placement.

to proceed with concreting, the inspector can include the checklist in the final report.

CHAPTER 9—CONCRETING OPERATIONS

Once contract document requirements have been met regarding site preparation and the location and condition of forms and reinforcement (as discussed in Chapter 8), concrete placement can begin. At this critical stage of the project, it is necessary for inspectors to see that the proper procedures are followed to convey, place, consolidate, finish, and cure the concrete. On smaller placements, inspectors may also be required to conduct field testing of fresh concrete. Typically, qualified field technicians will perform this function on larger placements. Methods for sampling and testing of fresh concrete are described in Chapter 19.

9.1—Placing conditions

Contract document requirements for placing conditions should be followed, particularly in regard to weather. For example, some contract documents prohibit concreting during periods of extreme heat or cold, wind, or rain unless protection of the work is provided. (See the discussion on hot and cold weather concreting in Chapter 10.) In hot weather, concreting at night may be required to reduce moisture evaporation and the temperature of the concrete, especially for structures in which cracking may be a problem, such as bridge decks, pavements, parking decks and slabs, and mass concrete. If work is to be performed at night, the contractor should provide adequate lighting. Some contract documents prohibit concreting at night to ensure visibility by placing crews.

In addition to observing provisions for weather, the inspector should:

- Review the planned placing sequence and the timing of deliveries. Concrete should be delivered, placed, and consolidated fast enough to prevent undue delays and the formation of cold joints. Delays also contribute to slump loss and slump variations. Delivery, however, should not be so fast that proper placement and consolidation are difficult or impossible; and

- Verify that all equipment is operating properly and that backup equipment is on hand to prevent interruptions of the placement if equipment fails.

9.2—Handling of concrete

When concrete is to be delivered to the job site in truck mixers, the revolution counters should be checked to ensure sufficient mixing. The minimum is normally 70 to 100 revolutions at mixing speed; additional revolutions should be at agitating speed only. Water tanks should also be checked to see that they are still full when concrete is discharged (minus any water used for initial slump adjustment) or kept completely empty, as required. The batch ticket of each truck mixer should be read to verify that mixture proportions are correct along with any other contract requirements, and the ticket should be properly signed. Expediency and job efficiency usually dictate that slump, air, and unit weight of the first load be measured, so that adjustments to the mixture can be made quickly if required. If the load was not chosen randomly, it should not be used in a statistical analysis of the concrete.

Most contract documents limit the time after mixing that the concrete can be used. As explained in Chapter 7, a time limit can be waived if the concrete can be properly placed, consolidated, and finished without the addition of water (retempering). Most contract documents prohibit such water additions. Retempering should not be confused with adding water initially to adjust slump during mixing or when truck mixers first arrive at the job. Such initial additions of water are acceptable if slump is below the maximum allowed and the maximum w/cm is not exceeded.

9.2.1 *Conveying*—Many types of devices can be used, either alone or in combination, to convey concrete from onsite mixers or delivery trucks to the forms. Options include buckets, buggies, chutes and belts, wheelbarrows, trucks, and pumps. For more details about conveying equipment, refer to ACI 304R.

When placing crews load concrete into conveying devices, the inspector should verify that they take the following measures to preserve concrete quality and uniformity:
- Dump or drop concrete vertically to avoid segregation of coarse aggregate. The best method of ensuring a vertical drop is to pass the concrete through a short section of drop chute. Baffle plates are not satisfactory because they may merely change the direction of segregation (Fig. 9.1);
- Avoid unconfined drops that can cause segregation. A short fall that results in stacking can cause more serious segregation than a longer fall that creates a bulging mass. Scattered individual pieces of coarse aggregate are acceptable because they will be re-embedded in the concrete. Piers and caissons are deep confined spaces that can be properly placed by allowing concrete to fall extended distances provided that freefall of concrete is unobstructed. Discharge of concrete through a hopper with a short downpipe carefully centered on the pier shaft is recommended. Per ACI 336.3R, freefall of concrete with a minimum 4 in. (100 mm) slump into piers produces adequate compaction up to the top 5 ft (1.5 m) of depth; and
- Do not dump concrete over reinforcement into deep forms, because segregation will occur as the coarse aggregate rattles past the bars.

9.2.1.1 *Buckets and hoppers*—Bottom dump buckets and hoppers permit placement of low-slump concrete, and they can be airlifted for delivery of concrete to remote job sites (Fig. 9.2). They should have side slopes of not less than 60 degrees and wide, free-working discharge gates that can be readily opened or tightly closed any time during discharge. Do not permit buckets to swing over freshly finished concrete.

Placing crews should load buckets and hoppers on platforms rather than directly on the ground to prevent contamination of the equipment. They should also avoid concrete buildup during use, and after each use, remove hardened concrete and lightly oil the bucket and gate control mechanism.

9.2.1.2 *Chutes and belts*—The slope of chutes should be steep enough to permit concrete to flow at the slump specified or required for placement. (The slope usually required is 1 vertical to 2 or 2-1/2 horizontal.) The use of round chutes will avoid accumulation of concrete in corners. Conveyor belts can be configured to deposit concrete into relatively inaccessible areas. To move concrete as close as possible to the placement location, the use of mobile conveyor belts or chutes should be considered.

Concrete should be deposited through hoppers onto conveyor belts in a continuous ribbon. Discharging the concrete through a suitable drop chute will help to avoid segregation. In hot, dry, or windy weather, it may be necessary to cover long lines of chutes or belts to prevent drying of concrete and excessive slump loss. Covering high-speed belt conveyors may not be required unless drying conditions are extreme. When chutes or belts are flushed with water for cleaning, the water and diluted concrete should not be allowed to drain into the forms or onto freshly cast or finished concrete.

9.2.1.3 *Pumps*—Some contractors prefer pumping to crane-and-bucket placement of concrete because it is faster and can reduce labor costs (Fig. 9.3). The contractor, however, needs to test the proposed equipment and line layout with concrete materials and mixtures before production. Pumping requires a continuous supply of medium-consistency concrete that is uniform, plastic, and workable. Because the force of pumping pressure will tend to reduce air content and concrete slump, the slump and air content of concrete discharged into the pump hopper usually needs to be greater than that required at the point of placement. When the concrete contains very low absorptive aggregates, the slump loss due to pumping will be minimal as long as the pump and line is not excessively long or subject to very hot temperatures and dry conditions. It is sometimes useful to know the extent of slump loss (if any) by conducting tests at the hopper and point of discharge. The inspector should verify that w/cm limits are not exceeded when slumps are adjusted before pumping.

Typically, a grout slurry is used to lubricate the pump system before introduction of the concrete mixture. This grout slurry should be wasted and not placed in any portion of the structure.

Pipelines are available in steel, aluminum, and plastic. Aluminum pipelines should not be used because aluminum abraded by the flow of concrete can form hydrogen gas, which causes damaging expansion of the concrete.

The inspector's responsibilities during pumping operations include:
- Inspecting the condition of the concrete at the end of the pipeline and the slump and air content at each end of the line;

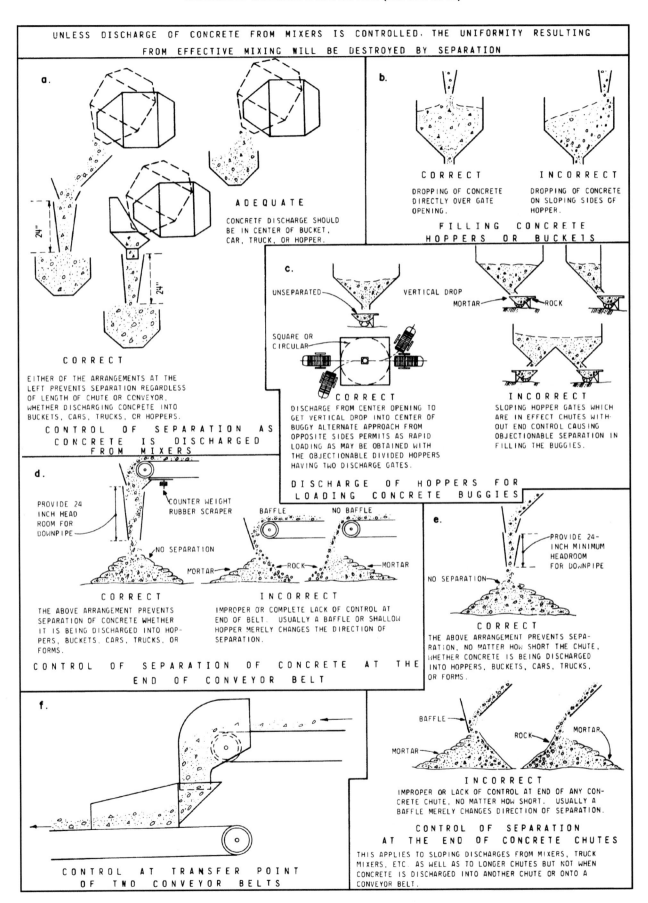

Fig. 9.1—Correct and incorrect methods of handling concrete (ACI 304R).

Fig. 9.2—Buckets allow for delivery of concrete work for above ground lifted by crane or, at inaccessible sites, by helicopter.

Fig. 9.3—Pumping enables placement of slab concrete without runways or a crane. Direct communication between the pump operator and placing crew is essential. Inspectors should verify that the contractor has supported the reinforcement so workers will not inadvertently move it.

- Verifying that no unauthorized water is placed in the pump hopper;
- Monitoring concrete flow to ensure segregation is not occurring. A smooth, even flow is desirable. If pumping is rapid, baffles may be required at discharge points to prevent spurting and segregation of the concrete. Pumped concrete is more uniform when discharge lines are horizontal or slightly inclined; and
- Seeing that placing crews thoroughly clean the pump and pipeline after pumping is done, dispensing the water used to clean the line outside the forms.

For more details about pumping practices and equipment, refer to Chapter 15.

9.2.2 *Placing*—Placing methods that keep the concrete uniform and free from obvious imperfections are critical to the success of the entire concreting operation. Proper placing methods will prevent segregation and porous or honeycombed areas, avoid displacement of forms and reinforcement, secure a firm bond between layers, and minimize shrinkage cracking (U. S. Bureau of Reclamation 1988; ACI 304R).

9.2.2.1 *Depositing concrete into forms*—Figure 9.4 shows correct and incorrect methods of introducing concrete into forms. Concrete should be dropped vertically, and drop chutes used, if necessary, to avoid striking reinforcement or the sides of the forms. This minimizes segregation and premature coating of form and reinforcement surfaces with mortar. (A dry mortar coating on reinforcement reduces bond.) The lowest section of a sectional drop chute should be held in a vertical position. If it is pushed or pulled to a considerable angle from vertical, serious segregation can occur. When concrete is placed near the top of the forms, dumping the concrete directly from buckets or other delivery equipment is acceptable.

Placing crews should deposit concrete near its final location and not allow it to flow laterally unless the whole mixture is moving without segregation. In slab construction, they should dump new concrete against the concrete in place, not away from it. In wall forms, they should place concrete directly in corners and ends so that flow is away from these areas rather than toward them.

To ensure proper consolidation, crews should deposit only as much concrete as can be consolidated efficiently. The depth of the first layer placed on hardened concrete or rock should be limited to 20 in. (500 mm).

9.2.2.2 *Deep lifts*—Near the top of a deep lift, concrete tends to become wetter because water in lower concrete migrates upward. It is possible to offset this by using a drier consistency of concrete as the level of concrete rises. This normally causes no placement problems because concrete near the top can be reached more easily for vibration. If excessive bleedwater rises to the surface, it is probably due to sand that is deficient in fines, a low-cementitious-content mixture, or a mixture having a high *w/cm*. Such bleeding, either of clear water or water and fines, produces a top surface of weak concrete unsuitable for a construction-joint surface to which additional concrete will be bonded or for exposure to weather or traffic. Air-entrained concrete has a significantly lower bleeding potential than non-air-entrained concrete.

The inspector should verify that placement of additional lifts of concrete in columns and walls are not placed until the specified period between lifts has elapsed. This precaution allows for settlement, hardening, or cooling of previously placed concrete. Before crews place concrete in floor slabs and beams on top of fresh concrete in walls and columns, they should allow ample time for settlement of concrete in the vertical placement to prevent cracking of the slab or beam concrete. Placement should continue, however, before a cold joint develops.

9.2.2.3 *Placing operations*—During concrete placement, continuity should be maintained without undue delays (except as noted previously). If equipment breakdowns, rain, or other unexpected events interrupt placing operations, the concrete should be shaded and protected by being covered

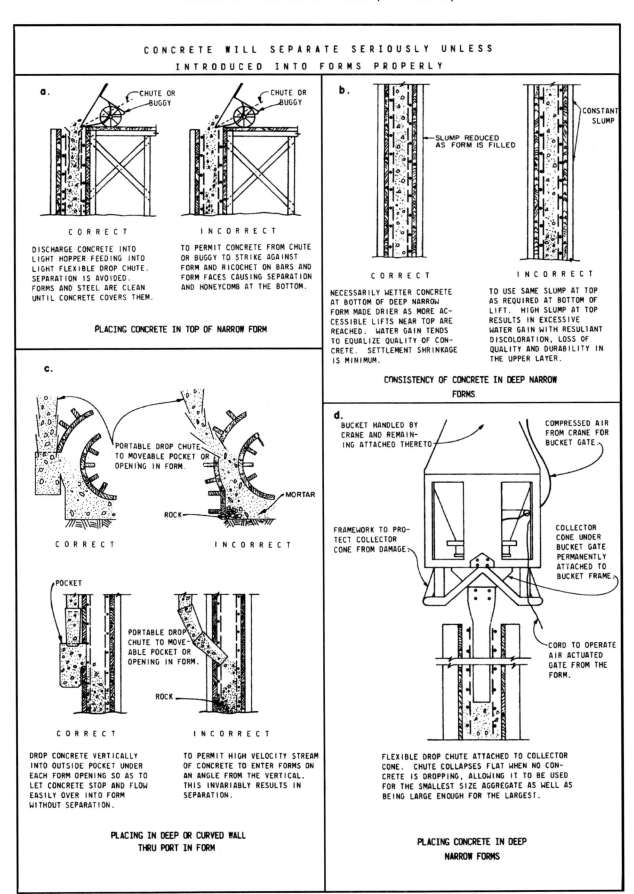

Fig. 9.4(a) to (d)—Correct and incorrect methods of placing concrete into forms (ACI 304R).

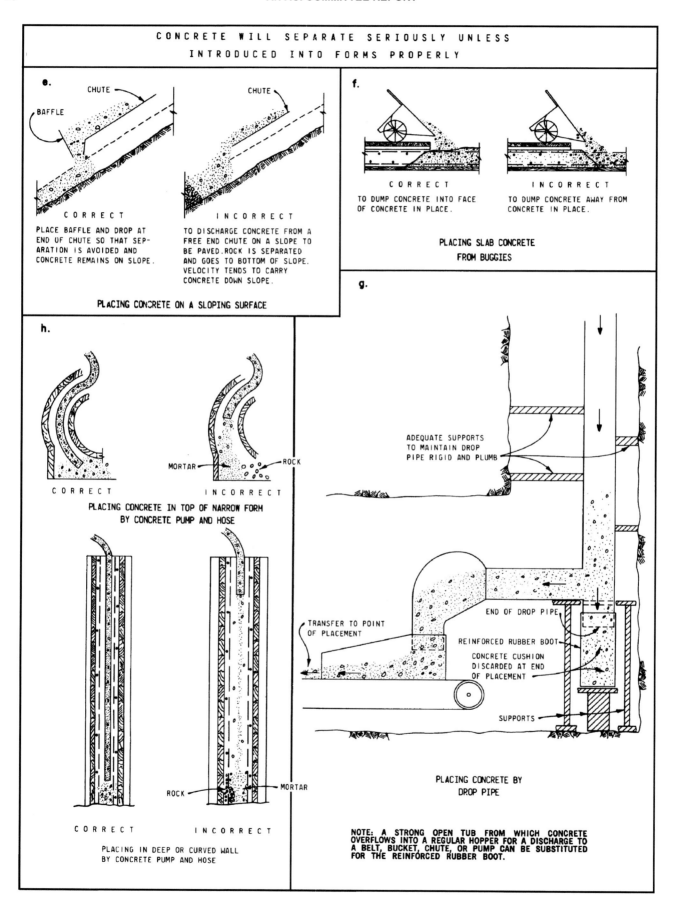

Fig. 9.4(e) to (h)—Correct and incorrect methods of placing concrete into forms (ACI 304R).

with plastic sheeting or wet (not dripping) burlap, particularly if conditions are hot, dry, and windy. A fog spray is another suitable protection method. If the concrete reaches final set before placing resumes, refer to the section on unplanned construction joints later in this chapter.

Workers should minimize disturbance to fresh concrete or reinforcement once placed and consolidated and should avoid activities that will affect concrete uniformity, finish, or bond. On wide slab placements, the contractor should provide working platforms that span the entire width of the slab to avoid disturbing the freshly placed concrete.

When spreading concrete, placing crews should only use solid-faced tools, such as shovels, and not toothed rakes or vibrators, which may promote segregation.

The contractor should measure surface elevations before, during, and after concreting. Slow settlements often go unnoticed, and excessive deflection or settlement of shores during or after concreting is a sign of trouble. Shores on ground are particularly susceptible to creeping settlement, which causes excessive deflections and cracking of green slabs. Workers may neglect to check and adjust forms, shores, and bracing after the inspector's preliminary approval (before the start of concrete placement). The contractor should include provisions for workers to make such adjustments during placing operations. (Refer to the discussion of forming and shoring in Chapter 8.)

9.3—Consolidation

Consolidating concrete thoroughly as it is being placed—using hand tools, mechanical vibrators, vibrating screeds, or finishing machines—is necessary to obtain a dense concrete, good bond with reinforcement, and smooth surfaces. The contractor should ensure that sufficient consolidation equipment and personnel are available for the planned production rate so the concrete can be consolidated fast enough to prevent delays and possible cold joints. If consolidation is slowed unexpectedly by congested conditions, equipment failure, poor workability of the mixture, or other causes, the contractor should reduce the rate of batching and mixing.

Crews should work the concrete well around reinforcement and embedded fixtures and into corners of forms. If the concrete mixture tends to segregate or stratify when worked or vibrated, its water content should be reduced or reproportioned it as necessary. Unconsolidated concrete should not be allowed to accumulate in the forms or to stand idle and stiffen in the mixer, hopper, bucket, or any other part of the conveying system.

9.3.1 *Vibration*—Generally, mechanical vibration is much more effective and less labor-intensive than hand tamping, especially for low-slump mixtures (Fig. 9.5). There are three general types of vibrators: internal, surface, and form. Vibration permits placing and consolidating of concrete with a slump of 2 to 4 in. (50 to 100 mm) in heavily reinforced members, and even lower-slump concrete in open placements. Low-slump concrete has less tendency to segregate than the wetter, higher-slump mixtures necessary when concrete is to be consolidated by hand. Vibration should not be expected to correct segregation that has already occurred because of faulty methods of handling and placing, nor to guarantee good results if the mixture proportions are incorrect.

Fig. 9.5—Internal vibration readily consolidates low-slump concrete, momentarily liquefying the mixture and removing entrapped air.

Detailed recommendations for consolidating concrete by vibration, including information on vibrator types and sizes, are given in ACI 309R. Refer to Chapter 13 for proper procedures for consolidating concrete pavements. To ensure compliance with required minimum vibration frequencies, vibrators should be checked with a tachometer or similar instrument before use.

9.3.1.1 *Internal vibrators*—For best results when using internal vibrators, follow the size, frequency, and amplitude recommendations of ACI 309R. The operator should insert the vibrator vertically through the full depth of the layer being placed and at close intervals so that vibrated areas of concrete overlap without omission. The vibrator should not be dragged through the concrete. Instead, it should be slowly withdrawn vertically while operating continuously so that no hole is left in the concrete. The operator should not use the vibrator to cause concrete to flow from one location to another because this practice usually causes segregation, with the larger coarse aggregate remaining behind. In thin slabs, it may not be feasible to use internal vibrators inserted vertically. Refer to Chapter 12 for a detailed discussion of consolidation of concrete in slabs.

For normalweight concrete, vibration should be continued until the concrete is thoroughly consolidated and the voids are filled, as evidenced by the leveled appearance of the concrete at the exposed surface and the embedment of surface aggregate. When in doubt as to the adequacy of the vibration procedure for normalweight concrete, more vibration should be applied. There is little chance of overvibrating a properly proportioned mixture.

It is possible, however, to overvibrate concrete that has a high water content or is susceptible to segregation. In such cases, the slump should be reduced or the mixture proportions modified rather than the vibration reduced. When the coarse aggregate is much lighter than the mortar, as with light-weight-aggregate concrete, overvibration tends to stratify the mixture. Because of the lower specific gravity of light-

Fig. 9.6—External vibrators are mounted on brackets attached to forms. Use only low-amplitude, high-frequency units, and make sure forms are rigid enough to withstand the vibration without distortion.

weight coarse aggregate, some large particles are likely to rise to the surface, even under careful vibration. When this occurs, a tamper grate can be used to drive the particles below the surface, and thus allow proper finishing. (Refer to Chapter 16 for further information on placing, consolidating, and finishing lightweight concrete.) For either lightweight or heavyweight concrete mixtures, vibration should be limited to that necessary for effective consolidation. For structural concrete, the desired results will usually be achieved within 5 to 15 seconds when vibration points are 18 to 30 in. (450 to 750 mm) apart.

When concrete is placed on hardened concrete or rock, the first layer requires more vibration than succeeding layers to ensure continuous tight contact at the cold joint. This is best accomplished by inserting the vibrator at about half the normal spacing for short periods of time. Succeeding layers should be thoroughly vibrated into the preceding layer while both are still plastic.

On sloped surfaces, crews should begin placing and consolidating concrete at the bottom of the slope and delay finishing operations to avoid sag. Heavy, nonvibrating power screeds are helpful in obtaining the proper slope.

Ordinarily, internal vibrators will not damage steel reinforcement or concrete in lower lifts. In fact, revibration of concrete is beneficial if the concrete will respond to the vibrator and again become plastic. Revibration can help eliminate horizontal checks and plastic shrinkage cracks caused by the settlement of concrete held up by reinforcement or irregular forms. It can also increase concrete strength, decrease the number of air-bubble holes in upper formed areas, strengthen bond under horizontal bars and embedments, and reduce leakage under form bolts. Revibration around built-in frames, such as windows, can help avoid cracking from uneven settlement relative to the finish quality stated in the contract documents. The effect of revibration on concrete-to-steel bond, however, is not clear. Revibration appears to improve bond strength to reinforcing steel placed in high-slump concrete but may severely damage bond strength in well-consolidated, low-slump concrete. Therefore, caution should be exercised in the use of revibration.

9.3.1.2 *Surface vibrators*—A surface vibrator, such as a vibratory screed or pan-type vibrator (Chapter 13), should consolidate the layer of concrete being placed to its full depth. If it does not, the depth of the layer should be reduced or a more powerful machine used. Successful use of a surface vibrator requires the concrete mixture to be properly proportioned and have a low slump. Otherwise, an undesirable amount of laitance can be brought to the surface.

9.3.1.3 *Form vibrators*—Form, or external, vibrators are used most frequently in precasting operations, but they also are suitable for thin sections of cast-in-place concrete and as a supplement to internal vibration at locations where it is difficult or impossible to insert an internal vibrator. In addition, these vibrators can reduce air voids (bug holes) on formed surfaces. Form vibrators must be powerful enough to effectively vibrate the forms. This, in turn, requires the forms to be rigid enough to withstand the vibration without distortion or leakage of mortar (Fig. 9.6). Thus, it is usually necessary to strengthen and stiffen the forms when external vibrators are used.

9.4—Finishing

The quality of concrete is largely judged by the condition and appearance of the finished surface. Exposed surfaces are subject to conditions—ranging from benign to severe—of wetting or drying, temperature changes, and mechanical wear. In addition, most concrete surfaces are subject to cracking from excessive drying shrinkage. To withstand these conditions, the concrete should be properly proportioned (without excessive water content), properly consolidated and finished, and properly cured for the specified time, according to ACI 302.1R. Some contract documents may require the use of ACI-certified flatwork finishers (or equivalent).

9.4.1 *Unformed surfaces*—Concrete proportions and consistency and consolidation methods should be selected so that only sufficient mortar for finishing is available at the surface. If the mixture is oversanded or has a high water content, or if the concrete is overworked during consolidation or finishing, the surface is likely to be covered with bleedwater or to contain a relatively deep layer of overwet mortar or laitance.

Concrete should be spread evenly ahead of the strike-off screed (Fig. 9.7). After screeding, finishers should use darbies or bullfloats (Fig. 9.8) to remove high and low spots and to produce a true plane surface. After darbying or bullfloating is complete, and while the surface is still fairly soft, the surface should be checked for alignment by using a straightedge or template. The inspector should have finishers correct high and low areas at once. They should also perform initial edging and joint grooving at this stage while the concrete is still fairly plastic.

Hand or machine floating should be delayed until all bleedwater has evaporated from the surface. Conducting finishing operations while bleedwater is still present will increase the w/cm of the surface paste, resulting in reduced strength and durability of the surface. Concrete is generally considered to be ready for floating once the surface water sheen has disappeared and when a person stepping on the surface leaves an imprint approximately 1/4 in. (6 mm) deep. In instances where excessive bleeding occurs, dragging a large loop of garden hose across the surface effectively removes excess water. Sprinkling of dry cement or a mixture of dry cement and sand directly on the surface of the fresh concrete to absorb water should not be permitted.

Troweling should be delayed as long as possible. The proper time interval varies with the cement, weather, and other conditions. Generally, the surface is ready for troweling when it just reaches the stage that it can no longer be indented with a finger. Troweling the surface too soon can trap late-rising bleedwater below the troweled surface and lead to scaling. If troweled too late, the concrete will become too hard to be troweled effectively. Sprinkling of water onto a stiffened surface should not be permitted because this will increase the w/cm of the surface paste and result in lower strength and durability of the surface. During troweling, the finishing techniques used should be observed (Fig. 9.9). Finishers should tilt the steel trowel at a slight angle and exert heavy pressure to compact the paste and form a dense, hard surface. If the surface will receive more than one troweling, both the angle and pressure of the trowel should be increased with each operation. When finishing edges and joints, finishers should use thin, small-radius edgers and jointers and avoid overworking the concrete at these locations.

For additional details on finishing slabs-on-ground, refer to Chapter 12. Refer to Chapter 15 for information on finishing vacuum-treated slab surfaces.

9.4.1.1 *Air-entrained concrete*—Because of lower bleedwater volume and slower rates of bleeding, successful finishing of air-entrained concrete can be more difficult. The surface tends to stiffen more rapidly than the interior portions of the concrete, so there can be a tendency to finish the surface before all bleedwater has risen, leading to trapped bleedwater beneath the finished surface. After evaporation of this trapped water, the resulting near-surface voids promote rapid scaling of the overlying paste crust. ACI 302.2R recommends that air-entrained concrete not be used in floor slabs requiring a steel-troweled finish unless exposure conditions require the use of air entrainment for durability. The use of a magnesium or aluminum float, instead of wood, facilitates finishing of air-entrained concrete.

9.4.1.2 *Plastic shrinkage cracking*—Cracks appearing on unformed concrete surfaces soon after placement (while the concrete is still plastic) are usually caused by high evaporation rates at the surface due to a combination of high temperatures, low humidity, and wind. These randomly oriented, unconnected surface cracks can be thin or wide, but are generally shallow in depth. The use of fog spraying, windscreens, sunshades, plastic sheets, application of monomolecular film (evaporative retarder), or other means

Fig. 9.7—A screed strikes off the concrete surface.

Fig. 9.8—A long-handled bullfloat removes highs and lows on a pavement slab.

Fig. 9.9—Troweling smooths and compacts fresh concrete.

to inhibit moisture loss between finishing operations will minimize plastic shrinkage cracking in flatwork placed under unfavorable conditions. Plastic cracking can sometimes be prevented or remedied by well-timed working of the surface, with somewhat later-than-usual floating followed by slightly earlier troweling. Cracks that form before troweling often can be melded together successfully with the float. If cracks are merely troweled over, they are likely to be visible later.

9.4.1.3 *Protection from rain*—Rain falling on freshly placed concrete can erode the surface and dilute the surface mortar, quickly damaging the newly finished work. During rain, therefore, the concrete should be fully sheltered, or work should discontinue until the rain has stopped. Preparations should be made in advance for protection from rain when work must continue under such conditions.

9.4.2 *Formed surfaces*—Contract documents should clearly state what finish is required or permitted on formed surfaces. Finishing can range from merely knocking off the fins and repairing imperfections to applying one of several decorative treatments such as sandblasting, bush hammering, grinding, brushing, bagging, and rubbing. For more information about finishes for architectural concrete, see Chapter 14 of this manual, ACI 303R, and *Color and Texture in Architectural Concrete* (Portland Cement Association 1995).

Regardless of the finishing method, uniformity of surface texture and color is of primary importance for good appearance. At the time forms are removed, the condition of the concrete surface should be observed to determine the need for repairs and to plan the timing of finishing operations. Many finishing operations (other than textures provided by the forms) should be applied as soon as possible because the concrete has hardened considerably by the time the forms are removed. If the surface will be ground, chipped, or bush hammered, operations should be delayed until the concrete has gained sufficient strength to prevent loosening of the coarse-aggregate particles. Plastering can be done any time after concrete has cured sufficiently.

Imperfections should also be repaired as early as possible (using approved methods) to make the repair more monolithic with the base concrete. Repair work, however, should not be allowed to interfere with immediate application of continuous moist curing on adjacent areas. Because the two operations tend to conflict, this matter requires special attention. Throughout the repair work, the surface should not be allowed to dry, nor should curing operations result in damage to the new repairs.

9.5—Construction joints

9.5.1 *Planned*—Construction joints do not necessarily allow movement across the joint. Generally, they mark the top of a lift, the end of a monolith, or the end of a day's work. Construction joint locations should be planned before placing operations begin, and those locations should be adhered to whenever possible. Many contract documents will require the design professional's approval for construction joint locations and details. Because construction joints frequently leak and degrade from exposure to weather, they should be used only where necessary, consistent with contract document limits and good practice on depth and extent of placements.

In floor slabs and beams, construction joints should be located near the middle of the span (where the shear is least), and be made vertical (normal to the axis of the slab or beam). In highway or airport pavements, construction joints should always be located at planned contraction or expansion joints. In addition to being properly located, it is also important for construction joints to be neat and well-bonded. Contract documents may require surface cleaning, key construction, or insertion of dowels or tie bars at construction joints. Recommended practice regarding location, design, and preparation of construction joints is given in *Concrete Manual* (U. S. Bureau of Reclamation 1988) and ACI 304R. Refer to Chapter 13 for a discussion of pavement joints.

9.5.1.1 *Preparing joint surfaces and edges*—In a wall, column, or other vertical lift, when the level of concrete reaches a horizontal construction joint, the exposed edge of the joint should be formed or trimmed to produce a neat line. For succeeding placements, holding and tying the form tightly against the joint edge, with minimum overlap, will prevent mortar leakage or offset of surfaces as additional concrete is placed.

The earlier the subsequent lift is placed, the better the chances for achieving satisfactory bond. If only a few hours elapse between successive lifts, it is not necessary to prepare the contact surface of the older concrete to achieve a good bond. If the surface is clean and damp, but not wet or covered with a layer of laitance, new concrete can be adequately bonded by vibrating it thoroughly over the area of contact. To ensure cleanliness, cleanup should be postponed until just before placement of the next lift of concrete, when crews should use wet sandblasting or high-pressure water jets to remove any surface film and contaminants.

In addition to being clean, it is essential that joint surfaces on the older concrete not have loose aggregate or shattered edges or corners. Cleaning by use of an air-water jet or wire brooming should be done only when the concrete is still soft enough that any laitance can be removed but has hardened enough that the aggregate will not be loosened. Until the new concrete is placed or the specified curing time has elapsed, the surface should be kept damp by ponding or sprinkling with water or by covering with damp sand or burlap.

These procedures are usually applicable to dams, but not widely used in building construction. Similar procedures, however, can be used for buildings when high-quality joints are desired.

Roughness of a joint surface is not required for a well-bonded joint, and it can interfere with thorough cleaning of the joint surface. To prevent rough joint surfaces, the coarse aggregate should be embedded in the concrete surface when the concrete is placed.

9.5.1.2 *Use of starting mixtures*—When new concrete is placed on a horizontal surface, a bedding layer of mortar of the same mixture as that in the concrete is sometimes broomed into the old surface after it has been kept continuously moist for several hours. Another common practice is to place a starting concrete mixture on a clean joint surface that is damp, but not wet.

If the regular concrete contains aggregate larger than 3/4 in. (19.0 mm), a suitable starting mixture can be made simply by omitting that portion of the aggregate that is larger than 3/4 in. (19.0 mm). If 3/4 in. (19.0 mm) or less is the specified maximum aggregate size, an extra bag of cement per cubic yard (cubic meter) in the first batches plus enough additional water to make a 6 in. (150 mm) slump will serve the purpose.

Enough of this starting concrete mixture should be produced to make a layer 4 to 6 in. (100 to 150 mm) deep. Dropping the mixture onto the joint surface at various locations will allow more even spreading. When the first layer of the normal concrete mixture is placed on this starting mixture, workers should vibrate both layers thoroughly and allow the vibrators to penetrate to hard bottom.

If the joint surface is clean, the starting mixture method is just as effective in bonding as the mortar method, but it avoids the possible concentrations of mortar that can occur with the latter method. With either process, however, a slight change in color may be noted at the joints. The mortar procedure is often the easier option when the contact surfaces are small, as in most building construction.

With vertical or steeply inclined joint surfaces, a good bond can generally be achieved by thoroughly vibrating the succeeding concrete placement so that mortar is pushed against the joint surface. If a strong bond is required, however, it may be necessary for workers to sandblast the surface of the old concrete before form erection and then later, as concrete is placed, thoroughly vibrate the concrete against the joint. To prevent weakening of the upper part of a vertical joint by bleeding and water gain, the concrete should be vibrated at the joint deeply and as late as the running vibrator will penetrate the concrete.

9.5.2 *Unplanned*—Breakdowns, delays in concrete delivery, and many other construction problems may make it necessary to stop placement of concrete at locations other than those previously planned. The contractor and inspector should contact the designer and receive tentative plans and details ahead of time for installing unplanned construction joints. When the need for such a joint arises, the designer should be advised so that a review of possible effects of the unplanned joint on the behavior and safety of the structure can be conducted. Because these joints are not indicated on the plans, it is important to locate and construct them so as to least impair the strength of the structure. During construction of unplanned joints, it should be confirmed that the contractor observes all of the precautions and methods discussed previously for planned joints.

CHAPTER 10—FORM REMOVAL, RESHORING, CURING, AND PROTECTION

Inspection does not end with the actual casting of concrete. After concrete is placed, it is necessary for inspectors to verify that proper procedures are followed for removal of forms and supports, reshoring, curing of concrete, and protection of concrete from physical damage and exposure to extreme weather conditions. Inspectors should familiarize themselves with contract document requirements for these post-placement activities before the work begins.

10.1—Removal of forms and supports

10.1.1 *Time of removal*—Detailed recommendations for removal of forms and supports and for reshoring are provided in ACI 301, 306R, 318, and 347. Timing of removal of forms and shores should be approved by the architect/engineer or be in accordance with the contract

Fig. 10.1—High wall with form ties. Removal of high wall support (ties, studding, wales, and bracing support) is dependent on time and strength gain of concrete.

documents (Fig. 10.1). Usually contract document requirements for time of form removal are based on tests of job-cured cylinders or in-place concrete. When test cylinders are used to determine the time to strip forms, they should be cured under conditions no better than those for the in-place concrete. Records of weather conditions and other pertinent information should be kept and used in conjunction with the test results to confirm form removal timing and sequencing.

Usually, the architect/engineer's approval is required before primary supporting forms can be removed. As a general rule, forms for columns and piers can be removed before those for beams and slabs. When the architect/engineer does not specify the minimum strength required of concrete at the time of form removal (stripping), ACI 347 provides minimum times that should elapse before forms are stripped, depending on the type of concrete member. These times are the accumulated number of days for air temperatures above 50 °F (10 °C), assuming that Type I portland cement is used. If other cements or accelerating admixtures are used, these times may be reduced with the engineer's approval. If temperatures fall below 50 °F (10 °C), the minimum times should be increased.

Early form removal is desirable for finishing, and usually desirable for curing. In warm, dry weather, it is especially beneficial to remove forms and start curing as soon as possible. Where protection is not needed, workers can remove nonsupporting forms (such as for walls, columns, and beam sides) as soon as they can without damaging concrete surfaces and edges.

It is imperative to allow forms supporting concrete to remain undisturbed until the concrete can carry its own weight plus construction live and dead loads within acceptable deflection limits. At no time should the construction load exceed that for which the member was designed. The inspector should make sure forms and supports are removed without impact or shock, and allow the concrete to assume load gradually and uniformly.

10.1.2 *Multistory work*—Multistory work presents special conditions for removal of forms and shores. The

Fig. 10.2—Multistory work requires careful design of shores and reshores to ensure adequate support of loads.

Fig. 10.3—Concrete should be reshored soon after form removal.

shoring that supports green concrete is supported by lower floors that may not have been designed for these loads. In such cases, shoring of the lower floors also is necessary to help carry the load of shores above. Shoring should be provided for a sufficient number of floors to support the imposed loads without excessive stress or deflection. ACI SP-4 (Hurd 2005) and ACI 347 provide detailed information on the design of forms and shores to support concrete.

For multistory construction, a structural engineer should design all shores, reshores, and other supports. Previous failures have indicated that past engineering practice regarding reshores has not been sufficiently conservative. Reshoring (Fig. 10.2) is one of the most critical operations in multistory construction. Reshores (shores placed under a stripped concrete slab or structural member after the original forms and shores have been removed) are needed when large areas of new construction are to support combined dead and construction loads that exceed their capacity. The structural engineer should plan for reshores to extend through a sufficient number of lower floors. Reshores should also be located in the same position on each successive floor, as indicated by the engineer's drawings. Where shores are not directly over reshores, an analysis should be made by the engineer responsible for shoring to determine if detrimental flexural stress or punching shear will be produced in the lower slab. ACI 347 provides detailed guidance for placing reshores.

Workers should install reshores for the new construction as soon as possible after stripping operations are complete, but no later than the end of the working day on which stripping occurs (Fig. 10.3). While reshoring is under way, construction loads should not be permitted on the new construction and live, impact, and vibratory loadings on the balance of the affected structure should be minimized. All shores and reshores should have adequate lateral bracing. The structural engineer should be consulted about any questions or doubts.

Do not allow final removal of reshoring until the engineer confirms that the supported concrete member has attained sufficient strength to carry all loads. Removal operations should be carried out in a sequence that does not subject the supported structure to impact or excessive loading.

10.2—Protection from damage

Unless precautions are taken, construction operations can overload, jar, or mar the concrete already in place. Occasional jarring or vibration, if not severe, is generally not detrimental. The most severe loads are typically those imposed by storage of construction materials, reshoring of upper floors, operation of construction equipment, and construction traffic. Inspectors should discuss provisions for these construction activities with contractors and subcontractors to minimize the risk of physical damage to the structure. Protective measures include:

- Making sure storage loads are spaced to avoid overloading of any portion of the structure. Additional reshoring may be required in congested areas where such loads cannot be spread out;
- Using protective coverings on floors on which construction activity is taking place. Curing membranes and coverings should be checked regularly for signs of damage; and
- Protecting inserts, piping, and projecting ornamentation from falling material and debris that can plug openings or damage the concrete. Special attention should be given to protecting architectural and ornamental concrete.

10.2.1 *Backfilling*—Placement and compaction of backfill on and against concrete should be permitted only when the concrete is strong enough to support active and passive soil pressures and surcharge loads. The contractor should take the following measures to avoid impact and damage to concrete during backfilling:

- Leave underpinning, protective sheet piling, and bracing in place until the concrete has developed adequate strength to resist loading;
- Carefully control backfilling and compaction against walls, particularly if they are tall, unbraced, or both. The use of heavy equipment close to any wall can cause cracking, abrasion, or other damage; hand compactors should be used instead. For some basement and retaining walls, the design may require interior floors or specific bracing and superimposed loads to be installed before exterior backfilling;
- Prevent damage to curing compounds, waterproof coatings, insulation, and similar materials by excluding large or sharp pieces of material in backfill placed near the concrete; and

- Monitor concrete wall or structure performance during backfilling for any cracking or deformation. Advise the contractor immediately if any damage is observed.

10.3—Curing

Most contract documents require keeping exposed surfaces of concrete containing standard Type I portland cement continuously moist for at least 7 days. Concretes containing high-early-strength cements (Type III) require less moist curing time (about half), and slow-hardening cements (Types II, IV, V, and IP and pozzolanic portland cement materials) need longer curing times for best results. Extensive tests indicate that the greater the amount of moisture retained within the concrete, the greater the curing efficiency and strength gain. Early and continuous curing can also reduce or eliminate plastic shrinkage cracking.

For more information on standard curing practices, refer to ACI 308R.

10.3.1 *Moist curing*—The preferred method of curing is moist curing by use of continuous sprays, flowing or ponded water, or continuously saturated coverings of sand, burlap, or other absorbent materials.

Fig. 10.4—Fogging minimizes moisture loss during and after the placement of concrete.

Water should be applied to unformed surfaces as soon as it will not damage the finish, and to formed surfaces immediately after forms are stripped (Fig. 10.4). Wood forms, kept wet, and metal forms provide some protection against moisture loss, as long as exposed top surfaces of the concrete are kept wet. Otherwise, forms should be removed as soon as possible so prescribed curing can begin with minimal delay.

Nonstaining water should be used for curing where appearance of the completed structure is important. Staining can be caused by water containing a high iron content, by ferrous pipes used to spread the curing water, and by other staining agents. Perforated plastic tubing or canvas soaker hoses are satisfactory alternatives to ferrous pipes for distributing curing water.

Wet burlap is inexpensive and can be applied without damage to the surface almost immediately after concrete is finished. Wet cotton mats and other highly absorbent materials can also be used. Burlap or other wet covers should be clean. New burlap or old burlap that has been contaminated can stain concrete.

More than one thickness of burlap should be used, and the burlap should be maintained in contact with the concrete surface, especially during hot, sunny weather. Burlap on formed surfaces should be kept wet with perforated plastic tubing or a soaker hose placed along the top of the work or by other means. If the burlap or other cover is used for the entire curing period, it should be allowed to dry before removal, particularly if weather conditions are dry. The concrete will then dry more slowly and be less prone to cracking.

If water-curing sprays are planned, moist burlap or other mats should be used first. The mats should be left in place and kept wet until there is no danger of surface erosion by the curing sprays.

If wet earth or sand is used for moist curing, it should be free of large lumps or stones (drying occurs more quickly at such points) and kept continuously wet. It should not contain organic matter or other substances that can damage or stain the concrete.

During curing operations involving flowing or ponded water, care should be taken to prevent damage to the concrete surface from erosion or washout of the cement matrix.

10.3.2 *Membrane curing*—Membrane-forming curing compounds applied to the concrete surface immediately after final set will retard evaporation of the mixing water if they meet ASTM C309 requirements and are used in strict conformance with the manufacturer's recommendations. These curing compounds are a satisfactory means of curing, particularly if preceded by wet curing. White or gray pigmented curing compounds are commonly used where future appearance of the concrete is not critical because coverage is easily verified and they reflect sunlight, minimizing surface temperatures. On areas where the appearance of a white or gray curing compound would be objectionable during its weathering-off period, a clear compound containing a fugitive (quick-fading) dye can be used to ensure complete coverage.

The surface receiving a curing compound should be moist and above 40 °F (4 °C) when the compound is applied. Workers should apply two coats smoothly and evenly using power sprayers. The second coat should be applied at right angles to the first to more effectively seal the surface (Fig. 10.5). The use of hand-operated garden-type sprayers should be permitted only on small jobs. It should be noted that windy conditions can significantly reduce effective coverage.

If the required coverage rate is not specified by the manufacturer, the maximum coverage for each coat should not exceed 200 ft^2/gal. (5 m^2/L) as stipulated in ASTM C309. Curing compounds should not be used on surfaces that will receive additional concrete, paint, coatings, or other materials that require positive bond unless it has been clearly demonstrated that the membrane can be removed satisfactorily before the subsequent materials are applied or that the membrane can serve as a base for the later application. Controlled, high-pressure water sprays that can remove curing compound without damaging the concrete are often

Fig. 10.5—Liquid membrane curing compounds should be applied with uniform and adequate coverage for effective curing of concrete.

used to prepare concrete surfaces before application of subsequent materials.

The completeness and uniformity of curing compound coverage and the amount of material used should be checked; the latter should be adequate for the area of surface covered. It should be noted that broom or other rough finishes increase the surface area and, therefore, require more curing compound than a smooth or troweled surface. If work resumes on an area that has been treated with a curing compound, the compound should not be damaged by foot or other traffic. If any spots become exposed, they should be resealed immediately.

In arid climates when the rate of moisture loss is accelerated due to a combination of high ambient temperatures, high concrete temperatures, low relative humidity, wind speed, and solar radiation, a membrane curing compound can be satisfactory if one of the wet curing methods discussed previously is used during the first 24 hours after finishing or form removal. Without ample initial water curing, sealing of formed concrete is of limited value in such climates.

If a membrane curing compound is to be applied to surfaces on which forms have been left for 24 hours or more, the surface should be soaked for several hours before application of the compound. Soaking is important, especially for rich mixtures, because concrete partially dries due to the self-desiccation (consumption of internal water due to the hydration of cement).

10.3.3 *Impermeable sheets*—Polyethylene film can provide effective moist curing if it is held close against the concrete surface and is either protected from or not subjected to construction activity that would tear or otherwise affect its performance as a moisture barrier. Waterproof (Kraft) paper also can be used successfully as a moisture barrier if seams between adjacent sheets are tightly sealed and the sheets are protected from damage. The use of dark-colored plastic film should be avoided during warm weather, except for interior work. In cold weather, however, the heat-absorption qualities of dark plastic can be advantageous. The concrete surface should be inspected beneath the sheets occasionally; if it is dry, the surface should be rewetted. Under certain conditions, the combination of plastic film bonded to absorbent fabric works more efficiently than plastic sheets alone by helping to retain and distribute moisture from the concrete that condenses on the curing cover.

Where appearance is important, concrete should be cured by means other than plastic sheets because moisture condensing on the underside of the plastic (particularly in wrinkled areas) creates an uneven distribution of water in the surface concrete. The resulting migration of soluble substances, as well as differential hydration, can cause mottling and discoloration of the surface. On decks and pavements requiring a textured surface, coverings should be kept from damaging the texture while the concrete is still plastic.

10.3.4 *Accelerated curing*—Most plant-precast and plant-prestressed concrete (other than decorative panels) is cured by accelerated curing procedures. Accelerated curing is achieved with the use of saturated steam or dry heat, which requires the concrete member to be sealed to prevent loss of mixing water. Refer to Chapter 17 for more information about accelerated curing methods.

10.4—Curing and protection during weather extremes

Although the curing requirements for concrete placed during cold or hot weather remain the same as during normal temperatures, the techniques used to achieve them become extremely critical. All members of the design and construction team should discuss curing and protection strategies for both types of weather extremes well in advance of their occurrence.

10.4.1 *Cold weather*—Whenever mean daily air temperatures are expected to drop below 40 °F (4 °C) soon after concrete placement, artificial heating should be provided and measures taken to protect the concrete, as outlined in ACI 306R. Fresh concrete that is allowed to freeze before gaining sufficient strength (500 psi [3.5 MPa]) will not regain strength after thawing. When the mean daily temperature is likely to be above 40 °F (4 °C), newly placed concrete should be protected from freezing during the first 24 hours. ACI 306R recommends that thin sections (less than 12 in. [300 mm]) of newly placed, air-entrained concrete made with Type I or II cement, without the addition of an accelerator, should be kept at a temperature of not less than 55 °F (13 °C) for 6 days to ensure adequate strength under partial load. Thin exposed sections of concrete made with Type III (high-early-strength) cement, an accelerator, or higher cement contents (100 lb/yd^3 [60 kg/m^3]) of additional cement) should be kept at not less than 55 °F (13 °C) for 4 days to ensure adequate durability and strength of members carrying a portion of their design load. Non-air-entrained concrete (although not recommended when freezing-and-thawing durability is necessary) requires protection periods approximately twice as long as those for air-entrained concrete. When there are early-age strength requirements, it is necessary to extend the protection period beyond these recommendations.

Tests of field-cured cylinders or other tests, such as maturity testing, are required to confirm strength of the concrete before removal of shoring and loading of the structure. The engineer should provide guidance or criteria to ensure that allowable stresses in the concrete are not exceeded.

At the end of the protection period, artificial heating should be discontinued and protective enclosures removed so the fall in temperature at the concrete surface will be gradual (not exceeding 20 °F [11 °C] in 24 hours for members 72 in. [1800 mm] thick; 30 °F [17 °C] in 24 hours for members 36 to 72 in. [900 to 1800 mm] thick; 40 °F [22 °C] in 24 hours for members 12 to 36 in. [300 to 900 mm] thick; or 50 °F [28 °C] in 24 hours for members 12 in. [300 mm] thick or less, as required by ACI 306R). If the temperature is allowed to fall too rapidly, excessive surface shrinkage will occur, resulting in cracking.

Wet curing should be stopped soon enough to allow the concrete to dry somewhat before protection from freezing is stopped. A record of the temperatures of the outside air, the enclosed area, and the concrete surface should be kept. Surface temperature is most accurately measured by thermometers or thermocouples embedded in the concrete, with 1/16 to 1/8 in. [1.5 to 3 mm] of cover.

Fig. 10.6—Concreting in cold weather requires special vigilance to see that no part of the concrete freezes or dries out.

10.4.1.1 *Protection methods*—The preferred method of protection at low temperatures is to completely insulate or enclose the fresh concrete (Fig. 10.6) and heat the enclosed area as necessary. On above-grade construction, it can be difficult to enclose the top surface of the concrete. One solution is to cover the concrete with polyethylene film and then with protective blankets and insulation designed for this purpose. Electric heating blankets are another option. In cases where top enclosures are impractical, partial enclosures or windbreaks should be provided to prevent chilling winds from freezing the concrete. It is critical to provide the specified curing temperatures.

The use of low-pressure steam or electric or fuel-burning heaters can maintain the air temperature within the enclosure. To prevent rapid drying and shrinkage of the concrete, excessively hot and dry air from heaters should not be allowed to blow directly onto the concrete. Fuel-burning heaters should be properly vented (Fig. 10.7) so rapid carbonation of the concrete surface will not occur, resulting in dusting problems later. Both drying and carbonation problems during cold weather curing and protection can be avoided by injecting saturated steam within the enclosure. It is important, however, that the enclosure remains tight and that the steam be maintained as near the saturation point as possible.

During curing and protection, the inspector should make sure that all concrete surfaces are at the proper temperature. If concrete is being artificially heated, the heat should be applied evenly so that no sections differ in temperature by more than 50 °F (28 °C) (ACI 306R).

10.4.2 *Hot weather*—ACI 305R defines hot weather concreting as conditions involving any combination of the following:
- High ambient temperature;
- High concrete temperature;
- Low relative humidity;
- High wind velocity; and

Fig. 10.7—An enclosure with heater provides cold weather protection. Combustion gases must be vented outside the enclosure.

Fig. 10.8—Early covering with mats, kept water-soaked, protects concrete while it cures in hot weather.

- Solar radiation.

These conditions can impair the quality of fresh or hardened concrete by accelerating the rate of moisture loss and cement hydration, which can lead to increased cracking and reduced long-term strength development.

Practices for offsetting the negative effects of hot weather concreting are also discussed in ACI 305R. They include:
- Keeping aggregate stockpiles well-sprayed and cooled;
- Reducing temperatures of freshly mixed concrete by using chilled water, ice, or liquid nitrogen;
- Placing concrete at night or during early morning hours when ambient temperatures and wind velocity are lower and humidity levels higher;
- Minimizing delays during transport, placing, and finishing of concrete;
- Shading and dampening ready-mixed transit drums during waiting periods;
- Providing shading or windshields or misting concrete surfaces when high evaporation conditions likely to cause plastic shrinkage cracking are present;
- Using a monomolecular film (evaporation retardant) during finishing to reduce evaporation rate;
- Providing fogging nozzles spaced above and around the work to raise humidity;
- Starting moist curing as soon as possible (Fig. 10.8); and
- Maintaining a strict curing regimen per contract documents and ACI 308R.

Coverings or membrane curing compounds that are dark in color should be avoided because they can increase the surface temperature of the concrete by absorbing radiant heat. Keep accurate records of the protection provided to the concrete, including measured surface temperatures and type and duration of curing.

CHAPTER 11—POSTCONSTRUCTION INSPECTION OF CONCRETE

In spite of suitable design, good workmanship, and efforts to produce and install flawless concrete, correction of defects may be required after forms are removed and curing has taken place. In addition, repairs or rehabilitation may be necessary to constructed or in-service concrete structures as a result of improper installation (such as poor materials, improper batching, and incomplete placement and curing), design deficiencies, exposure to excessive structural loading or fire, or degradation (from conditions such as abrasion, freezing and thawing, aggressive chemical attack, corrosion of reinforcement, deleterious chemical reactions between the aggregate and cement).

Contractual documents typically define the roles and responsibilities of the inspector in post-construction inspection of concrete structures. Often, the concrete inspector is involved in later stages of "acceptance inspection" (ACI 311.4R), where the field testing records from the construction phase are reviewed and confirmed as adequate and a final visual inspection of the post-constructed structure is completed to verify the absence of defects or nonconformities to design documents. The inspector may also be requested to complete a visual inspection (condition survey), coordinate nondestructive testing, or complete the same for repairs or rehabilitative measures on a concrete structure that has been in service for a number of years. The type and extent of services provided by the inspector are often linked to capabilities and experience of both the inspector and inspector's employer, status of the structure, and needs of the owner.

The remainder of this chapter provides an overview of the inspector's role in acceptance inspection, visual inspection (condition survey), roles, and responsibilities with respect to other forms of inspection, testing, and repair inspection.

11.1—Acceptance inspection

Typically, the concrete inspector is used during the construction phase of a project for tasks including verification of materials (aggregate, cement, and admixture) conformance, review of batch plant operations, and on-site field testing (Chapters 3, 6 through 10, and 19). As described herein, field testing may include fabricating compressive test cylinders; performing air content, unit weight, slump, and other field tests; as well as inspecting formwork and shoring, placement equipment, and the post-constructed structure for conformity to design documents. Associated laboratory work includes determining unconfined compressive strength (compression testing of field-cast cylinders) and other testing (such as tensile strength, air void analysis, and petrographic review). The inspector's role, responsibilities, scope, and communication methods should all be specified in contract documents and dictated by design specifications and drawings.

Acceptance inspection may also include on-site inspection of the post-constructed structure to confirm that as-built dimensions conform to the design documents and allowed tolerances and to confirm the absence of defects in exposed surfaces. Defects such as plastic shrinkage cracks, honeycomb, surface voids, and excess laitance may be observed at this stage and corrected before the structure is placed in service. More extensive defects may require repair/rehabilitation (refer to Section 11.4).

An important final task in acceptance inspection is preparing a final written report for the structure that summarizes the results of field/laboratory inspection and testing and identifies any follow-on actions for consideration by the owner or design professional. Should there be any anomalies remaining (for example, compressive strength derived from cylinders below specified design strength), commentary on the specifics should be provided in the report.

11.2—Visual inspection (condition survey)

Visual condition survey and inspection encompasses a variety of techniques including direct and indirect inspection of exposed surfaces, crack and discontinuity mapping, physical dimensioning, environmental surveying, and protective coatings review. Inspection may provide significant information on the current condition of an accessible concrete structure, including the absence or presence and cause of degradation, material deficiencies, performance of coatings and cover concrete, and response of a structure under load or operating condition (inherent deflection, vibration, strain, or similar). Visual inspection may be completed to gain an understanding of general condition, effects of aging, or to assist in planning for modifications, alterations, or rehabilitation of the structure.

Typically, visual inspection is an initial technique employed to gain a general knowledge on the overall condition of the post-constructed structure. Structures, or components thereof, that are primarily inaccessible without removal of soil or neighboring structures, may require other preliminary efforts to locally characterize the current physical condition. These efforts may include qualification of the surrounding environment by sampling the air, soil, and groundwater chemistry, examination using other test methods (for example, nondestructive testing), inspection after local disassembly (concrete or tendon removal), or sampling via concrete coring and laboratory testing. Other testing and analysis methods may also be selected and implemented after review of visual inspection results to gain additional insight on structural condition.

Commonly used practices and checklists for the visual inspection and condition survey of existing concrete are contained in ACI 201.1R, ACI 207.3R, and ASCE Standard No. 11 (ASCE 2000).

The scope of the visual inspection should include all exposed surfaces of the structure, joints and joint material, interfacing structures and materials (for example, abutting soil), embedments, and attached components such as base plates and anchor bolts. These components should be directly viewed (maximum 24 in. [600 mm]) focal distance) if possible, with photographs or video images taken of any discontinuities and pertinent findings. Comprehensive direct viewing may require the installation of temporary ladders, platforms, or scaffolding. The use of binoculars, fiberscopes, and other optical aids (indirect visual inspection) is recommended if needed to gain better access, to augment the inspection, and to further examine any discontinuities. Such equipment should have similar resolution capabilities, under ambient or enhanced lighting, as gained for direct viewing. The condition of surrounding structures should also be observed to better assess the aggressiveness of the local operating environment. Documentation of physical condition and alignment may also be enhanced through the use of close-range photogrammetry. This technique may provide a computer file of the mapped surface or geometry for comparison to repeated data gathering in the future.

Visual inspection also requires the use of physical measuring equipment for dimensioning and measuring the size of degraded areas. This equipment should be in good working order and properly calibrated. Calibration may be achieved through statistical review of equipment accuracy via multiple measurements or through formal calibration performed by a certification body such as the National Institute for Standards and Technology. For crack investigations, a feeler gauge, optical crack comparator, or crack width meter should be used for quantifying the width and depth if possible. For crack length measurement and general dimensioning purposes, a standard retractable metal tape should provide the desired accuracy. Overall structural dimensions should be checked against the design documents and, where not specified, against ACI 117.

Limitations with visual inspection include accessibility problems with all surfaces of a structure. The inability to detect fine or internally generated defects, and lack of quantitative bases for certain damage observations. Inability to access all surfaces of a structure reduces the ability to completely verify physical condition and absence of degradation. For structures that are largely inaccessible (for example, foundations and lower walls), this represents a primary concern for use of direct visual inspection as the main inspection tool. In addition, certain degradation mechanisms, such as fatigue, may manifest and propagate within a structure before any visible signs are displayed. For structures exposed to thermal effects and time-varying or vibratory loads, consideration should be given to supplementing any visual inspections with nondestructive or destructive testing to examine subsurface condition.

Documentation of inspection results should include a general description of observed surface conditions, location/size of any significant discontinuities, noted effects of environmental exposure, and presence of degradation. Sketches, photographs, and other means should be used to supplement text descriptions. In the event that additional testing is needed, any limitations on use, such as "access to one side only," should be noted on the visual inspection report. Upon completion of a visual inspection, the results should be assessed by comparing any findings to standard acceptance criteria. Such criteria should be included in contract documents and inspection plans as they are necessary to provide a consistent means for interpreting the condition and integrity of a structure.

11.3—Other roles and responsibilities

The concrete inspector may also be requested to participate in or coordinate additional testing on the post-constructed structure, depending on experience and qualifications. As described in ACI 228.2R, nondestructive testing may be an effective tool to supplement visual inspection in the evaluation of the postconstructed structure. Destructive sampling and testing, as well as load testing, are also tools available for evaluation. This chapter provides an overview of nondestructive and destructive techniques available to aid in the inspection and evaluation of the post-constructed structure.

Nondestructive evaluation (NDE) or testing techniques commonly employ specialized equipment and procedures to obtain specific data about the structure in question and, in certain instances, the aggressiveness of the surrounding environment. The heterogeneity of concrete, coupled with varying cross-sectional thickness and presence of reinforcing steel, limit the effectiveness of applying some of the nondestructive testing tools used for steel structures. The goal of this type of testing is to provide quantitative information about a structure and its constituents without destructively removing any material. In addition to ACI 228.2R, ASCE Standard No. 11 (ASCE 2000) also provides useful nondestructive testing techniques and application data for concrete structures. The following methods have been cited as having merit (technology and primary function are noted in parentheses) (refer to Table 11.1):

1. Load testing (structural integrity);
2. Sonic and ultrasonic methods (subsurface concrete cracking/degradation);
3. Microseismic and radar techniques (subsurface concrete degradation);
4. Infrared testing (subsurface concrete voids and large cracks);
5. Vibration and structural motion monitoring (structural integrity);
6. Acoustic emission and impact methods (subsurface concrete degradation);
7. Magnetic methods (reinforcing steel location and evidence of corrosion);
8. Electrical potential/resistance measurements (reinforcement corrosion);
9. Radiographic methods (reinforcement placement/corrosion);
10. Adhesion and holiday testing (coating durability and pin hole voids);
11. Modal analysis (structural integrity);
12. Surface strength testing (uniformity and location of questionable concrete); and
13. Tomography (subsurface voids, cracks, and degradation).

Many of these nondestructive techniques are oriented toward providing specific information about the material constituents and their condition, or the presence of voids and other internal defects, as opposed to verifying overall structural integrity. As an example, surface-strength testing may be coupled with visual inspection to examine structures for general condition, but will likely not identify structural cracking associated with internal degradation mechanisms.

11.3.1 *Nondestructive evaluation (NDE)*—Table 11.1 provides a comprehensive summary of NDE methods, their principal uses, capabilities, limitations, and potential application to nuclear structures. Key aspects of these methods are assessed in the following paragraphs. In general, the use of several methods, including ultrasonics and radar, have been successfully used to identify voids and cracks and review subsurface integrity, both before and after repairs. As continued experience is gained with these and other methods applied to concrete structures, their usefulness will grow.

There are a number of limitations associated with the application of NDE techniques to concrete structures due to their size, quantity and size of reinforcing steel, expense and reliability of equipment, and interpretation of results. Many test methods also require a representative calibration standard comprised of similar materials to accurately conduct the work. Only limited published and industry-accepted procedures exist for many of the methods listed (especially Methods B through I in Table 11.1). Interpretation of the results of many nondestructive tests on concrete also requires significant background and experience with the methodology. In spite of these limitations, NDE can be a valuable tool in the evaluation of post-constructed structures.

Load testing, as defined in ACI 437R, and modal analysis may provide information about the remaining structural integrity. These nondestructive tests involve the application of a test load representing design requirements and measurement and assessment of measured response. Practical applications for load testing can consist of individual elements or entire structures.

In the event that NDE is used, the equipment type(s), calibration records, and complete description of application should be noted in the inspection report. The referenced test method (for example, ASTM designation and title) should also be identified, along with the names of testing personnel, date, and complete description of techniques used and the results.

11.3.2 *Destructive testing*—Destructive sampling and testing focuses on the removal of concrete, tendons, and embedded items or reinforcing steel specimens from the structure for laboratory testing to determine physical, chemical, microstructural, and mechanical properties or other information relative to condition. Generally, this technique is limited to a controlled number of specimens to minimize any impact on remaining structural performance. Included in this category are concrete core sampling and follow-up laboratory testing (for example, unconfined strength tests and petrographic analysis), concrete cover removal and reinforcing steel evaluation, and removal of coatings for laboratory testing. Destructive testing also involves the testing of concrete exposed to soil, groundwater, or other environmentally significant materials to determine the effect of exposure on its condition. The presence and concentration of aggressive chemicals such as chlorides, sulfates, and other salts in addition to the pH and conductivity of the contacting fluid or solid is of particular interest. Table 11.2 summarizes destructive testing techniques and their uses.

Destructive testing also provides information needed to determine the remaining durability of the cover concrete, structural concrete, and reinforcing steel system by testing of exposed or removed samples. Factors important to concrete durability include permeability, porosity, reactivity between constituents, chloride ion content, degree of carbonation, and chemically induced degradation. Factors important to the performance of the reinforcing steel system include electric potential and resistance, degree of surface corrosion, and bond characteristics with surrounding concrete (monolithic

Table 11.1—Nondestructive evaluation methods

Concrete

A. **Surface hardness and penetration resistance techniques**: Tests including the Windsor probe (ASTM C803/C803M) and Schmidt rebound hammer (ASTM C805) provide relative uniformity and hardness information for the tested structure; this information is useful in establishing the overall condition and areas of possible weakness of concrete (strength can be estimated from results when properly correlated with compressive strength tests). These methods have limited ability to detect damage. (ACI 228.1R and 437R).

B. **Ultrasonic methods**: Tests, including the pulse-echo, pulse-velocity (ASTM C597), and through-transmission methods, can identify the presence of internal discontinuities and cracks, provide relative strength and soundness, and give the approximate concrete strength of a tested member. Sound waves are transmitted into the structure, and the reflection of waves and transit time response are used to identify internal discontinuities. The detectability success of these methods is highly variable and dependant on configuration; pulse-echo and pulse-velocity have suitable detectability of voids and cracks in thin, lightly reinforced members (ACI 228.2R).

C. **Stress wave, resonant frequency, and impact-echo testing**: These test methods provide improved understanding of internal concrete structural condition from a single surface. Equipment used ranges from a simple impactor and sound receiving-and-processing device to sophisticated spectral and waveform analyzers. Similar in principle to ultrasonic methods, the impact-echo method has been successfully used in concrete structures and other civil structures and pavements to find voids and other discontinuities. Impact-echo is one of the more promising techniques for detecting minor internal damage with relatively high defect detectability (ACI 228.2R).

D. **Microseismic and radar testing**: Short-pulse radar or microseismic electromagnetic waves have been shown to identify shallow subsurface discontinuities in concrete structures such as delaminations, the extent of cement hydration, location of reinforcement, member thickness, and water content with reasonable accuracy. The equipment is expensive and detectability for other forms of damage is still questionable (ACI 228.2R).

E. **Acoustic emission testing**: This method involves monitoring the noise and release of strain energy created by the formation and propagation of defects, such as cracks, in structures. Its use has been limited to primarily laboratory testing because of difficulties in measuring signals and interpreting their cause in the field. This method can also be applied during load testing. The equipment currently available is also expensive and difficult to transport.

F. **Infrared thermography**: Infrared waves may be used to identify volumetric subsurface discontinuities by measuring differences in temperature (heat flow). It is particularly useful for finding delaminations and shallow voids that contain air or moisture at different temperatures than the concrete, although this testing cannot easily identify damage depth or thickness dimensions. Equipment is becoming more affordable and is relatively simple to use. The method has reasonable detectability, especially for external structures exposed to moisture, which will display greater thermal differences. Minimum damage size detected is greater than for other methods.

G. **Tomography**: Testing consists of using acoustic wave or x-ray transmission into a structure to examine for presence of subsurface discontinuities; analysis of the wave form and response is managed via computerized imaging. This technique has demonstrated good resolution and accuracy in the laboratory, but has had little field application. The acoustic method is judged to be reasonable for relatively small internal voids and discontinuities to depths of 18 in. (450 mm).

H. **Physical response monitoring and testing**: The response of a structure to load and vibration may introduce behavior such as high strain rates, deflections, settlement, or resonance. Structures that demonstrate sensitivity in response to loading may be analyzed through the use of a variety of instruments, gauges, motion detectors, and measuring devices. This testing is analogous to static load testing. Detectability of damage and loss of integrity is greater for smaller, lightly reinforced concrete structures.

I. **Radiography**: Passage of gamma, neutron, or x-ray radiation through a structure may be used to locate internal defects, such as voids, and reinforcing steel through differences in radiation intensity passing through the structure that is captured on photographic film placed on the opposite side of the structure from the source. The equipment is expensive, however, and requires special licensing, and this method is not suitable for partially accessible, thick, and heavily reinforced members. This method has reasonable detectability for larger internal damage in thin members.

J. **Magnetic resonance imaging**: This method is based on the interaction between nuclear magnetic dipole moments and a magnetic field, as developed in the medical industry. This interaction is used as the basis for determining the amount of moisture present in a material. Although a prototype system has been developed, only limited results are available, and damage detectability is uncertain.

K. **Surface topography**: This method involves the examination of a concrete crack and fracture surface profile through the use of enhanced scanning with a scanning electron microscope and use of Fourier spectral analysis. Although the technology behind this method has a long history, its application to concrete has been slow primarily due to the nonhomogeneous nature of concrete and difficulty with correlating damage observed to cause and significance. Only limited experience has been gained with this method, and damage detectability is uncertain.

Embedded steel (conventional and prestressing steel)

A. **Half-cell potential testing**: This test involves measuring the electrical potential difference between a locally exposed and connected reinforcing steel bar to the neighboring embedded bars through the use of a copper-copper sulfate electrode and high-impedance voltmeter (ASTM C876). This method, however, only indicates the presence of corrosion activity and not the rate of attack. The damage detectability of the method is greatly enhanced when concrete resistance measurements are also taken and used with reinforcing steel potentials to define corrosion cell locations. Detectability is limited by factors including varying moisture contents and operator inexperience. This method has the highest detectability for active corrosion, short of destructive cover concrete removal.

B. **Linear polarization testing**: This method is similar to the half-cell method. Only the cyclic potentiodynamic polarization resistance of the embedded steel is measured, and an impressed current in the embedded steel is used to compute the shift in potential, corrosion current, and degree of corrosive activity. Lack of field experience limits the relative detectability of this method; it is more intensive to conduct than the half-cell method.

C. **Magnetic location testing**: Using the principles of magnetic induction, flux leakage, or nuclear magnetic resonance, the location, depth, and size of embedded reinforcing steel may be determined. Simple, inexpensive equipment is available for this testing, which is restricted in accuracy by the presence of large size or quantity of steel.[1,51] The damage detectability of this method is limited, as the minimum damage size identifiable is relatively large (concrete surface damage may already be present).

---

behavior). Sampling and chemical analysis of tendon grease or surrounding grout for presence of water, chlorides, or other contaminants is also considered a destructive test.

The removal of cores from concrete is the most common of destructive testing/evaluating methods. Core samples can be used primarily for: strength determinations, visual/condition examination (such as cracks, color, aggregate size and distribution, voids, and reinforcing cover), length (as for slab-on-ground thickness), petrographic analysis (see following paragraph for explanation), permeability tests, and durability assessment.

Petrography involves the study of hardened concrete samples through the use of detailed physical examination (primarily through microscopy, spectroscopy, and x-ray diffraction techniques), usually combined with various chemical analyses. Petrographic techniques are most

Table 11.2—Destructive sampling and testing methods

The following sampling and testing methods are often employed when additional information, unavailable from visual inspection or NDE techniques, is needed to better characterize the condition of a concrete structure. These techniques generally have a high degree of defect/degradation detectability.

A. **Core sampling and laboratory mechanical, petrographic, and chemical testing** (ASTM C42/C42M, C823; refer to summaries on petrographic analysis and material testing techniques)

B. **Local static or dynamic load testing (to failure)** (ACI 437R)

C. **Reinforcing steel excavation and laboratory mechanical and metallurgical testing**

D. **Powder sampling and sulfate or chloride ion testing** (powder samples taken via drilling into concrete at varying depths from surface for determination of deleterious ion content) (ASTM C1152/C1152M)

E. **Miniature core sampling and depth-of-carbonation testing** (phenolphthalein etching of miniature core into cover concrete identifies depth of carbonation from surface)

F. **Tendon removal and laboratory mechanical and metallurgical assessment**

G. **Tendon grease or grout removal and laboratory chemical testing** (chemical screening for presence of water, chlorides, nitrates, sulfides, and reserve alkalinity) (ASTM D95, D512, and D974)

Petrographic testing on concrete sampling

Petrographic analysis includes a variety of laboratory tests that have the purpose of indicating the durability, physical condition, and soundness of postconstructed concrete. These methods may also be used to characterize original mixture proportions, type of aggregate, and expected durability. The following list identifies specific tests and their uses:

A. **Light microscopy** (air-void system, aggregate soundness, cement hydration, w/c, stress level, degradation type)

B. **X-ray diffractometry** (extent of hydration, type of aggregate)

C. **X-ray spectroscopy** (approximate cement content, impurity content)

D. **X-ray radiography** (cracking and stress patterns)

E. **Wet chemical analysis** (cement content, chemical constituents)

F. **Electro microscopy** (type of aggregate, presence of degradation or distress)

G. **Infrared spectroscopy** (presence of organic material)

Laboratory material tests on concrete and metals

The following tests are typically used to quantify in-place material condition, mechanical property, or other relevant data from samples taken from the subject structure. This list is not exhaustive.

A. **Concrete**
1. Compressive strength
2. Splitting tensile strength
3. Modulus of elasticity
4. Permeability/porosity
5. Chemical testing (chlorides, sulfates, etc.)

B. **Metals**
1. Ultimate or yield strength
2. Elongation
3. Toughness (brittle behavior)
4. Corrosion susceptibility

frequently used to estimate w/c, degree of hydration, depth of carbonation, air void distribution, aggregate-cement reaction, aggregate soundness, presence of deleterious materials (chlorides, sulfates, and foreign substances), freezing-and-thawing damage, and depth of fire damage (see also ASTM C295 and C856).

Limitations to the use of destructive testing include the cost for retrieving samples and performing testing, impact on the sampled structure, accuracy of various laboratory tests, and interpretation of results. Similar to NDE, the number and location of samples and tests taken in destructive testing is important to the accuracy of the results and their use in analyzing the structure. Destructive testing is especially useful, however, for reviewing the ultimate impact of degradation and establishing the presence of microstructural and internal damage. Destructive mechanical testing and petrography provide the greatest insight on future structural performance of the as-sampled structure.

11.3.3 *Summary*—When responsible for coordinating or executing further NDE or destructive testing on a concrete structure, the inspector should ensure that such work is in accordance with contract/design documents and that the results are documented and submitted to the owner or responsible party.

11.4—Observations leading to repair/rehabilitation

Visual inspection, other testing, or results from construction-phase testing of a completed structure may all identify conditions where additional evaluation and repair are required. A thorough review and evaluation by qualified professionals is mandatory to determine the cause(s) for the defects or degradation observed. If required by the owner, written specifications and repair procedures should then be prepared by the design professional or qualified repair specialist to correct the condition and causes for the problems identified during the evaluation. Inspection of the repair should be completed to verify that the acceptance criteria (methods, materials, and equipment) have been met, or, if not met, to document the details of nonconformance for the record.

An inspector should not be expected to prescribe how repairs are to be made and then pass judgment on the acceptability of such repairs.

The following paragraphs describe conditions that typically require repair, either as a result of improper construction or in-service degradation, to maintain compliance with minimum building code, insurance carrier, or owner interests. In each instance, design documents consisting of a specification, repair procedure, drawing set, or combination thereof, should be used by the organization responsible for making the repair. Such documents should contain the acceptance criteria to be used by the inspector to verify and document conformance or nonconformance.

11.4.1 *Minor defects*—The following conditions are typically the result of problems occurring during the batching, placement, or curing of concrete, and generally do not result in any significance to the structural integrity of the structure. These conditions are typically manifested in the cover concrete. As such, repair of the condition may use industry-proven procedures and materials established for this purpose, without any further evaluation or engineering analysis.

1. Blemishes;
2. Small holes (for example, form tie holes or bug holes);
3. Stains, discoloration, or other undesired surface finishing problems;
4. Honeycombing and voids (should be explored for extent and depth to ensure that they are not significant);
5. Spalls;
6. Fractured edges and corners; and
7. Shallow, passive cracking.

The inspector's role in repair of nonstructural defects includes verification that the organization making the repair has proper design documents in hand, follows such documents and procedures, and uses materials conforming to repair design, and confirmation that the repair meets its published acceptance criteria. For example, if repair of a passive crack is completed using epoxy injection and the procedure calls for epoxy filling for its entire length, the inspector should confirm both the existence of sufficient ports and the presence of epoxy at the surface for the length of the crack.

11.4.2 *Structural defects*—Structural defects are those conditions that could affect the load-carrying capacity of the concrete structure. All repair and rehabilitation work associated with structural defects repair requires the review and approval of the design professional (for example, Engineer of Record). Before conducting these repairs, it is important to conduct an evaluation of existing concrete conditions and to analyze root causes of damage and deterioration that may exist. To accomplish this, as well as to establish the scope and nature of repairs needed, the design professional may direct the inspector, a qualified testing agency, or a concrete repair specialist to conduct various examinations, measurements, and analysis of in-place materials, complete load tests

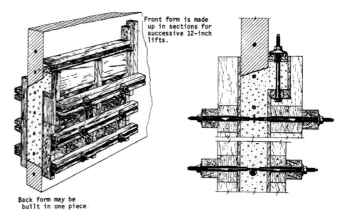

Fig. 11.1—Forms for concrete replacement in walls.

and nondestructive testing and structural condition. Detailed instructions on conducting evaluations of concrete structures before repair are provided in ACI documents, including ACI 201.1R, 207.3R, 224.1R, 349.3R, 364.1R, and 437R. Following the evaluation, design documents consisting of specifications, repair procedures, drawing set, or a combination thereof should be developed and then approved by the design professional.

Typical structural defects include:

1. Large voids, honeycombs, and spalls that may result in insufficient strength of the structural element (foundations, walls, floors, columns, beams, and stairs);
2. Form failures resulting in dimensional deviations from the design drawings;
3. Cracking into the structural concrete, with or without active movement;
4. Degradation due to improper curing or lack of protection from freezing or excessive heat of hydration;
5. Inadequate bond between concrete and reinforcement (conventional or prestressed);
6. Inadequate structural support before the development of sufficient concrete strength;
7. Fire, earthquake, overload, or mechanical damage (for example, impact loading from falling objects); and
8. Conditions leading to improper strength of the finished structure (for example, improper mixture proportioning, batching, placement, or curing).

Many variables should be considered by the design professional in establishing a program for structural repair of concrete, including:

- Methods to be used in removing damaged and deteriorated concrete;
- Reinforcement evaluation and repair;
- Surface preparation of the existing concrete;
- Special forming considerations (Fig. 11.1);
- Selection of repair materials;
- Sequencing of repairs;
- Inspection and testing requirements and acceptance criteria for the repair; and
- Curing and protection.

The design professional may involve the concrete inspector in the preparation of repair design and procedures, particularly with respect to inspection steps. Preparation of the repair design documents, however, is solely the responsibility of the design professional.

The inspector's role in repair of structural defects includes verification that the organization making the repair has proper design documents in hand, follows such documents and procedures, and uses materials conforming to repair design, and confirmation that the repair meets its published acceptance criteria. Given that such repairs often require removal of concrete, addition of structural elements, and new formwork or shoring, the inspector may be assigned additional responsibilities during the repair cycle. For example, if a beam in a composite beam/column/slab system is found to possess insufficient concrete compressive strength, the inspector may be involved in the following tasks:

1. Verification that temporary shores for the slab and columns are in place per design drawings;

2. Confirmation of reinforcing steel condition and position after removal of existing concrete;

3. Inspection of batch plant preparing replacement concrete, and field testing of the delivered material;

4. Observation of the repair process; and

5. Acceptance inspection of the completed repair.

Numerous resources are available for guidance in the preparation of repair specifications and the selection of materials, application procedures, and specialized equipment, as well as minimum post-repair acceptance criteria. ACI committee publications addressing concrete evaluation, repair, and durability include: ACI 201.1R, 201.2R, 207.3R, 224.1R, 303R, 349.3R, 364.1R, 437R, 503.4, 506R, 515.1R, 546R, 546.1R, 546.2R, and 548.1R.

In addition, the following associations offer publications, technical guides, and other resources on various aspects of concrete evaluation and repair:

International Concrete Repair Institute (www.icri.org)
Precast/Prestressed Concrete Institute (www.pci.org)
Portland Cement Association (www.cement.org)
National Precast Concrete Association (www.precast.org)
ASTM International (www.astm.org)

CHAPTER 12—SLABS FOR BUILDINGS

Because of their large, unformed surfaces, concrete slabs demand close attention to construction and inspection to ensure a high-quality finished product. Some aspects of construction may be more important than others, depending on the location and intended use of the slab. In all cases, however, the primary objective is to conform to contract document requirements for finished design elevation and surface tolerances, serviceability, durability, and appearance.

Before becoming involved with slab construction, inspectors should review the guidelines presented in ACI 302.1R. They should also verify that the finishers working on the project are experienced and qualified. ACI provides a certification program for concrete flatwork technicians and finishers.

12.1—Positioning of reinforcement

Specifications for nonstructural slabs-on-ground usually require crack control reinforcement, typically welded-wire fabric. The Wire Reinforcement Institute recommends that welded-wire reinforcement be placed 2 in. (50 mm) below the slab surface or in the upper 1/3 of the slab section, whichever is closer to the surface. Proper placement and support of welded-wire reinforcement is often not achieved, resulting in reinforcement at the very bottom of the slab or so near the top that cover is inadequate. Structural slabs usually contain two layers of reinforcing bar mats—one placed near the top and one near the bottom of the slab—or a system of post-tensioned strands.

Before concreting begins, it should be checked that reinforcement conforms to drawing requirements and is rigidly supported at its designed elevation. Chairs are the support method usually specified. Crews should never place welded-wire fabric or other reinforcement on the subgrade and then later attempt to pull it into place after placing the concrete. Special attention should be paid to placement of reinforcement around corners of slab openings, where cracking often occurs. For composite metal deck slabs, the inspector may also be required to verify the installation, size, location, and number of shear studs before concrete placement.

Some industrial floors subject to heavy traffic may have special joint details such as large, closely spaced dowels that require careful installation (Fig. 12.1). The inspector should verify that the dowels are set at the specified height, depth, and spacing and are set level and perpendicular to the joint.

12.2—Mixture requirements

General requirements for concrete mixtures, as discussed in Chapter 6, are usually satisfactory for slab concrete, although slab concrete mixtures should also meet minimum cementitious contents given in ACI 301. Good finishability characteristics are especially important, and should meet the requirements of ACI 302.1R. If not otherwise specified, use a water content to produce slumps conforming to the provisions of ACI 301. The concrete should have satisfactory plasticity for finishing and be sufficiently cohesive to minimize segregation. The use of water-reducing admixtures can help overcome problems of placement in large, flat structures. For exterior slabs, concrete should be air entrained where required by exposure conditions. For interior slabs, ACI 302.2R cautions against using air-entrained concrete when slabs are to receive a steel trowel finish unless the air entrainment is required for protection from freezing-and-thawing exposure during construction. (Refer to previous comments in Chapter 9). Concrete made with lightweight aggregate may also require air entrainment to improve workability. Refer to Chapter 6 for a discussion on how temperature, slump, and water reducers affect air entrainment.

If excess bleeding (presence of free water on the slab surface) occurs at recommended slump limits, the sand may contain insufficient fines passing the No. 50, 100, and 200 (300, 150, and 75 μm) sieves. This can sometimes be corrected by the use of admixtures, blending sand to improve the gradation, adding a supplementary cementitious material

(SCM), or increasing the cement content, with an appropriate reproportioning of the mixture. If the mixture causes difficulties in workability and finishing, the mixture or admixtures should be adjusted to overcome the problem. Before allowing the mixture proportions to be changed, the designer's approval should be obtained.

Mixtures should contain the maximum amount of coarse aggregate that can be used without causing difficulty in placing and finishing. The nominal maximum size of coarse aggregate should not exceed 1/3 the depth of the slab and generally not exceed 1-1/2 in. (37.5 mm).

12.3—Slabs-on-ground

12.3.1 *Subgrade and subbase*—Before slabs-on-ground are placed, the subgrade should be inspected to see that it is prepared and compacted in accordance with the contract documents. Slab design requirements usually require that a layer of crushed-stone subbase be placed on top of the subgrade to provide free drainage beneath the slab and to prevent capillary rise of moisture through the concrete. The bottom of the granular base courses should be drained so they do not become reservoirs for water. Concrete for slabs-on-ground should never be placed on frozen ground.

In many cases, particularly for floors of enclosed buildings, specifications require placing impervious sheeting or similar material over the subbase as a vapor retarder to minimize moisture migration through the concrete and vapor emission transfer that can affect floor covering adhesion. If a vapor retarder is used, it should provide full coverage, and each sheet should be adequately lapped and preferably taped to adjoining sheets. According to ACI 302.1R, drying shrinkage and curling of slabs can be reduced by placing several inches of compactible granular fill (not sand) above the vapor retarder. Considerations, however, for meeting vapor emission standards before installing floor coverings may dictate placement of concrete directly onto the vapor retarder. The engineer needs to weigh these considerations and specify requirements accordingly.

To prevent puncturing of the vapor retarder, construction crews should take precautions during installation and concreting. If no vapor retarder is specified, crews should lightly dampen the subgrade by spraying with water before concreting.

12.3.2 *Placing and consolidation of concrete*—Concrete for slabs can be placed by crane and bucket, pumps, truck chutes, conveyors, or hand or power buggies. Whatever the placing method, workers should exercise care to maintain proper location of reinforcement. Measures should be taken to not damage epoxy-coated reinforcement and post-tensioning sleeves and anchorages during placement. As concrete is deposited, the drop onto the subbase or decking should be vertical and as low as possible. The concrete also should be deposited close to its final location to avoid excessive horizontal movement and segregation. Workers should distribute the concrete horizontally with square-edged shovels, not with vibrators.

To ensure that the designed slab thickness is attained within specified tolerances, the concrete should be struck off

Fig. 12.1—Installation of dowels after first placement of concrete.

uniformly by checking the depth to grade at sufficient points using grade stakes or alternative methods. Bulkheads or forms can serve as screeds, but they should be set at proper elevation and have no projections. Modern construction techniques for controlling finish elevation often employ the use of lasers to check alignment and elevation of screeds and finishing equipment.

Another method for obtaining proper elevation of the concrete is with the use of a concrete screed, sometimes called a wet screed. Strips of low-slump concrete are cast between grade stakes and struck off at the proper level. The grade stakes are then removed, and the remainder of the slab is placed and finished using these strips of concrete as screeds. The use of wet screeds for placing concrete in slabs is difficult to control properly. Their use should be discontinued immediately if there is doubt about consolidation of the concrete in the screeds or in the slab concrete alongside the screeds.

Internal vibrators provide the best means of consolidating thick reinforced concrete slabs. They should be inserted and removed vertically, at short spacings, over the entire area. It is especially important that crews do not overvibrate the concrete, which can bring excess fine material and water to the surface. Close attention should also be paid to vibration along bulkheads and at corners. If internal vibrators are not available, the concrete should be thoroughly spaded as it is placed to compact the concrete and eliminate voids.

Concrete in thinner slabs (4 in. [100 mm] or less) can be consolidated by the sawing motion of rigid strikeoffs, by vibratory screeds, and by rollers. Pan-type surface vibrators can also be used, but care should be taken to prevent working too much paste to the surface. Grate tampers (jitterbugs) can be used with low-slump concrete, but should not be used with higher-slump mixtures or with lightweight-aggregate concrete. Crews should never use vibratory-grate tampers because they are likely to overvibrate the concrete, resulting in segregation and excessive paste worked to the surface.

12.3.3 *Finishing*—Immediately after the concrete has been consolidated and screeded to grade, finishers should remove surface irregularities using a bullfloat or darby. This should be done before any water appears at the surface.

Fig. 12.2—A power float finishes the surface of a concrete slab.

After strikeoff and initial leveling, floating is performed to remove remaining minor irregularities and begin densification of the surface. Floating should not begin until the water sheen has disappeared from the surface and the concrete has stiffened sufficiently so that a person stepping on the surface leaves an imprint approximately 1/4 in. (5 mm) deep. Premature finishing will bring excessive fines to the surface, which will tend to dust in service. If the amount of water or laitance is excessive, it should be scraped off before the surface is floated. A large loop of garden hose pulled across the surface is effective for removing excess surface water.

Floating can be done by hand or power float (Fig. 12.2). To produce the best results, however, finishers should apply the proper amount of pressure. If the concrete is air entrained, the use of magnesium or aluminum tools, rather than wood, will prevent tearing of the surface. Finishers should not overfloat or overfinish the surface; the operation should continue only to the extent necessary to ensure that a layer of mortar will cover the coarse aggregate. This precaution in particular should be observed when power floats are used.

Finishers should perform edging only if required by the contract documents, using edgers with a radius no greater than specified. If the radius is not specified, tools with a radius of 3/16 to 3/8 in. (5 to 10 mm) should be used. Edges at construction joints can be lightly stoned, rather than edged, after the forms are stripped and before placement of the adjacent slab.

Workers should not apply additional water to the concrete surface during finishing or edging operations by dashing with a brush, sprinkling, spraying, or using finishing machines with water attachments. With proper scheduling of the finishing operations, taking into account prevailing weather conditions, no additional water should be necessary.

Floating may be the final finishing operation or may be followed by troweling. Many slabs require a troweled surface. To prevent working of excess fines and water to the surface, troweling should be delayed as long as possible, and certainly until the surface moisture remaining after floating has disappeared. Troweling, however, should not be delayed so long that the surface becomes too hard to permit compacting the fines at the surface.

On most jobs, power troweling is performed first, followed by finish troweling by hand. Hand trowels are made of spring steel, and for second and succeeding trowelings, the finisher should tilt them slightly to increase the pressure of the contact area, thus producing a dense, hard surface. Hand troweling requires a high degree of skill to produce a uniformly dense surface free of chatter marks or other imperfections. Where a hard-troweled or burnished surface is required, finish troweling should be continued until the trowel makes a ringing sound.

For some walks or ramps, particularly exterior walks, it is more desirable to use only a float finish, without troweling, to produce a rougher surface with improved traction. Traction can also be improved by first applying a steel-troweled finish and then, just before the concrete sets, lightly brooming the surface with a fine-haired push broom or similar texturing tool.

Requirements for flatness and levelness of slabs are often specified using the F-number system. Measurements of F-numbers are conducted using equipment and methods specified in ASTM E1155. Additional information on finish tolerances and other requirements for flatness and levelness can be found in ACI 117.

12.3.4 *Hardened surfaces*—Some industrial floors, particularly those for warehouses, are designed to receive a hardened surface produced by use of a dry shake of metallic or natural hard mineral aggregate. The dry-shake material is usually made by blending aggregate with dry cement. Whenever a dry-shake hardener is used, it is imperative that the manufacturer's recommendations are followed.

Proper installation of these surfaces requires finishers to evenly distribute the dry shake over the slab surface after floating it once. Typically, concrete should be non-air-entrained to minimize the possibility of finishing problems such as blisters and scaling. Mechanical application is superior to hand methods, and should be used whenever possible. Finishers first apply 2/3 of the dry shake in one direction, and then float it into the surface without the addition of water. The remaining third is then applied at right angles to the first and again floated to ensure uniform application. Finishing operations are then conducted as described previously. Abrasive aggregate often is added to slab surfaces in much the same way to produce a nonslip surface.

12.3.5 *Two-course construction and special toppings*— Two-course floors are constructed by applying a thin integral topping to a base course of concrete that has not completely hardened or by applying a bonded topping to a base course that has already hardened. Procedures for base-course construction are similar to single-course methods except that only a bullfloat or darby is used for finishing to produce a surface to which the topping will bond. Workers should apply integral toppings when the base-course concrete has hardened to the point that their footprints are barely perceptible. Before applying a bonded topping to hardened base slabs, it is necessary to roughen or profile the surface and to thoroughly clean the base slab of all loose material, laitance, scale, oil, paint, or other contaminants that can affect bond. Sandblasting or high-pressure water can be used to clean exterior slabs. For indoor slabs, it may be necessary for workers to use cleaning agents that will not leave a residue or to scrub the contaminated area with a 10% solution of muriatic acid, followed by thorough flushing. Acid cleaning should be used only as a last resort; the acid should be stored and handled carefully.

Before placing the top course, workers should thoroughly moisten the bottom course, making sure no pools of free water remain. A neat cement grout mixed to the consistency of thick paint or a specified bonding agent is sometimes applied to the surface just before concrete placement. These materials should not be allowed to dry or set before the top course is applied.

Bonding of two-course floors is an operation requiring meticulous attention to procedure. Complete bonding is almost never achieved, and post-construction sounding of slabs will sometimes indicate large areas of debonded topping. As a result, some flooring contractors no longer use regular bonded two-course floors, preferring instead to install the overlaying second course thick enough to be serviceable by itself, thus removing the need for bonding between the layers.

The mixture proportions, thickness of the topping, and spacing of control joints are important considerations for two-course floors. Mixture proportions usually contain smaller-sized aggregates, low water contents, are flowable, and have high cement contents. Water content should be carefully controlled using the minimum amount necessary to permit easy placing, consolidating, and screeding (usually a slump of less than 1 in. [25 mm]). Heavy-duty topping mixtures should have no more than 31 lb (3.75 gal. [14.2 L]) of water per 100 lb (45 kg) of cement and a slump not to exceed 1 in. (25 mm), unless a high-range water-reducing admixture is used to increase slump.

To control random cracks, control joints should be placed above existing joints in the base course, and quite possibly at additionally reduced spacings, depending upon the thickness of the top layer. Top layers are usually a minimum of 3/4 in. (20 mm) thick. Thinner toppings tend to debond more frequently and have a greater tendency to develop random shrinkage cracks and excessive curl.

The finishing methods for two-course slabs are the same as for single-course slabs. Disk-type power floats equipped with an integral impact mechanism can be used to consolidate and finish heavy-duty toppings.

12.3.6 *Curing and protection*—Curing is one of the most important factors in attaining durable slab concrete, and should begin immediately after finishing is completed.

The use of membrane-forming curing compounds to seal the concrete surface, thus preventing moisture loss, is the most common, but not the best, method of curing concrete slabs. The contract documents usually list acceptable types and brands, along with approved application methods and rates. Application is typically by power sprayer. As long as the compounds are applied immediately after finishing and according to the manufacturer's directions, they provide reasonably good results. During windy weather, the application rate may need to be increased, depending on the application method and wind velocity. The application rate should also be increased when the compounds are to be applied to broomed or similar rough surfaces, which have greater exposed surface areas.

Where practical, the best method of curing slabs is by ponding the entire surface with water after finishing has been completed and the concrete has achieved initial set. In many cases, however, this method will not be used because of the cost and complexity of retaining the water.

Other effective curing methods include covering the slab with wet burlap, curing paper, plastic sheeting, or similar materials that can maintain a moist surface condition during the entire curing period. Refer to Chapter 9 for a discussion on plastic shrinkage cracks, and Chapter 10 for a detailed discussion of curing methods, including protective measures to take during hot and cold weather.

12.4—Structural slabs

Satisfactory construction of structural slabs involves most of the same requirements as those for slabs-on-ground, with the following additional requirements:

- Screeds for reinforced structural slabs should be placed over the form-supporting members. In all cases, the deflections of the supporting members caused by imposed concrete loads should be taken into account, and such deflections in setting the screeds should be allowed for. This may require shoring of the support members. If shoring is used, the vertical loads should be traced and additional shoring provided at lower floors, if required, to prevent possible deflections of lower floor members that support the shoring;
- The final top-of-slab elevations should be checked around columns carefully, as there is a tendency for workers to overfill at these points; and
- Curing and protection should be applied to both the bottom and top surfaces of structural slabs.

12.5—Joint construction

Types and locations of joints for building slabs, either on ground or structural, are usually shown in the contract documents. No deviation should be allowed except for allowable tolerances. Joints should be carefully planned by the designer to serve the intended purpose, such as isolation of columns (Fig. 12.3) or control of expansion or shrinkage (Fig. 12.4). (Refer to Chapter 8 for descriptions of the various joint types.)

Casting of concrete for the slab should be planned in conjunction with joint locations. If possible, construction joints should coincide with planned joints. No unspecified joints should be placed in a structural slab without approval of the designer.

Depending on the purpose of the joint, fillers or armored protection may be specified. The material used should be specified and properly placed and treated.

Contraction joints for control of drying shrinkage often are sawed. (Refer to Chapter 13 on pavements for the optimum conditions for successful joint sawing.) In building slabs-on-grade, the joints should be sawed to at least one-fourth of the slab thickness. To be effective, joints should be installed as soon

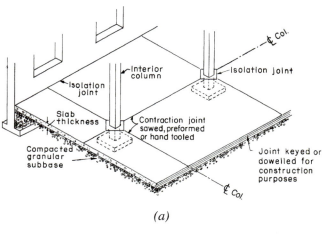

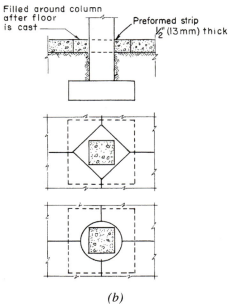

Fig. 12.3—(a) Locations of isolation and control joints for concrete floor slabs (ACI 302.1R); and (b) detail shows isolation joints at columns.

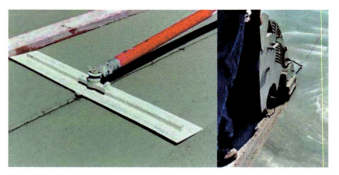

Fig. 12.4—In slabs-on-ground, a contraction joint formed by tooling fresh concrete or saw-cutting after final set provides a weakened plane that prevents random cracking caused by shrinkage or thermal effects.

as the concrete will permit cutting without raveling of edges (usually 4 to 12 hours after finishing). ACI 302.2R recommends that reinforcing bars or welded-wire reinforcement should be discontinued at any joints where the intent of the designer is to let the joint open and to reduce the possibility of shrinkage and temperature cracks in an adjacent panel.

CHAPTER 13—PAVEMENT SLABS AND BRIDGE DECKS

Concrete pavements vary widely in thickness and reinforcement requirements, depending on expected traffic loads. Primary streets and roads and highways designed to carry high-speed, heavy-load traffic typically require thicker slabs than those used for parking lots and residential streets with light traffic. Runways designed to carry aircraft with gross loads up to 750,000 lb (340,000 kg) may need slab thicknesses exceeding 20 in. (500 mm). Each of these pavement types may be unreinforced, have distributed steel only, be heavily reinforced (such as a continuously reinforced concrete pavement), or even be prestressed.

In general, the thicker the pavement or the higher the load rating, the higher the cost per unit area and the greater the priority given to inspection, testing, and sophisticated construction control measures. Regardless of slab thickness, load requirements, and cost, however, the purpose of inspection and control of pavement construction should be the same: to provide a functional pavement that meets the performance criteria for the design life. A little more money spent on inspection and construction control of even lighter-duty pavements can return great benefits by upgrading their quality and lengthening their service life.

Inspectors and testing technicians have an important role in the production of satisfactory pavements, whether employed by the owner for quality assurance (acceptance) inspection or by the contractor for quality-control inspection. ACI provides certification programs for concrete transportation construction inspectors and concrete field-testing technicians. This chapter gives an overview of recommended practices for producing satisfactory concrete pavements and bridge decks, but it should serve as additional guidance only on items not fully covered by the contract documents. For more detailed information, refer to ACI 325.9R and 325.12R.

13.1—Foundation (subgrade and subbase course)

The degree of pavement smoothness depends, to a large extent, on the quality of the foundation. Grading and compaction of subgrades and construction of subbase courses should be in strict accordance with the contract documents. Regardless of the type of subgrade or subbase required, uniform support is essential to achieving good pavement performance. Close attention to uniformity in materials, thorough compaction, and base thickness can result in reduced maintenance and longer pavement life.

Low-traffic-volume pavements, such as residential streets and secondary roads, may not require a subbase. The use of a subbase course, however, can facilitate construction, and may be permitted by contract documents at the contractor's option.

13.1.1 *Fine grading*—Fine grading for slipformed concrete pavements is usually done by machines with automatic grade-control devices that operate from preset string lines or wires or from adjacent, completed, fine-graded subgrade (Fig. 13.1). Grade wire positioning is based on accurate measurements from the designer's survey stakes followed by careful sighting along the wire, correcting for discrepancies. For fixed-form paving jobs, fine grading is done by machines operating off of grade stakes or the set forms. In both cases, the inspector's job is to see that the fine grading is at the correct elevation based on the elevation of the grade stakes. Fine grading to the correct elevation will reduce concrete yield losses and conflicts over pavement depth following coring.

If permitted by the contract documents, concrete transit mixers or agitators are sometimes allowed to operate on the prepared subgrade. In this case, any subgrade deformations should be corrected and recompacted before concrete placement.

13.1.2 *Stabilized base*—Some heavy-duty highway and airfield pavements are placed on a granular subbase or base course that has been treated with chemical stabilizers such as cement, asphalt, lime, or combinations of these materials. It is important that such bases be finished to grade within specified tolerances at the time of original construction because they will be difficult to rework later. Equipment for placing the subbase or base course should have automatic grade control. A typical surface tolerance for stabilized bases under concrete pavement is ±1/4 in. (6 mm) when checked by a 10 ft (3 m) straightedge. Excessive manipulation of the surface should be avoided, as this results in a weakened top layer susceptible to deterioration during service. This can lead to joint faulting and other pavement problems. A stabilized base should be protected from damage after placement.

The use of lean concrete bases—often called econocrete—has increased. The same equipment already on the job for constructing the concrete pavement can place these bases, and normally no hand finishing of the base behind the paver is needed. A membrane-forming curing compound sprayed on the base surface will retain the moisture required for cement hydration and provide a bond breaker between the subbase and pavement. The lean concrete base should contain entrained air for durability. Testing of strength and fresh concrete properties, such as air content and consistency, should be done in the same manner as for concrete pavements.

13.2—Forms

Early pavement construction required the use of fixed forms to contain the concrete in the paving lane. In the late 1950s, slipform paving equipment was introduced, and its use increased rapidly. Today, most large paving projects use this method (refer to the discussion of slipform paving later in this chapter). Although fixed forms are still used for some large jobs, they are most often used for paving of small areas. Spreading and finishing of the concrete is typically done by hand, but for some applications, such as variable-width ramp or street paving, the concrete is placed and finished by equipment that rides on the forms.

Fixed forms are normally made of steel. If they are to support paving equipment, they should be strong enough (at least 5/16 in. [8 mm] thick) to support the machines without excessive deflection (Fig. 13.2). The forms should be straight within a tolerance of 1/8 in. (3 mm) in 10 ft (3 m)

Fig. 13.1—Subgrade trimmer operating from string line and grade.

Fig. 13.2—Paving equipment for street paving rides on fixed forms. After form removal, the lanes between are paved.

along the top rail and 1/4 in. (6 mm) in 10 ft (3 m) on the sides. Flexible or curved forms can be used to form curves with a radius of 100 ft (30 m) or less.

The form depth should be equal to the thickness of the concrete. For small areas of variable-thickness pavement, full-width wood boards securely attached to the form bottoms can be used to increase the depth not more than 25% of the original form depth. Do not add to the top of the form. To ensure stability, the base width of the forms should be at least 3/4 of the form depth.

When inspecting fixed forms, it should be verified that:
- No bent, dented, or twisted forms not meeting the above requirements are used;
- Rusty forms or forms caked with hardened concrete are cleaned before use;
- The forms are securely staked to the grade and are in full contact at all points. Filling and compacting of low spots under the forms should not be allowed after they are in place. Low spots in the grade should be corrected before forms are set;
- Shims are not used to correct uneven support;
- Forms are the full depth of the pavement. They should never be placed in depressions or on mounds to compensate for improper height;
- Abutting form sections are locked together tightly; and

Fig. 13.3—Male keyway formed by slipform paver. For maximum effectiveness, keyways should be located at mid-depth of slab.

- Inside form faces may be lightly oiled or treated with a releasing compound before concrete placement.

The contractor may need to make final adjustments to fixed forms after checking grade from grade stakes and checking smoothness with a straightedge.

13.2.1 *Keyway forms*—Keys often are required for load transfer across longitudinal joints. When fixed-form paving is used, the keyway form is attached to the paving forms for the first lane (or lanes) of pavement placed (Fig. 13.3). When slipform paving is used, the keyway form is attached to the sliding form, usually to form a female key (although the reverse has also been used successfully). On the adjoining or fill-in lane, the other half of the key is formed as the concrete in this lane is molded against the first placement.

To prevent movement of the keyway form, secure attachment to the steel paving form, usually by tack welding, is required. For maximum effectiveness, the keys should conform to the dimensions required by the contract documents and be located at mid-depth of the slab. The use of wood keys should be avoided, which can swell after contact with water and crack the keyway.

13.3—Steel reinforcement

Welded-wire fabric, sometimes referred to as distributed steel, is used in concrete placements having a high tendency for cracking. Distributed steel usually occupies approximately 0.05% of the slab's cross-sectional area. It acts to keep any cracks that occur tightly closed so that aggregate interlock will be sufficiently maintained and provide load transfer at these locations. Normally, transverse contraction joints are constructed at close enough intervals to prevent transverse cracking, in which case distributed reinforcing steel is not provided, except for odd-shaped slabs and areas where joints are mismatched.

Structural slabs and continuously reinforced concrete pavement (CRCP) use high percentages of reinforcing steel. Reinforcement may be welded-wire fabric (plain or deformed), bar mats, or individual bars.

13.3.1 *Storage*—When the steel is unloaded and stacked at the paving site, rust protection is needed only for long-term storage. This protection can be provided by placing the steel on a layer of polyethylene sheeting and covering the pile with another layer of sheeting.

Light rusting of reinforcement is not harmful, and actually improves bond with the concrete. Any loose or flaking rust, however, should be removed. If the rust is so deep that it reduces the cross section of the steel, the reinforcement needs to be replaced.

13.3.2 *Installation methods*—There are several methods of installing steel reinforcement in concrete pavements. In two-course construction, the usual procedure is to strike off the first course to the specified depth of the steel, place the steel on the concrete, and then place the top course of concrete. The top course should be placed before the lower course begins to harden to avoid a cold joint, which can result in pavement failure.

Another method of positioning steel is to set it on fabricated chairs (supports) staked in the grade before paving begins. The concrete is then placed at full depth in one pass. If this method is used, it should be checked that the steel is positioned correctly in the slab to ensure adequate concrete cover.

A third method is to place the concrete at full depth, lay the steel on the surface, and then depress it to its specified position by machine. Crews should never lay reinforcing steel, either welded-wire fabric or reinforcing bars, on the subgrade and attempt to pull the reinforcement up into position after placing the concrete.

If contract documents for continuously reinforced pavement call for only longitudinal bars, the bars can be preassembled and fed through the bell-shaped entrances of tubes attached to the front of the paver. The tubes are adjusted to install the bars at the correct spacing and elevation in the completed pavement.

Regardless of the method used to position reinforcing steel, the fresh concrete behind the paver should be probed to verify that the steel is in its proper position within specified tolerances.

13.4—Concrete

13.4.1 *Materials*—On large paving projects, cement will be furnished in bulk. Documentation should be checked with each shipment to ensure compliance with contract documents. Temperatures of cement should also be checked upon arrival before transfer to job-site silos and use in production. Close control of aggregate conditions should be maintained. Variations in aggregate grading make it difficult to maintain uniform concrete consistency. In many areas, coarse aggregates, particularly gravels, can contain significant amounts of low-density particles that become unstable when exposed to freezing and thawing or even wetting and drying. When used in pavement concrete exposed to severe weather, these unstable particles can cause popouts in the pavement surface. This is particularly undesirable in runways for jet aircraft because the jet engines can pick up the loose particles from the surface and suck them in, causing major damage. For these pavements, contract documents may have very tight limits on such deleterious materials.

Some limestone coarse aggregates containing unsound particles can also be the source of pavement problems. As such unsound particles deteriorate, D-cracking, a progressive crack pattern roughly parallel to pavement edges and joints, can develop. To minimize or prevent D-cracking when such coarse aggregates must be used, the maximum size of the aggregate should be limited to 1/2 or 3/4 in. (13 to 19 mm) and the percentage of coarse aggregate limited to 40 to 50% at most. It is best to use aggregates that are not susceptible to D-cracking to allow the use of larger aggregates.

Admixtures commonly used in pavement concrete include:
- Air-entraining admixtures to provide resistance to freezing and thawing;
- Water reducers to improve workability;
- Retarders to prevent premature stiffening of the concrete, particularly in hot weather; and
- Accelerator to promote faster setting times of concrete in cold weather.

Most specifications discourage, or even prohibit, the use of calcium chloride as an accelerator in pavement concrete. ACI 318 provides specific limits for the amount of chlorides that may be present in reinforced concrete. Chapter 7 gives more information about the use of admixtures in concrete.

13.4.2 *Mixture proportioning*—Flexural strength (or modulus of rupture) is the basis of most pavement thickness design, and is therefore commonly used for developing mixture proportions. The use of standard cylinders for compressive strength rather than flexural testing of beams is becoming more widespread, however, particularly on smaller projects, for quality control and owner testing. On smaller projects, the use of cylinders to determine compressive strength often is the method of control. These cylinders should be correlated against flexural beam tests. In either case, standard sampling, curing, and testing procedures should be strictly followed to obtain meaningful results. It is particularly important that the upper exposed surface of test beams are protected to avoid drying and that transportation and handling do not induce cracking, both of which can produce beam strengths lower than the actual pavement strength.

13.4.3 *Batching and mixing*—Generally, pavement concrete for large projects is batched at a central plant set up on or near the job or batched at a commercial ready mixed concrete plant. The concrete is either mixed in stationary mixers at the central plant, mixed in transit-mix trucks en route to the paving site, or shrink-mixed (partially mixed at the central plant and finished in truck mixers). Central-mixed concrete is hauled to the paving site in agitating trucks, concrete trucks, special nonagitating trucks, or even in regular dump trucks (when well-proportioned low-slump mixtures are used).

Contract documents usually specify time limits for concrete delivery, allowing longer times for agitating units and truck mixers. What is most important, however, is the condition of the concrete at the time of placement. It should meet all requirements for proportions, uniformity, consistency, temperature, air content, and strength. Visible changes in the concrete and its handling characteristics may indicate significant changes in essential properties. If there are variations among the batches, differential settlement can occur during hardening and result in a rough pavement. Uniformity tests should be made when specified or when there are doubts as to the within-batch uniformity of the concrete. Uniformity samples should be taken after discharge of approximately 15 and 85% of a batch and tested as described in ASTM C94/C94M (see Chapter 19) or as otherwise specified.

It also is important to ensure that the concrete contains the specified air content and is of the proper consistency (slump). In areas where the concrete will be subjected to freezing-and-thawing and deicing salts, the proper amount of entrained air is essential for durability. In all climates, air entrainment improves concrete workability and reduces bleeding. Slump of the concrete is important because variations will affect the operation of the finishing equipment. This is particularly true for slipform pavers. Contract documents often limit the slump of concrete to 2 in. (50 mm) maximum for fixed-form paving, and 1-1/2 in. (40 mm) maximum for slipform paving. Only specified test equipment and test procedures should be used for acceptance or rejection of the concrete.

13.5—Paving operations

13.5.1 *Concrete placement*—To provide moisture that will later aid in curing the concrete and to help hold down concrete temperatures on hot days, the subgrade should be moist, but free of standing water, when the concrete is placed. It is good practice to keep concrete hauling equipment off the prepared subgrade in the lane being paved, particularly where subgrades are too soft to support haul units without deformation. Instead, equipment such as belt spreaders or moving hoppers should be used to transfer the concrete from trucks on the shoulder to the front of the paver. If the subgrade or base course has been suitably stabilized by treatment with lime, cement, or asphalt, hauling equipment may drive in the paving lane and discharge the concrete immediately in front of the paver.

Paving crews should deposit the concrete evenly across the width of the lane. If placed in piles or windrows, it can consolidate unevenly, resulting in rough pavement.

13.5.2 *Vibration*—Adequate vibration is essential to producing high-quality concrete pavements. Often, a combination of internal spud vibrators and surface tamping bars are used (Fig. 13.4). The vibrators should be rigidly gang-mounted on the front of the paver so all are at the same depth and angle. Spud vibrators are usually oriented parallel to the longitudinal direction of paving and should be spaced at intervals of not more than 2 to 2-1/2 ft (500 to 750 mm) across the width of the paver. A minimum vibration frequency usually is specified; a minimum amplitude may be specified as well. To ensure compliance, vibration frequency should be measured using a tachometer. Variables affecting vibrator frequency and depth include concrete consistency, materials, and weather conditions.

During paving operations, the crew should maintain a sizeable head of concrete over the vibrators to achieve effective consolidation. Interlocks should be provided for automatic cutoff of vibration when the paver stops moving. In addition to gang vibrators, the paver should have a knock-down spreader at the front to spread the concrete uniformly across the lane, with the surface leveled slightly above finished

Fig. 13.4—Types of paver-mounted vibrators including gang-mounted spud vibrators shown above.

Fig. 13.6—Spreader and finishing machine supported on previously paved lane (background) and on form rail (foreground). Oscillating screed produces specified crown in pavement.

Fig. 13.5—Slipform paver guided by string line.

grade. The spreader mechanism is usually of an auger- or paddle-type design.

13.5.3 *Slipform paving*—In slipform paving, the concrete is extruded or formed by a paver equipped with moving side forms that mold and retain the stiff plastic concrete. The pavers range in size from those designed for curb and gutter construction to those that can pave 50 ft (15 m) in one pass. They can handle depths from as little as 2 in. (50 mm) on resurfacing projects to 20 in. (500 mm) or more for airfield pavements.

Slipform pavers are self-propelled and ride on crawler tracks outside of the paving lane. Sensors and electronic or hydraulic controls guided by string lines, lasers, or wires control grade automatically (Fig. 13.5). It is essential for the string lines or wires that guide the control devices to be carefully set to line and grade, with sufficient tension maintained in the string line to prevent sag.

Some slipform pavers are equipped with oscillating transverse screeds to strike off and finish the concrete surface. Others have a broad extrusion plate that forms the surface. In either case, it is important for the paver to keep moving forward at all times without stopping because each stop can produce a ripple in the surface.

With properly proportioned concrete, slipform pavers can extrude exact male and female keyways where additional lanes are planned. They can also install tie bars to the specified position and dowels for transverse and longitudinal contraction joints. Care should be taken, however, to ensure proper positioning of dowels in thick pavements.

Edge slump (sagging of the plastic concrete at the pavement edge) is sometimes a problem with slipform paving, especially with thicker pavements. Excessive edge slump can often be corrected by adjusting the moving forms of the paver to squeeze in more and by varying the consistency and proportions of the concrete mixture, particularly changing the sand content. A commonly used limit for maximum allowable edge slump is 1/4 in. (6 mm). Many projects do not permit repair of excessive edge slump, and instead require removal and replacement of the entire slab.

Uniform concrete is a large factor in successful paving, particularly slipform paving. Testing is one of the tools inspectors can use to ensure acceptable and uniform concrete, but being alert to changing conditions during paving operations provides additional assurance that the desired results are being achieved. An understanding of the subtle effects of mixture proportions on finishing and rideability of the pavement, as well as edge slump, comes with experience.

13.5.4 *Fixed-form paving*—Pavers for fixed-form paving also are self-propelled, but they ride on top of the forms with flanged wheels or on the adjoining pavement slab with hard rubber-tired wheels. Wheels riding on the forms should be equipped with scrapers that can be adjusted tightly against the wheels to keep them and the top of the forms free from concrete. Most fixed-form pavers have oscillating transverse screeds to strike off and finish the concrete (Fig. 13.6). Some have an extrusion plate.

13.5.5 *Finishing*—Mechanical finishing operations behind the paver are performed by wide transverse floats, rollers, and nonrotating pipe floats operated at an angle to the centerline of the lane. The use of rotating-type pipe or drum floats or screeds is usually undesirable because they work too much mortar to the surface.

Pavers are designed to produce pavements of specified cross section and riding quality with a minimum of hand finishing. The only hand-finishing operations usually required behind the paver are occasional smoothing with long-handled floats and straightedges, edging, and correcting minor surface defects. If significant hand finishing is required, the paver should be stopped and properly adjusted. Vibrators, screeds, extrusion plates, and pan floats can quickly be adjusted to regulate the amount of concrete surge behind the paver so that the completed surface is at the specified elevation with the required crown (Fig. 13.7). Quick adjustments to screeds and mechanical floats permit smooth changes in crown, as necessary.

Even if the screeds and floats are initially in correct adjustment, further adjustments will be required as paving progresses due to changes in weather, mixture proportions, and other variables that can affect the finishing characteristics of the concrete. Such changes should be watched for and take immediate measures taken to correct any problems. If the surface begins to tear, for example, the paver or concrete mixture proportions should be adjusted so that the need for hand finishing to close the surface is reduced or eliminated.

Excessive finishing, either by mechanical equipment or by hand, creates a weakened layer at the pavement surface that tends to deteriorate with exposure to traffic and weather. Additions of water should only be permitted to the surface in emergencies, such as when rapid drying conditions cause tearing of the surface and the formation of plastic shrinkage cracks. Under such conditions, finishing crews should add water only in the form of a pressurized fog spray, using just enough to restore the surface sheen. They should not add water during hand finishing because it can wash cement and entrained air out of the surface layer. A good entrained-air system at the surface is essential for durability of pavements exposed to freezing climates.

13.5.6 *Texturing*—Producing a satisfactory surface texture that will last under traffic is extremely important to the skid resistance of a concrete pavement. Surface texture determines how fast water escapes from between vehicle tires and the pavement and how fast water drains from the surface during rain. Water on the pavement can result in loss of contact between tires and the pavement, causing vehicles to lose control and skid. This phenomenon, called hydroplaning, often occurs during high-speed vehicle travel when the depth of water on the surface is at a critical level in relation to vehicle speed.

The two major factors affecting the initial skid resistance of a concrete pavement are the fine aggregate in the surface mortar and the texture formed in the surface. The fine aggregate should have a high percentage of siliceous particles, and the proportion of fine aggregate in the concrete mixture should be near the upper limit of the range that permits proper placing, finishing, and texturing. For low-speed streets and parking lots, dragging burlap over the plastic concrete can provide adequate texture. At least 3 ft (1 m) of burlap should be in contact with the surface for the full width of the pavement. The operation should begin as soon as possible behind the finisher, and certainly before the water sheen has disappeared from the pavement surface. Dragging a stiff-bristled broom (either a hand broom or machine attachment) transversely across the pavement surface can also provide good texture

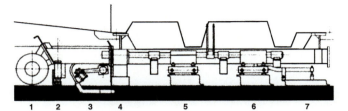

Fig. 13.7—Components of typical slipform paver that extrudes concrete to exact width, depth, and crown: (1) primary concrete spreader screw; (2) primary concrete feed meter; (3) vibrator mounting arm; (4) secondary concrete feed meter; (5) primary oscillating extrusion finisher; (6) final oscillating extrusion finisher; and (7) floating fine surface finisher.

Fig. 13.8—Brooming produces a brushed texture in freshly paved concrete.

(Fig. 13.8). This procedure should be done at a time when it will not cause excessive tearing.

For high-speed pavements, a burlap-drag texture may not provide adequate skid resistance. A deeper texture that permits faster drainage and better tire contact with the pavement can be produced by forming grooves in the surface using combs with wire tines (Fig. 13.9). The grooves can be transverse or longitudinal. Transverse grooves provide better drainage; longitudinal grooves result in a quieter ride. For highways, the grooves should be about 1/8 in. (3 mm) wide and 1/8 in. (3 mm) deep and at spacings of 1/2 to 1/4 in. (6 to 13 mm). For airfield pavements, groove depths and widths should be about 3/16 in. (5 mm), with spacings up to 2 in. or more. The timing for forming grooves in plastic concrete is

Fig. 13.9—A comb with wire tines forms deep grooves in freshly paved concrete.

Fig. 13.10—Grooves for skid resistance are sawed into hardened concrete pavement by multiple diamond-tipped blades. Sawed grooves improve skid resistance by serving as escape channels for water between vehicle tires and the pavement.

Fig. 13.11—Application of a liquid membrane-forming compound to ensure adequate curing of concrete pavement.

extremely critical, and should be determined by an experienced operator.

Another method of grooving concrete pavement is to saw the grooves after the concrete has hardened, using a machine equipped with multiple diamond-tipped blades. The grooves prevent hydroplaning by providing an escape path for the water between vehicle tires and the pavement (Fig. 13.10). Sawing grooves in the longitudinal direction improves the directional control of vehicles. This procedure is often done on highways, especially on curves, and can dramatically reduce wet-weather skidding accidents. Sawing the grooves transversely increases the coefficient of friction, and is often done at locations where vehicles stop or reduce speed. Grooves in airfield runways are sawed transversely.

For detailed procedures for texturing of concrete pavements, refer to ACI 325.6R.

13.5.7 *Curing and protection*—Adequate curing is essential to obtaining the concrete strength and durability for which the pavement is designed. The most common method of curing concrete pavements is by spray application of a liquid membrane-forming compound (Fig. 13.11). It is important to verify that coverage is complete, but no further attention is required during the curing period, except to see that areas are resprayed if the membrane is damaged by construction traffic. Curing compounds for pavements generally contain white pigment to aid in identifying areas that have been sufficiently covered. The white color also helps to reduce concrete temperatures by reflecting sunlight.

Curing compound should be applied by power sprayers to all exposed concrete surfaces, including edges, as soon as pavement texturing is completed. If fixed forms are used, it should be applied to the edges as soon as the forms are removed. The specified rate of application of curing compound may vary with pavement type and surface texture. Usually, two coats are applied at a typical application rate for each coat of 300 to 400 ft^2/gal. (7.4 to 9.8 m^2/L); slightly heavier coats are applied to heavily textured surfaces. Check that the specified amount of compound is being applied and that the compound is agitated before and during application to ensure uniform dispersion of pigment. For more information about the use of membrane-forming curing compounds, refer to Chapter 10.

Other methods of curing concrete pavement include waterproof paper, plastic sheeting, and wet burlap. A disadvantage of using these methods to cure pavements is that all require constant attention during the specified curing period. Waterproof covers are hard to keep in place on windy days, and burlap covers should be kept constantly wet during the curing period. On large paving projects, such as those averaging a mile or more per day, using loose curing covers requires keeping large amounts of material on hand. Plastic covers, however, should always be on hand to protect the pavement surface in case of rain before hardening. If exposure conditions are likely to produce plastic shrinkage cracking, see that immediate measures are taken to protect the concrete.

13.6—Final acceptance

Final acceptance requirements contained in contract documents for concrete pavements vary with the class of the project. It is more important to specify and enforce closer tolerances for a runway at a major airport than for a residential street. To ensure fairness to all parties involved and to avoid expensive legal disputes, inspectors should follow specified sampling and testing methods for determining concrete pavement properties required for acceptance.

A common acceptance requirement is strength, which is usually determined from beams or cylinders cast from the concrete being placed. Nondestructive test methods (Chapter 19) may also be employed to verify strength. Another requirement can be slab thickness, which is determined from cores removed from the hardened concrete. Specified sampling plans vary with the type and volume of traffic the pavement is expected to carry. Sophisticated and statistically based plans are specified for many larger projects.

Surface evenness may be specified for pavements serving large volumes of high-speed traffic. Traditionally, this was determined by measuring vertical deviations of the hardened concrete from the underside of a 10 ft (3 m) straightedge; generally, pavement evenness is considered acceptable for high-speed traffic if the deviations do not exceed 1/8 in. (3 mm). The use of profilometers (devices on wheels that measure surface deviations in inches or millimeters per mile), however, has become the most common method used for measuring surface evenness. When profilometer measurements are specified for the pavement, it is a good practice to also check the subbase surface with a profilometer before paving.

Pavement evenness is critical for freeways and airfield runways. Contract documents may be less strict for low-speed pavements, such as intersection ramps, residential streets, parking lots, and airfield parking aprons, where evenness is not as important. Tolerances of 3/16 or 1/4 in. (5 or 6 mm) in 10 ft (3 m) are typically specified for such applications.

Another requirement being enforced more often for high-speed pavements is average texture depth, an indication of skid resistance. This is determined by spreading a known volume of fine, dry sand on the pavement surface in a circular area and measuring the area covered. The test is described more fully in ACI 325.6R.

13.7—Joints

Pavement design engineers use a number of different joint types in concrete pavements, each intended to serve a special function. If these joints are not constructed as shown on the contract documents and installed at the proper locations, pavement performance will suffer. To ensure that pavement joints will function as intended, it is necessary for inspectors to understand their purpose and see that they are constructed properly.

13.7.1 *Transverse contraction joints*—Normally, concrete pavement shrinks upon hardening and drying, and will never again be as long (occupy as much volume) as when first constructed. This shrinkage results in cracking, particularly transverse cracking, because the long axis of paving lanes normally runs between joints. Contraction joints are weakened planes installed to predetermine the locations of transverse cracks and to ensure that they occur in straight lines. In plastic concrete, the weakened planes can be formed by special tools or removable or permanent inserts; in hardened concrete, they can be formed by sawing with diamond- or carbide-tipped blades.

The vertical fin on the bottom of a hand float is sometimes used to form contraction joints in parking lot and residential pavements. The use of the tool should be delayed until it is determined that the depression it forms is permanent so that a crack will form below it. If temporary inserts are used to form the weakened plane, they should not be removed until the danger of damage to the concrete has passed. These inserts are sometimes removed by sawing after the concrete has hardened. Permanent inserts should be made of material that will not deteriorate under traffic and weather. Asphalt-impregnated fiberboard is economical and has performed satisfactorily.

When contraction joints will be sawed in the hardened concrete, timing is critical. Sawing can begin as soon as it does not cause excessive raveling of the concrete, but it needs to begin before the concrete starts to crack. This timing can vary widely—generally from 4 to 24 hours after concrete placement—depending on weather conditions, concrete materials, the type of sawing system employed (soff-cut or water-cut), and the type of pavement. Thus, crews should be on hand day or night to do the work. Adequate lighting needs to be available for night sawing and a standby saw available in case of equipment breakdown.

13.7.2 *Transverse construction joints*—Transverse construction joints are formed by transverse forms—commonly called headers—at the end of a day's run or wherever concrete placement is interrupted long enough that there is danger that the concrete in place may begin to harden. Usually, wood forms with holes drilled for insertion of dowels or tie bars are used.

Where possible, transverse construction joints should be at planned contraction joint locations. If the last delivered concrete does not quite reach the header, do not allow excess grout carried over from normal paving operations to extend the pavement up to the header. Concrete at transverse construction joints should be of the highest quality. Extra concrete should be delivered to finish the day's run, if necessary, or the joint should be moved back. The paver should run past the header, carrying a roll of concrete with it. Careful hand vibration is required on both sides of transverse construction joints to ensure adequate concrete consolidation.

13.7.2.1 *Mechanical load transfer*—For heavy-duty pavements, mechanical load transfer may be specified for transverse joints. This is provided by smooth round or square steel dowel bars or by plate dowels. Dowels in transverse joints are usually installed in wire baskets staked in the grade ahead of paving (Fig. 13.12). The purpose of the dowels is to transfer loads across the joint and reduce deflections at the joint. They also need to permit unrestricted horizontal movement as the joint opens and closes. Therefore, they should be completely smooth and free of burrs and painted or lightly oiled on one end to prevent bonding to the concrete. Some dowels are manufactured with plastic coatings to reduce bonding and to prevent corrosion.

To function properly, dowels should be installed parallel to the centerline and surface of the pavement. Specified

Fig. 13.12—Dowels installed in wire dowel basket for mechanical load transfer at transverse joints.

Fig. 13.13—Sawing a longitudinal contraction joint.

tolerances often require that misalignment shall not exceed 1/8 to 1/2 in. (3 to 13 mm). Levels, measuring tapes, and templates are used to check dowel alignment. Stiff concrete can shove baskets and dowels out of alignment unless precautions are taken as the concrete is placed.

When a construction joint occurs at the normal location of a contraction joint, smooth dowels are usually installed, even if they are not provided at the other joints on the project. If a construction joint is required at other than the normal contraction joint spacing, deformed tie bars should be installed to prevent subsequent joint movements. This is done to prevent cracking in pavement abutting previously completed lanes.

In continuously reinforced pavement, the longitudinal steel extends through the transverse construction joint for a specified length and is lapped by new steel when paving resumes. Maintaining the specified lap lengths is extremely important. Extra steel is often specified at construction joints in continuously reinforced pavement.

13.7.3 *Longitudinal contraction joints*—Where concrete pavement is placed in one wide pass, a weakened plane is required to prevent uncontrolled longitudinal cracking. Such cracking is due to warping stresses caused by the tendency of concrete to curl when temperature or moisture differences exist between the top and bottom of the slab. Because of this, these joints in two-lane pavements often are called hinge joints.

Unlike transverse joints, longitudinal contraction joints are not intended to open and close. In road pavements and in outer lanes of wide aprons, deformed steel tie bars are normally installed across these joints to keep them tightly closed. The depth of the weakened plane for longitudinal joints is extremely critical in preventing longitudinal cracks, and should be closely monitored. The specified depth is usually 1/4 of the pavement thickness plus 1/4 in. (6 mm).

One way to form the longitudinal joint is with a strip of polyethylene tape of the specified thickness and width, installed automatically from a reel through a special attachment on the paver. The tape should be installed vertically, with the top at or slightly below the pavement surface. Another common way to construct longitudinal joints is by sawing after the concrete has hardened (Fig. 13.13). Timing is not as critical as for sawing transverse contraction joints because shrinkage in the transverse direction is less with the shorter length of concrete. Longitudinal joint sawing, however, should be completed within 24 hours and before any traffic, including construction traffic, is allowed on the pavement.

13.7.4 *Longitudinal construction joints*—When one or more pavement lanes are constructed separately, longitudinal construction joints are required where new concrete abuts previously placed pavement. Dowels are commonly used for load transfer in heavy-duty pavements. For slipformed pavements, deformed tie bars bent at right angles can be mechanically inserted into the slab keyway and straightened before the adjoining slab is placed.

Machines are available that can mechanically insert dowels at transverse and longitudinal joints from the pavement surface after the concrete has been placed. Some types of dowel inserters do not function properly for longitudinal construction joints in thick slipformed pavements. If the inserters cannot be used for these pavements, dowels can be installed by cementing them with epoxy resin into holes carefully drilled into the hardened concrete.

13.7.5 *Expansion joints*—Expansion joints permit movement of adjoining slabs and isolate slab movement from fixed structures, such as bridge approaches or concrete walls. They are formed by vertical, compressible premolded expansion joint filler inserts extending the full depth of the pavement. The premolded expansion joint filler material should be in contact with the grade throughout its length and extend the full width of the lane to exclude plugs of concrete from the expansion area. The filler should also be kept vertical during concrete placement and hardening, otherwise, when movement occurs, one slab will tend to ride up over the other, possibly causing joint faulting, spalling, or a blowup. To provide space for joint sealant, the expansion joint filler should extend to approximately 1/2 in. (13 mm) below the pavement surface. Fillers often come with removable inserts on top for this purpose.

When dowels are used at expansion joints, special caps designed to provide space for concrete expansion should be installed on the free-moving ends of the dowels. This space provides for concrete movement without the buildup of critical stresses that could shatter the concrete.

13.7.6 *Joint sealing*—Joint sealing prevents the entrance of incompressible material, which can cause extreme stresses in the concrete during expansion cycles, resulting in pavement damage. Joint sealing also reduces the penetration of water into the foundation. Joint sealing is a critical operation on heavy-duty pavements, but is often not required on low-volume streets and parking lots if joints are closely spaced so that movements are minimal.

Materials commonly used to seal joints in concrete pavements include hot-poured rubber asphalt and cold-applied mastics (single- or multiple-component types). On pavements with high traffic volumes, long-lasting preformed neoprene compression seals often are installed in sawed joints. The dimensions and configurations of preformed seals are designed for individual joint widths so that the seals are always in compression to maintain a tight seal. For more information about the inspection and testing of joint materials, refer to Chapter 3.

Joints need to be clean and completely dry at the time of sealing. No curing compound should get into the joints, as it prevents bonding of the sealant to the joint faces, and all foreign material should be removed by compressed air or sawing before joint sealing. If preformed seals are to be installed, all spalls at the joints should be patched so the seals will remain in place.

Hot-applied liquid sealant should be heated in a double-boiler-type kettle with positive temperature control to prevent chemical breakdown by overheating. Reheating of material should never be allowed. Workers should apply liquid sealants through a nozzle inserted into the joint, filling the joint from the bottom up to prevent voids. A backer rod is sometimes installed in the joint to a specified depth before sealant application to control sealant shape and to prevent the material from bonding to the bottom of the joint (Fig. 13.14). The use of jute or rope for this application should be avoided because these materials can rot.

The closer the installation temperature is to the mean annual temperature, the less strain there will be in the sealant in service and the better it will perform. The manufacturer's recommendations should be referenced for hot and cold weather installation. Preformed seals should be installed mechanically, after a liquid adhesive-lubricant is applied to the joint faces, to prevent excessive stretching of the material. Contract documents usually limit stretching of the material to 1% during installation to ensure satisfactory performance.

13.8—Bridge decks

The quality of bridge decks depends on the same factors as those for concrete pavements on grade, and the same principles of concrete quality and uniformity apply. Construction traffic and wind loads can cause vibration of the bridge deck, resulting in random cracking. Because of potential bridge deck vibration, storage of concrete test cylinders on bridge decks is strongly discouraged. Special attention, however, should be given to concrete placement, consolidation, and

---

SPECIAL PRECAUTIONS

To meet project completion dates, contractors can rarely afford to wait for ideal weather to pave. They need to use all potential paving time, including days when weather conditions may require special precautions to produce an acceptable product. After the pavement is placed, they may find it difficult to keep traffic off the surface. Before concreting begins, inspectors should alert contractors to be prepared to take precautions, such as those given below, to deal with these situations. Detailed protection measures for concreting during hot and cold weather are described in Chapter 10.

Hot weather
- Verify that the hot weather plan that was submitted and approved is used;
- Before concrete placement, use water to cool forms, stabilize subbases, and existing pavements being resurfaced;
- Verify concrete delivery temperatures;
- Use set-retarding admixtures to facilitate finishing. Another option is to limit pavement construction to the cooler hours of the day or night; and
- It is sometimes difficult to saw joints soon enough to prevent uncontrolled shrinkage cracking. Surfaces subject to high evaporation should be kept moist with fog sprays or evaporation-retarding admixtures to prevent plastic shrinkage cracks from developing.

Cold weather
- Verify that the cold weather plan that was submitted is used;
- Before concrete placement, verify that subbases are not frozen and forms are free of snow and ice;
- Verify concrete delivery temperatures;
- Use effective combinations of blankets and temporary heat to prevent freezing and facilitate curing;
- Use set-retarding admixtures to increase the rate of hydration; and
- Monitor strength development by maturity method.

Rain
- When newly placed concrete is exposed to unexpected rain, apply protective covers immediately. Keep materials, such as plastic sheeting or burlap, on hand at all times. A roll of plastic sheeting often is carried on the curing machine for such emergencies;
- If the rain stops before the concrete has hardened, repair any damage to surface texturing caused by the protective covers, and apply additional curing compound. If the concrete has hardened, leave it undisturbed until after the curing period. Texturing can then be restored by sawing grooves in the surface; and
- It is sometimes specified that temporary side forms be installed on slipform paving projects when it rains. This can delay the placement of protective covering, however, allowing surface water to flow to the pavement edge and down between the concrete and the forms, damaging the edge. Any edge damage should be repaired after the rain stops.

Premature traffic
- Keep all public and construction traffic off the new pavement until the concrete has attained its specified strength. Only equipment for sawing and sealing joints should be allowed on the pavement; and
- Use adequate barricades and warning signs on projects involving improvements of existing roads on which traffic is maintained during construction.

---

curing to ensure durability. The depth of the top reinforcing steel is critical to bridge deck performance. If sufficient cover is not provided, cracks can develop in the concrete over the steel, allowing water and deicing chemicals to penetrate to the steel. A minimum cover of 2 in. (50 mm) is normally specified where deicing salts are to be used. Refer to ACI 345R for standard practices for bridge deck construction.

13.8.1 *Concrete placement*—In bridge deck paving, grade control is based on the screed rails that support the paver. These are positioned on adjustable supports for correcting the elevation. The elevation of the rails should be checked to make sure they are positioned to produce an accurate grade line, taking into account deflections from dead loads. Before paving begins, the finishing machine should be adjusted to produce the specified crown. The screeds and floats should match the crown of the dam plates at the ends of the deck sections.

Concrete for bridge decks is normally placed by crane and bucket or by concrete pump (Fig. 13.15). When cranes are used, two buckets are usually provided so concrete can be placed in one while the other is discharging. The concrete is placed as close to its final position as possible to reduce segregation and differential settlement.

The close spacing of reinforcing steel in bridge decks requires special attention to concrete placement and consolidation to prevent the formation of voids and to provide adequate bonding of the concrete and steel. Spud vibrators are usually used to consolidate the concrete around the reinforcement, and vibrating screeds are used to finish the concrete. Vibrators should not be permitted to operate in one location long enough to cause segregation of the concrete materials.

13.8.2 *Finishing*—The finishing machine should move slowly at a constant rate, timed to concrete delivery, with an excess of concrete provided for the full length of all screeds and floats (Fig. 13.16). Because of limited space, hand finishing is done from work bridges; workers should not walk in the concrete behind the finishing machine. If much hand floating becomes necessary to close the surface, adjustments should be made to the finishing machine or concrete mixture proportions. Nonuniform distribution of entrained air bubbles is attributed to disturbance of the air-void system during final finishing operations. Therefore, it is important to avoid excessive manipulation of the concrete surface during finishing. In particular, the addition of water to the surface should be prohibited during finishing.

Additional skid resistance on bridge decks can be applied by use of metal tines to form grooves in the plastic concrete, as described previously. Just as for pavements on grade, the timing of the texturing operation is critical for achieving optimum results.

Bridges are often completed ahead of other paving operations so that paving equipment can cross the bridges, expediting project completion. The concrete should reach its specified strength before traffic is allowed on the bridge deck.

Fig. 13.14—Expansion joint with preformed neoprene compression seal.

Fig. 13.15—Placing concrete by pump for a bridge span.

Fig. 13.16—Mechanical finishing of concrete bridge deck. The finishing machine should move at a constant rate, timed with concrete delivery.

Fig. 14.1—A stamp with a stone pattern formed this rough stone surface finish.

Fig. 14.2—Textured surface applied by sandblasting concrete after form removal to provide a rough appearance.

Fig. 14.3—Assembled precast concrete panels from a mural.

CHAPTER 14—ARCHITECTURAL CONCRETE

Architectural concrete refers to elements for which unusual care is taken to produce unblemished surfaces or in which the exposed surfaces have been treated to enhance their visual appearance. Architectural concrete also includes elements designed with unusual, sometimes complex, shapes for aesthetic appeal.

Many architectural surface treatments are possible, including:
- Aggregate exposure, ranging from slight to deep;
- Surface abrasion, or removal of part of the surface;
- Color, either by the addition of coloring agents to the concrete mixture, by broadcasting of coloring agents onto the surface before finishing or stamping, by the selection of aggregates for a specific color, or by the use of a specific color of cement; and
- Patterns or textures, produced in unformed surfaces by tooling, stamping, and other treatments and in formed surfaces by the use of textured or patterned sheathing or liners (Fig. 14.1) or by various treatments after form removal (Fig. 14.2).

Architectural concrete can either be cast-in-place or precast, and either load-bearing structural elements or non-load-bearing facing units (Fig. 14.3). Except for appearance, there is no basic difference between architectural concrete and conventional cast-in-place or precast concrete. Consequently, the general procedures of inspection covered elsewhere in this manual also apply to architectural concrete, with additional attention given to achieving the desired appearance.

On projects involving architectural concrete, as with all other concrete construction, the requirements of the contract documents should serve as the inspector's primary guide. Inspectors of architectural concrete should also become familiar with the guidelines provided in ACI 303R.

14.1—Determining requirements for acceptability

In addition to the requirements applying to all concrete work, visual appearance of exposed surfaces is a major consideration in the acceptability of architectural concrete. Unfortunately, visual appearance is difficult to describe or measure in exact terms.

Ideally, the architect establishes acceptability criteria for appearance as the structure is conceived. To clearly convey those criteria to the contractor, demonstration panels or sections (described later in this chapter) often are necessary to supplement written descriptions in the contract documents.

Because evaluation cannot be based on precise measurement, general acceptability requires the cooperation of all parties. Although perfect uniformity in color and texture may be desired, materials in concrete have an inherent degree of nonuniformity. Even uniform placing and finishing operations cannot guarantee uniform results, especially over a large area. Nevertheless, the contractor and inspector should realize that the owner is entitled to a surface appearance as close as possible to the desired quality.

Fig. 14.4—Sample panel with a formed rusticated surface.

Fig. 14.5—Close-up view of a fluted rusticated surface.

Fig. 14.6—Preconstruction mockup with various finishes.

It is vital that early in the construction project, the architect, engineer, contractor, and inspector develop a common understanding of what constitutes acceptable work on each aspect of the project. This understanding is best accomplished by having all parties simultaneously inspect finished work of adequate quantity to permit the architect to point out acceptable areas, minimally acceptable areas, and unacceptable areas, if present. Any dispute between the architect and contractor regarding acceptability should be settled before work continues. These identified acceptable areas can be used for comparison in judging future work. Trials of approved methods and finishes can be demonstrated on less important parts of the structure, such as basement walls that will not be visible after project completion.

14.1.1 *Preconstruction samples*—The contract documents for architectural concrete can describe the desired surface only in general terms. The color of the cement may be specified, as well as the color, size, shape, and grading of the aggregates. In some cases, contract documents may also specify the source of these materials. Surface texture may be described in general terms such as "light sandblasting," or "heavy exposure of coarse aggregate." Although the specifications should be as precise as possible, it is difficult for a written document to fully communicate the desires of the architect. Thus, design reference samples or full-scale mockups often are required.

14.1.1.1 *Design reference sample*—The design reference sample should be at least 18 x 18 in. (450 x 450 mm) in area and 2 in. (50 mm) thick to adequately indicate the desired color and texture. The sample can be prepared for the architect and included as part of the contract documents, or the specifications can require the contractor to prepare various samples based on specified requirements, and the architect can select one to be the design reference sample. Not every area of the finished structure, however, will precisely match the reference sample, which can usually be cast with greater precision than is possible with a larger mass of concrete.

14.1.1.2 *Full-scale mockup*—The contract documents may require the contractor to construct a full-scale mockup section before construction begins (Fig. 14.4 and 14.5). Even if such a mockup is not required, the contractor should consider constructing one. The mockup should represent a typical portion of the structure and be built using the same procedures and equipment intended for the structure including formwork (along with form-tie holes and joints between form panels), placement of reinforcement, mixture proportioning, concrete placement and curing, and surface treatment. It should also include deliberate variations in the finish to demonstrate a range of acceptability (Fig. 14.6) as well as imperfections requiring repair so that repair operations can be demonstrated.

The primary purpose of the mockup is to provide a large sample of work that can be judged fairly by the architect. Approval of the mockup by the architect is usually required before work can begin. The inspector should reach an understanding with the architect and contractor regarding the acceptability of the mockup and how much variation is tolerable. For structures where architectural finishes are intended to be viewed at a distance rather than close up, specifications should state the appropriate viewing distance from which the mockup is to be evaluated.

The contractor should build the mockup at or near the job site so it is readily available for comparison with future site work. It can even be an inconspicuous part of the structure at the basement level.

14.1.1.3 *Precast concrete finish demonstration*—A mockup can also be constructed for precast architectural concrete, or a typical unit can be fabricated for approval by the architect and inspector. For inspection at the plant, this unit should be among the last scheduled for erection on the job so it will be available to the inspector for comparison with day-to-day production. If there is no plant inspection, the typical unit should be among the first delivered to the job for future comparison.

14.2—Importance of uniformity

Uniformity is the key to acceptable architectural concrete: uniformity of materials, uniformity of equipment, uniformity of operations (including scheduling), and uniformity of workmanship. Although an inspector representing the owner should not tell the contractor what equipment to use or how to distribute the work force, it is the inspector's duty to insist on uniformity of the finished product.

Where feasible, each material should be obtained from a single source, and for manufactured or processed materials, from a single production run. If materials are stockpiled or stored at the site before construction begins, they should be protected from deterioration, contamination, intermingling, or, in the case of aggregates, segregation. Changes in quality or appearance of materials as work progresses, as revealed either by test results or visual observation, should be constantly monitored.

Uniformity of product is best attained by using the same equipment and workmanship throughout the job. Changing the vibration equipment, for example, can affect the distribution of aggregate near the surface. Similarly, using two different workers to operate the same vibrator can cause variations in aggregate distribution if they have not been trained to use the vibrator in an identical manner. Best results are achieved when workers of equal ability perform tasks affecting the appearance of the finished product, such as vibration, sandblasting, and bush hammering.

The operations of concrete batching, mixing, transporting, placing, vibrating, form removal, finishing, and curing should be performed in the same manner and with the same timing from day to day, except as necessary to accommodate varying weather conditions or variations in structural elements. Variations in procedure and timing for any of these operations can cause variations in the appearance of the concrete surface.

14.3—Forms

Generally, structural design of the forms should meet the same requirements as for forms for other concrete construction. Tolerances for setting forms and limits for form deflection, however, are typically more restrictive than for ordinary exposed building concrete. Thus, workmanship in fabricating forms is critical (Fig. 14.7). Detailed information on forms is given in ACI SP-4 (Hurd 2005).

14.3.1 *Form sheathing or lining*—The materials, textures, and patterns of the form sheathing or lining are governed primarily by specification requirements or by requirements to match a preconstruction mockup. Form sheathing and

Fig. 14.7—An exposed-aggregate panel retains its appearance after more than 40 years.

lining can be made from many materials including wood, plywood, metal (aluminum, steel, or magnesium), plastic, plaster waste molds, and rubber liners. Each material has advantages and limitations.

Wood or plywood used for sheathing or lining can affect the color of the concrete surface because of variations in moisture absorption of different portions of the board. The more pervious portions will absorb more water from the fresh concrete and thus lower the *w/cm*, causing a darker surface color. Organic substances in the wood can also result in a dark-colored concrete surface, and can sometimes cause dusting. During the first use of wood forms, forming crews can reduce the effects of these variations by treating the wood surface with a limewater solution or a cement slurry that will react with and neutralize the organic substances and fill porous surfaces. After the first use, wood forms produce much less variation in concrete color. Form sealers and release agents discussed in this chapter can also prevent variations.

14.3.2 *Textures and patterns*—The textures and patterns possible with different form sheathing and liner materials vary widely. Some liners are factory-cast from plastics or rubbers to produce a specific texture, some are textured by treatments applied on the job, and some sheathing and liners are job-built to produce specific large-scale patterns on the concrete surface. However the surface texture or pattern is produced, it is important to ensure that it is the same for all sets of forms and that it is not altered from one form use to the next. Form surfaces should be thoroughly cleaned and forms retightened between uses, without damage to the pattern or texture.

14.3.3 *Form joints*—Joints between form sheathing or lining panels should be completely tight and sealed to prevent all leakage of water or paste, which can produce sand streaks on the surface and, in extreme cases, rock pockets. Even slow leakage or absorption of moisture can produce hydration discoloration (dark streaks), particularly on sandblasted surfaces. It is nearly impossible to remove these imperfections, although hydration discoloration may become lighter and barely visible after several years of weathering.

Leakage can be prevented by staggering joints of form liners from the joints in the sheathing by use of special rubber gaskets and, when further treatment of the concrete surface will occur, by use of pressure-sensitive tape. Caulking of joints, together with batten backing strips, can also be used if installation procedures are carefully inspected. No gaps should exist in the caulking or backing strips.

14.3.4 *Form sealers and release agents*—Form sealers often are used on wood or plywood forms to seal the surfaces to correct nonuniform absorption, prevent emphasis of grain pattern, and prolong form life. Release agents (form coatings) are used to prevent forms from sticking to concrete surfaces. Many types of release agents are available, but the product used should be compatible with the form material or sealer and not stain the concrete surface. The release agent should also be compatible with any joint sealants and caulking compounds to be used.

All release agents should be tested on mockups or unexposed surfaces to determine performance before using them on exposed surfaces. Release agents usually are fluid materials applied in thin coatings. Thick coatings of viscous materials should be avoided. If oily coatings are used, they should not be applied to the forms too far in advance of concrete placement; dirt can collect on the coating and cause problems.

14.3.5 *Form ties*—Holes left by removal of form ties should be filled or otherwise treated as required by the contract documents. Sometimes form-tie holes are designed to be plugged with manufactured plugs, usually plastic, that are driven into the hole. Even if filling of tie holes is not required, it is highly desirable to have form ties installed in a uniform pattern so that the holes in the concrete surface are less distracting.

14.3.6 *Form removal*—Form removal is especially critical for architectural concrete, and should be done with great care to prevent surface marring and damage to any intricate surface design. Normally, forming crews should not use wedges to remove forms, but if they must be used, they should be nonmetallic. For white or colored concrete, workers should remove forms for similar surfaces at the end of similar time periods, except when necessary to vary the time of removal due to weather conditions. During cold weather, form removal can cause a sudden drop in concrete surface temperature, which can result in surface cracking. The surface temperature should drop gradually, no more than 40 °F (22 °C) in 24 hours, and the drop should occur evenly over the 24-hour period. For additional information on form removal, refer to Chapter 10.

14.4—Reinforcement

Reinforcement should be located where shown in the contract documents. If it is positioned too close to the surface, coarse aggregate will be unable to work its way between the bars and the forms. As a result, the reinforcement pattern will mirror onto the concrete surface and will be particularly noticeable in smooth, light-colored surfaces. Reinforcement too close to the surface also tends to produce rust staining. Sometimes coated reinforcement is used in critical areas to prevent rust stains.

Rust staining can also occur if reinforcement supports are located adjacent to exposed surfaces. If supports need to be used near exposed surfaces, the supports should be made of plastic, plastic-coated metal, precast concrete block, or stainless steel. Workers should always bend the ends of tie wires back toward the interior of the concrete. Refer to Chapter 8 for additional information on installing reinforcement.

14.5—Concrete materials and mixture proportions

Some contract documents may specify that the materials to be used—particularly cement, aggregates, and pigments—come only from specific sources. Other contract documents may simply require the contractor to locate suitable materials to produce concrete with an appearance matching the design reference sample.

Mixture proportions (other than for gap-graded mixtures) usually are selected in the same manner as for ordinary concrete. It is common to keep the w/cm low (maximum of 0.46) and to limit the slump to a maximum of 4 in. (100 mm). Slightly drier mixtures are sometimes used for the top part of walls to prevent the color variation that can be caused when excess bleedwater from the concrete below rises into the concrete above. When pigments are used, trial mixtures should be made to verify that the required color is attained after the concrete hardens and dries.

14.5.1 *Cement*—Cement should typically meet the quality requirements for ordinary concrete construction, although greater importance to color is often required so that the color of the concrete matches the color of the design reference sample. To ensure color uniformity, all cement should come from one mill, and preferably from one grind.

Presently, there are no reference specifications controlling the color of white or colored cements. Some colored cements, particularly those in the buff-yellow-brown range, are produced by special grinding and burning operations using normal raw materials; other colored cements are produced by intergrinding mineral pigments at the cement mill. Cement produced by the first method is sometimes specified because it results in better color uniformity. Some of these special cements, however, have unusually high water demands that can result in low concrete strengths.

14.5.2 *Aggregates*—Aggregates for architectural concrete are required to meet the same quality requirements as those for ordinary concrete. Additional limits often are placed on the maximum permissible quantities of particles (principally iron compounds) that can cause staining on the concrete surface and of unstable compounds that can produce popouts. The color of the fine aggregate has much more effect on concrete color than does the color of the coarse aggregate.

Where concrete surfaces are to be treated to expose the aggregate, gap-graded mixtures often are specified to provide a higher percentage of coarse aggregate with better distribution in the concrete surface and thus a more pleasing appearance as the aggregate is exposed. A relatively large coarse aggregate with a narrow size range is used, along with intermediate-sized aggregates and fine aggregates consisting of concrete sand or masonry sand.

14.5.3 *Admixtures*—For air-entraining, water-reducing, and retarding admixtures, quality requirements usually are the same as for ordinary concrete.

Many contract documents prohibit the use of calcium chloride in architectural concrete. Even if not prohibited, its use should be discouraged. Calcium chloride may cause mottling or surface checking.

Pigments for coloring concrete include mineral pigments and organic dyes. No more pigment than necessary should be used to achieve the required color, because excess pigment can reduce concrete quality.

14.6—Batching, mixing, and transporting

Uniformity of concrete materials and uniformity of mixing are extremely important, especially when pigments are used. Achieving this uniformity requires close attention to the following procedures for batching, mixing, and transporting:

- All batching, mixing, and transporting equipment used for architectural concrete, particularly white or colored mixtures, should be thoroughly clean before the start of production. Preferably, separate equipment should be reserved for such mixtures;
- Aggregate storage piles should be controlled to prevent contamination, intermingling, or segregation. The fine aggregate and the smallest coarse aggregate should be uniform in moisture content to prevent variations in the water content of the concrete and in the consistency of the batches;
- Close control should be maintained over air content and slump of the concrete to provide uniformity;
- The temperature of the fresh concrete should be kept reasonably uniform and, if possible, in the range of 65 to 85 °F (18 to 29 °C) for optimum color uniformity. Concrete at higher temperatures is harder to handle properly and tends to have a shorter setting time, which could result in visible flow lines and cold joints;
- Scheduling of operations should be carefully monitored to prevent any delays in the period between charging of the mixers and depositing of the concrete in the forms. When such delays occur, the concrete should be held in mixers, transporting equipment, buckets, pump lines, conveyor lines, or elsewhere, which can lead to nonuniformity in the placed concrete; and
- Segregation of concrete should be guarded against at all stages of operations.

14.7—Placing and consolidation

Achieving acceptable architectural concrete also requires close attention to the procedures used for placing and consolidation. The rate of placement should be slow enough to permit proper vibration, yet rapid enough to prevent cold joints. All vibrating should be done by workers specifically trained to use the equipment correctly. They should lower the vibrators rapidly through the bottom of the lift and then raise them slowly to the surface, keeping the vibrators moving in the concrete at all times. When the vibrator is raised slowly and steadily, air bubbles dislodged from the form surface have time to rise ahead of the vibrator to the surface of the concrete. Removing these air bubbles is important because they are the source of bug holes in formed concrete surfaces.

Other measures for minimizing bug holes include:

- Placing the concrete in relatively shallow layers of not more than 15 to 18 in. (375 to 450 mm);
- Vibrating the concrete 50% longer than otherwise necessary;
- Double-vibrating drier batches of concrete. Two vibrator insertions are much more effective than a single insertion for twice the time;
- Revibrating the lower layer with each new layer placed (preferably to a depth of at least 6 in. [150 mm]); and
- Revibrating the top of the placement as late and as deep as the running vibrator will sink from its own weight, then withdrawing it slowly.

If forms are sufficiently rigid, external vibrators usually are satisfactory, but internal spud vibrators may also be required to remove air bubbles from the formed surface. Keep internal vibrators away from the form to prevent damage to form surfaces. If the forms are not sufficiently rigid, external vibrators can cause nonuniform distribution of the coarse aggregate in the vicinity of the vibrator.

Arriving at an acceptable vibration technique often requires experimentation early in the job. A technique that works well on one job may not be satisfactory on the next. Once established and approved, however, the technique should be applied uniformly for the remainder of the job. For the uniformity and consistency of concrete required for architectural concrete, overvibration is better than undervibration. Chapter 9 provides additional information on consolidation methods and equipment.

14.8—Surface treatments

Surface treatments typically applied to formed concrete consist of various degrees of abrasive blasting (using sand or other abrasives such as steel slag, corn cobs, walnut shells, or rice hulls), water-jet blasting (with or without the use of surface retarders), acid etching, bush hammering, and manual tooling. As with other operations, uniformity of workmanship is critical.

14.8.1 *Sandblasting*—Sandblasting is a common surface treatment and can achieve different finishes, depending on the degree of application. The various degrees of sandblasting include:

- *Brush*—Removes coatings and exposes fine aggregate. Produces no reveal (projection of coarse aggregate from the matrix after exposure);
- *Light*—Exposes fine aggregate and some coarse aggregate. Maximum reveal is 1/16 in. (1.5 mm);
- *Medium*—Generally exposes coarse aggregate. Maximum reveal is 1/4 in. (6 mm); and
- *Heavy*—Produces a rugged and uneven surface, exposing coarse aggregate to a maximum projection of 1/3 its dimension. Reveal is 3/8 to 2 in. (10 to 50 mm).

Fig. 14.8—One method of providing an exposed-aggregate surface is to seed specially selected coarse aggregate onto the concrete at the end of the finishing operation and then float it into the surface.

Whenever possible, the contractor should use the same sandblasting crew and equipment throughout the job. Changes in either tend to produce variations in the appearance of the finished surface. Be aware that the lighter degrees of sandblasting emphasize visible defects, particularly bug holes, and reveal defects previously hidden by the surface skin of the concrete. Sandblasting seldom removes hydration marks or defects in surface texture. For the heavier degrees of sandblasting, the concrete should be strong enough (at least 2000 psi [14 MPa]) to prevent dislodging of coarse-aggregate particles. Although sandblasting can be used to produce an exposed-aggregate finish, it results in a frosted appearance on the surface of the coarse aggregate. High-pressure water-jet blasting (1500 psi [10 MPa] or greater) is generally more effective for exposing aggregate (refer to the discussion on exposed-aggregate finishes). Experience has shown that sandblasting of slipformed concrete surfaces does not produce a desirable finish.

14.8.2 *Bush hammering*—This procedure, which is done with pneumatic tools fitted with a bush hammer, comb, or multipoint attachment, will remove concrete to a depth of approximately 3/16 in. (5 mm). The concrete should have strength of at least 4000 psi (28 MPa) before bush hammering is performed. Tool operators should be especially careful when working next to edges and corners to prevent damage.

14.8.3 *Grinding*—Grinding is used to produce both smooth and textured finishes on concrete surfaces. The procedure is normally done with power grinders. To avoid dislodging coarse aggregate, however, workers should not use these grinders until the concrete has reached a strength of at least 3000 psi (21 MPa). Hand grinding with a mason's stone can be done on green concrete as long as it goes no deeper than barely contacting the coarse aggregate. Deeper hand grinding, as well as power grinding, should be delayed.

14.8.4 *Manual treatment*—A common manual treatment consists of breaking off the tops of ridges in heavily fluted or similarly textured formed concrete to produce a broken surface appearance on the flutes, as shown in Fig. 14.2.

Although all the treatments just discussed are commonly used on formed surfaces, they can also be used on unformed surfaces. More common treatments for unformed surfaces are to impress a pattern in the plastic concrete or to give it an exposed-aggregate finish.

14.8.5 *Exposed-aggregate finishes*—Exposed-aggregate finishes are produced by removing, typically by water-jet blasting and brushing with a fiber brush, the cement-sand paste from the concrete surface to expose the coarse aggregate (Fig. 14.7). Usually, an attractive coarse aggregate is used in the concrete, and the mixture is gap-graded to increase the amount of coarse aggregate available for exposure. Another option is to seed specially selected coarse aggregate onto the concrete at the end of the finishing operation and float it into the surface (Fig. 14.8).

For unformed surfaces, aggregate exposure is relatively simple. The key is to watch the timing: the aggregate should be exposed as soon as the concrete has hardened enough that the coarse-aggregate particles will not be dislodged. Sometimes retarders are sprayed onto the surface of the fresh concrete to allow more flexibility in the timing of exposure.

For formed surfaces, the process is more involved. A surface retarder is sometimes applied to the inside of the forms immediately before concrete is placed. This retards setting of the concrete surface so that workers can expose the coarse aggregate by washing and brushing the surface after they remove the forms. Another option is the aggregate-transfer method, which involves gluing the attractive aggregate to the inside of the forms with water-soluble glue before the concrete is placed. Cement-sand paste surrounds the particles and can be removed from the surface after form removal. A third technique is the patented Arbeton method, which uses small-mesh wire fabric to hold the aggregate against the forms so that only the paste from the concrete seeps through the mesh to anchor the aggregate. For tilt-up panels, coarse aggregate can be spread on a bed of sand. For additional information on exposing aggregate, refer to *Color and Texture in Architectural Concrete* (Portland Cement Association 1995).

14.9—Curing and protection

Curing methods and timing of curing operations should be consistent to produce concrete of uniform color. To ensure color uniformity and to prevent marring or staining of architectural concrete surfaces, contractors should take the following precautions:

- If protective coverings are used, they should not mar the surface of immature concrete;
- Water or steam used for moist curing should be applied evenly to avoid blotching. The water should be nonstaining and not delivered through iron or steel pipes;
- Plastic sheeting used for curing should fit tightly against the concrete at all points, without wrinkling; otherwise, mottling will occur due to uneven moisture condensation;
- For membrane curing, colorless material should be used; it should be tried first on an unexposed surface to make sure it does not cause staining a few days after application; and

- When wood forms are used, curing in the forms (keeping the forms wet) is a good method. The form surfaces should be sealed to prevent staining.

Following completion of curing, contractors should take precautions to protect exposed architectural surfaces from the work of other trades. Acceptable architectural concrete can easily be ruined by these activities. For example, temporary erection of unprotected steel above the concrete can produce unsightly rust stains, and weld spatter can produce pockmarks.

14.10—Repairs
Some damage or blemishes are inevitable in architectural concrete. Therefore, specifications should include provisions for repair, and the inspector should be able to distinguish poor repairs from good ones. A poor repair in architectural concrete can look worse than the initial defect, and should be rejected immediately. Skilled artisans are available who can blend cements, aggregates, epoxies, or other polymers so that the repaired area cannot be detected with the unaided eye.

14.11—Precast members
14.11.1 *Storage and transportation*—Sunlight and weathering can affect the color and appearance of concrete surfaces. When practical, precast units should be stored so they will obtain equal exposure. If color variations do occur, however, they will diminish with time after the units have been placed on the structure.

When stacking precast units, blocking made of wood or other nonstaining materials should be used to prevent damage. Improper blocking and support during hauling is one of the most common causes of cracked members.

Road dirt can cause unsightly stains on precast concrete during transport. Because such stains can be difficult to remove, it should be requested that precast members be covered.

14.11.2 *Erection*—Workers should erect precast units carefully to prevent damage to exposed surfaces. Location and use of lifting inserts should be clearly indicated on shop and erection drawings and approved before unit production and erection. If concrete or grout will be placed above precast architectural concrete, sealing the surfaces of the precast units before placement will permit easier removal of grout leaks and drips.

14.12—Final acceptance
The most common problems found at the time of final inspection of architectural concrete are cracks, surface defects, bug holes, and color variations. There is a misconception that sandblasting cancels out these defects, but experience has shown that sandblasting actually magnifies them. Sandblasting intended for surface improvement, such as grime removal, should be very light and used with care.

Cracks in cast-in-place concrete are often caused by drying shrinkage. Although cracks can be patched by a skilled artisan, it is best to control them by providing sufficient crack-control joints in the architectural design or by the use of sufficient crack-control reinforcement to minimize potential crack widths.

Acceptability of surface defects such as honeycombing, bug-hole groupings, aggregate popouts, and local damage from handling depends largely on the distance of the defect from the viewer's eye. For example, a surface defect at the third story of a building is usually tolerated more easily than the same defect located adjacent to the entrance of the structure. Consequently, the inspector should evaluate defects from the vantage point of a potential observer.

14.12.1 *Bug holes*—Proper vibration techniques, as discussed previously, can reduce bug holes, but not completely eliminate them. As with other surface defects, bug holes should be evaluated from the vantage point of a potential observer.

14.12.2 *Color variations*—Even if all reasonable precautions are taken, some variation in color should be expected. In general, disputes concerning color variations are more likely to occur with precast units than with cast-in-place work. Often, selective placement of precast units on the basis of color or texture, rather than on the schedule of loading and delivery, can minimize any problems. As pointed out previously, color variations are often caused by variable curing conditions. Therefore, time and exposure to sunlight and weather may even out the variations. Color variations often can be reduced by using a hose to subject the surface to a series of wetting and drying cycles.

CHAPTER 15—SPECIAL CONCRETING METHODS
Preceding chapters cover inspection of more common types of concrete construction under normal conditions. This chapter briefly defines special types of concrete work that may have unique inspection requirements. To ensure satisfactory production of these special works, inspectors should study the requirements of the contract documents, which will serve as the inspector's primary guide.

15.1—Slipforming vertical structures
Slipform construction is basically an extrusion process. Concrete is placed at the top of the form, which is lifted continuously at a controlled rate so that the fresh concrete sufficiently sets by the time the form moves upward enough to expose it. A form must move continuously to be a slipform. A form that is stationary during concrete placement and moved upward in steps is called a climbing form, jump form, or lift form.

Crews can perform vertical slipforming continuously around the clock or stop at desired elevations and resume work the next day. Typical applications for this procedure include silos, chimneys, storage bins, bridge piers, water tanks, shaft liners, and service cores and bearing walls for buildings (Fig. 15.1 and 15.2).

15.1.1 *Formwork*—Forms for vertical slipforming are usually approximately 4 ft (1.2 m) tall and slightly tapered so that the concrete will detach itself from the forms as set occurs. A taper of 1/16 in./ft (1.5 mm/300 mm) of form depth is typical. The width of the forms at approximately midheight should equal the desired wall thickness.

As crews place the concrete, they slowly jack the forms upward at a rate based on the actual concrete setting rate. The jacks climb on smooth rods or structural tubing embedded in

Fig. 15.1—Slipformed service core of an office building.

Fig. 15.2—Slipforming the core of a high-rise building.

the hardened concrete. Typical slip rates are approximately 6 to 15 in./hour (150 to 375 mm/hour). It is important for the forms to stay level as they move upward to prevent binding, scoring, and lifting of the concrete from occurring.

Blockouts for openings in the walls are formed by attaching—usually to the reinforcement—polystyrene, cardboard, precast concrete, or wood frames so that they remain in place as the forms move by.

15.1.2 *Reinforcing steel*—Simple reinforcement details are essential because of time constraints on concrete placement and inspection during vertical slipforming. It is particularly important that all reinforcement be carefully placed as shown on the drawings. Templates mounted on the forms, or spacer bars attached to the top of the forms, are normally used to position vertical bars. Vertical laps should be staggered. For horizontal steel, workers place bars one layer at a time as the work progresses. The bars should be tied or firmly held in place, and lap splices should be staggered rather than in a vertical line to avoid "zipper" action failure. Using relatively short bars (10 to 12 ft [3 to 3.7 m]) permits easier handling.

Because the horizontal steel is continually disappearing into the concrete as the forms rise, monitoring horizontal bar spacing is difficult. To facilitate placing and inspection, scribe marks, crayon marks, tie wire, or other spacing indicators should be placed on selected vertical bars.

15.1.3 *Control of concrete placement*—Concrete is placed at the top of the forms in layers approximately 6 to 9 in. (150 to 230 mm) deep. Keeping the forms as full as possible gives the concrete enough time to set before it is exposed.

Vibration of concrete is normally restricted to the most recently placed layer of concrete, with penetration into the layer below, and is done no longer than necessary. Extra or deeper vibration can retard the essential early stiffening of the mixture and, even more serious, cause sagging or fallout below the forms. The setting time of the concrete should be monitored frequently, adjusting the slip rate as necessary.

To check on the progress of concrete set, a 1/2 in. (13 mm) diameter rod is forcefully pushed vertically into the concrete. If hard concrete stops the rod above 2/3 or below 3/8 of the form height, the rate of form lift should be increased or decreased, respectively. Vertical alignment and other dimensional aspects of the slipformed structure should be monitored frequently for compliance with ACI 117 tolerances.

15.1.4 *Finishing and curing*—Workers usually perform finishing and curing from a secondary (trailing) platform. From this work platform, they also can readily correct any defects in the plastic concrete leaving the forms, using a sponge float to fill small holes and achieve a uniform appearance.

A curing compound is normally applied immediately after finishing. Water curing also is possible (by means of a wet skirt of suitable length carried by and wetted from the finisher's platform), but wind and variations in water pressure can make this method impractical.

15.1.5 *Mixture requirements*—Concrete for slipforming may require a greater percentage of fine aggregate than conventionally placed concrete. The maximum aggregate size should be less than half the thickness of the cover over the reinforcement. Obtaining smooth-finish, slip-formed surfaces using silica fume concrete or other mixtures that might be described as "sticky" is often difficult to achieve.

The setting time of the mixture is extremely important. A penetration resistance of 50 to 200 psi (0.35 to 1.4 MPa), measured in accordance with ASTM C403/C403M, is normally required at the trailing edge of the formwork. Concrete with a penetration resistance of 15 psi (0.10 MPa) or below is susceptible to sagging or fallout. Concrete with a penetration resistance of 500 psi (3.5 MPa) has reached initial set. As initial set is approached, friction makes slipping difficult, and the forms tend to lift the concrete, resulting in horizontal checks and cracks and scoring of wall faces. Even worse, the forms could bind, and the operation must then be stopped. To measure the precise setting rate for the chosen mixture, laboratory tests should be conducted at the temperature expected to exist in the forms in the field.

Mixture uniformity may be monitored by observing the slump of the concrete as it is placed in the forms. For a given mixture, increases in slump imply a slower set, and decreases in slump imply a faster set. Concrete slumps should be kept within specified limits.

15.2—Slipforming cast-in-place pipe

Cast-in-place pipe is constructed in a previously excavated trench with near-vertical sides and a circular bottom. The circular trench bottom forms the outside bottom portion of the pipe, and a specially designed slipform forms the inside. Diameters of slipformed pipes generally range from 12 to 120 in. (300 to 3000 mm).

15.2.1 *Forms*—Because the trench is a portion of the form, its shape, line, and grade should be checked frequently to ensure correct wall thickness of the pipe. The circular bottom of the trench, which is the forming surface, should also be checked frequently to ensure that it is dense, clean, and free of serious irregularities. If standing water is in the trench, the contractor should remove it before concrete placement. If the trench is very dry, it should be dampened for better curing of the concrete.

Metal forms or a specially designed inflated tube can be used to slipform the inside of the pipe. Ordinarily, the top of the pipe is finished by hand.

15.2.2 *Control of concrete placement*—Concrete placement can be a two-stage manual operation or a one- or two-stage mechanical operation. In a two-stage operation, the lower portion of the pipe is placed first, followed shortly thereafter by the top portion. The condition of the top of the walls of the lower portion should be checked frequently before placement of the upper portion. Laitance or foreign matter, which can reduce watertightness of the construction joint, should be removed.

The concrete pipe should be continuously inspected as it leaves the forms to detect and resolve problems as soon as possible. To properly inspect the pipe, it should be viewed from both the outside and inside. Verification of inside dimensions will require coordination with construction activities and adherence to site safety regulations. For additional information on slipforming cast-in-place concrete pipe, refer to ACI 346 and 346R.

15.3—Tilt-up construction

In tilt-up construction, crews cast walls of a building horizontally on the job site, tilt or lift them into place, and then connect them to form an integral structure (Fig. 15.3). Walls up to three stories high are commonly constructed in this manner. The walls can be architecturally finished (Fig. 15.4), insulated, load bearing, or simply closure walls.

15.3.1 *Casting platform*—A surface called a casting platform provides a base for casting tilt-up panels. The building floor slab often serves as the casting surface. Tolerances for floor slabs used as casting platforms should also satisfy wall-panel tolerances, normally 1/4 in. (6 mm) or less in 10 ft (3 m).

Because surface imperfections in the casting platform will show on the wall panels, the slab may require some patching or topping. Temporarily filling cracks and joints with a

Fig. 15.3—A crane lifts a tilt-up panel. Note pipe bracing already attached to stabilize the wall panel until the entire structure is tied together.

nonabsorptive material will prevent reflection into panel surfaces. Platforms for architecturally treated wall panels may require a high-quality finish to achieve the specified surface. Where surface appearance is not critical, soil is sometimes used as the casting surface.

15.3.2 *Forms*—The only forms normally required for tilt-up construction are around the wall edges and at openings. For architectural finishes, textured or treated liners may be positioned on the casting platform before concrete placement. Edge forms should be sufficiently stiff and well-braced to keep the edges in alignment, particularly those forming the top and bottom edges of the wall (Fig. 15.5).

15.3.3 *Bond prevention*—Liquid membrane-forming curing compounds are commonly used to prevent bonding of tilt-up walls to the casting slab. At least two coats should be used, with the first coat applied soon after slab placement to seal the surface, and the final coat applied shortly before placement of the wall concrete to prevent bonding. It should be confirmed that both coats of the membrane are placed uniformly and that they completely cover the casting surface. The slab surface should be carefully inspected for defects that might mechanically bond the wall to the slab and inhibit lifting of the panels.

15.3.4 *Concrete placement*—Concrete should be placed and worked carefully to avoid damage to the casting platform. Particular care should be taken to obtain dense, homogeneous concrete along all panel edges. The use of jitterbugs to compact concrete surfaces that will have exposed aggregate, especially lightweight aggregate, is generally not permitted. For some architectural surface treatments, such as a pebble finish, casting the panels face up allows for better control of uniformity.

15.3.5 *Panel erection*—It should be verified that the contractor uses the erection procedures specified in the contract documents. Walls should develop sufficient strength before erection to prevent cracking. Testing of field-cured specimens, either cylinders or beams, or cast-in-place specimens are sometimes specified to confirm concrete strength prior to panel lifting. More recently, nondestructive

Fig. 15.4—Tilt-up construction permits a variety of architectural finishes by using textured forms.

Fig. 15.5—Wood edge forms and reinforcement are in place for casting a tilt-up wall panel.

Fig. 15.6—In lift-slab construction of a high-rise building, post-tensioned concrete slabs are supported by steel columns and slipformed concrete towers.

techniques such as the maturity method (ASTM C1074) have provided successful strength determination. Cracking can also be caused by lifting panels at points other than those indicated by the design, by using too much force to lift panels when it is difficult to break bond, or by jerking or impacting the panels.

Erection crews can lift tilt-up walls into position before, during, or after erection of the structural frame. If they set the walls before or during frame erection, temporary bracing is required. Be alert that workers use adequate bracing. Walls are typically anchored to the structural frame by bolting, welding, or keying to precast or cast-in-place columns. Because of the extremely critical nature of the transfer of stresses through precast concrete wall connections, not even minor deviations from the design documents should be permitted without the design engineer's approval.

For additional guidance on tilt-up construction, refer to ACI 551R.

15.4—Lift-slab construction

In lift-slab construction, the floors and roof slabs of a building are cast in layers, one on top of another, at or near ground level. Openings are left around the building columns. When the slabs have cured sufficiently, they are successively jacked up into position and connected to the columns (Fig. 15.6).

15.4.1 *Forms*—The only forms normally required for casting lift slabs are around the edges and at openings. The bottom form is the top of the previously placed slab. If the ceilings in the building are to be exposed, slab finishing requires special care because any imperfections in the top of a slab will be transferred to the underside of the slab above. Rough finishes can also inhibit lifting by developing a mechanical bond between slabs.

15.4.2 *Bond prevention*—As with tilt-up construction, membrane-forming curing compounds are commonly used to prevent bonding of lift slabs. It should be confirmed that the membrane is placed uniformly and that it completely covers each slab; poor application of bond breakers can be a major problem in lift-slab construction.

Polyethylene sheets also are excellent materials for preventing bond, but they can wrinkle during concrete placement, causing a poor finish. Their use will require special care to prevent wrinkling if the surfaces are exposed.

15.4.3 *Slab erection*—When the concrete has gained sufficient strength, jacks mounted on top of the building's columns successively lift the slabs into their final position. This is an extremely sensitive and potentially dangerous operation. It should be verified that the contractor uses lifting procedures for which the slabs were designed and pays close attention to construction-site safety.

15.5—Preplaced-aggregate concrete

Production of preplaced-aggregate concrete involves placing well-graded coarse aggregate, and then injecting grout to fill the voids between aggregate particles. Preplaced-aggregate concrete contains a higher percentage of coarse aggregate than does conventional concrete, and thus experiences approximately half the drying shrinkage.

15.5.1 *Aggregate placement*—Well-graded coarse aggregate (from approximately 3/4 in. [19 mm] up to the largest size) that can be placed without excessive segregation should be used. The void content of the coarse aggregate after placement in the forms ranges from approximately 38 to 48%. The aggregate should be washed and screened immediately before placement so that it will be free of undersized material and be surface-moist at the time of grout injection. Dry aggregate of less than SSD condition can cause premature initial setting of the grout and inadequate filling of voids in the coarse aggregate.

15.5.2 *Grout materials and mixing*—The grout is composed of portland cement, sand, and water, and may also contain admixtures such as pozzolans, fluidifiers, expansion agents, air-entraining agents, or coloring materials. Cement-sand ratios commonly range from 1:1 to 1:2 by weight. Although ratios as lean as 1:3 have been used for structural grout, it is usually not desirable to exceed 1:2. Sand for the grout should be well-graded, with 95% passing the No. 16 (1.18 mm) sieve, and have a fineness modulus of 1.2 to 2.0. A screen with openings not smaller than 3/16 in. (4.75 mm) or larger than 3/8 in. (9.5 mm) should be located ahead of the pump to remove oversized material.

Mixing grout requires the use of a high-speed mixer, such as a vertical-spindle paddle-type mixer or horizontal-shaft mixer (similar to a large plaster mixer). Conventional revolving-drum concrete mixers are not recommended because the mixing action is less effective. A grout-agitator tank can optimize use of the mixing equipment and provide storage capacity.

To control the consistency of grout containing fine sand, a standard flow cone should be used in accordance with ASTM C939. Cylinders should be made and tested according to ASTM C942.

15.5.3 *Grouting operations*—Grout is typically injected through insert pipes 3/4 to 1 in. (20 to 25 mm) in diameter and spaced 6 to 8 ft (1.8 to 2.4 m) on center. The insert pipes can extend horizontally through the formwork or vertically from above. Vertical insert pipes should extend to at least within 6 in. (150 mm) of the bottom of the form.

Grouting crews can inject material in horizontal layers or by advancing-slope techniques. With either system, they should start grouting from the lowest point in the forms. For horizontal layers, workers inject the grout through each insert pipe in sequence to raise the grout at each point from 3 to 5 ft (0.9 to 1.5 m), or as necessary to ensure that the next layer will go on while the one below is still soft. For the advancing-slope method, injection starts at one end of the form, and pumping continues through rows of inserts until the surface of the grout assumes a gentle slope.

It should be verified that grout flow is occurring at each insert pipe and that the flow is uniform throughout the placement; otherwise, rock pockets and severe honeycombing can result. Workers should clean all caked grout from equipment and all grout lines after each shift. Vent pipes should be provided at all locations where water or air may be trapped under form surfaces by the rising grout. Sounding wells, horizontal insert pipes, or electronically calibrated detector wires should be used to constantly monitor the rise of the grout. Sounding wells can be left in place.

For additional guidance on preplaced-aggregate construction, refer to *Concrete Manual* (U. S. Bureau of Reclamation 1988) and ACI 304R and 304.1R.

15.6—Underwater concrete construction

Underwater placement of concrete is used primarily for cofferdams, caissons, bridge piers, and dry-dock walls. Concrete should not be placed underwater, however, unless explicitly permitted by the contract documents or architect/engineer. It is seldom suitable for small or thin sections. Underwater placement can be successful in a highly controlled environment and in mass quantities. Quality control of the process and careful design are critical to successful placement and performance.

15.6.1 *Equipment and methods*—Concrete can be placed underwater by the preplaced-aggregate method, pump (using anti-washout admixtures), or tremie (Fig. 15.7). With the preplaced-aggregate method, previously discussed in this chapter, the rising grout displaces the water in the forms.

Tremie concrete is placed underwater by gravity feed through a vertical pipe, with one end above the water for charging of concrete and the bottom immersed in the concrete being placed (Fig. 15.8). The tremie pipe diameter should be approximately eight times the maximum aggregate size. First, workers plug or seal the tremie and lower it to the bottom of the placement. After the tremie is filled with concrete, they release the seal or, by lifting the pipe, allow the plug to discharge along with a flow of concrete. This concrete forms a mound around the end of the pipe to seal it. The pipe should be lifted slowly during concrete placement, with the bottom of the pipe always embedded in the concrete. Placing rates normally range from 1-1/2 to 10 ft (0.5 to 3 m) of height per hour. Refer to ACI 304R for additional information on tremie placement.

In difficult locations or where flowing water cannot be avoided, it may be necessary to use coarsely woven cloth bags partially filled with concrete. A diver carefully places the bags in a header and stretcher system so that the whole mass is interlocked. The bags should be free of harmful contaminants such as sugar, fertilizer, and organic materials.

15.6.2 *Mixture requirements for tremie concrete*—Tremie concrete should contain at least 650 lb/yd^3 (385 kg/m^3) of cement. The fine aggregate is usually 40 to 50% of the total aggregate by weight. Water-reducing retarders, air-entraining admixtures, and pozzolans often are added to improve flow. Anti-washout admixtures may also be added. The mixture should be sufficiently plastic to flow readily into place without vibration. Typically a maximum *w/cm* of

Fig. 15.7—A tremie installed for underwater concreting (top). Concrete being placed into top of tremie (bottom).

Fig. 15.8—The top of the tremie remains above water for charging with concrete. Water will be displaced when concreting begins (top) and concrete will overflow toward end of concreting (bottom).

0.44 by weight is recommended. A slump of 6 to 9 in. (150 to 230 mm) is necessary for the concrete to move properly underwater, where it weighs less than in air. For additional guidance on underwater concrete construction, refer to *Concrete Manual* (U. S. Bureau of Reclamation 1988), *Standard Practice for Concrete for Civil Work Structures* (U. S. Army Corps of Engineers 1994), and ACI 304R.

15.7—Vacuum dewatering of concrete

Vacuum dewatering is a method of extracting excess water from concrete shortly after it is placed. This permits concrete to stiffen rapidly by reducing the water content. The resulting reduced w/cm can also increase the 28-day compressive strength of the concrete as much as 25%, reduce shrinkage cracking, and improve wear resistance. The method is used almost exclusively on slabs, although it can also be used on vertical surfaces and on forms for precast concrete.

15.7.1 *Equipment and methods*—Vacuum mats should be arranged to begin dewatering of the concrete soon after it is placed, struck off, and bullfloated (preferably within 30 minutes in normal weather conditions). Most specifications require vacuum processing to continue for 1 to 3 min/in. (1 to 3 min/25 mm) of slab thickness. The effectiveness of water removal decreases with slab depth. It is seldom practical to reduce the w/cm below a depth of 12 in. (300 mm). Check the extent and uniformity of water extraction by observing the vacuum and the timing and duration of extraction. Final finishing operations should begin as soon as processing is complete.

Vacuum mats should be kept clean and in good condition. If a mat tends to pick up surface concrete, wetting it down thoroughly at the start of work, washing it halfway through the procedure, and thoroughly washing it at the end of the day will help prevent retention of cement film in the pores of the muslin. If the mat shows considerable wear, the muslin should be replaced, and the wire-screen backing scrubbed clean.

For additional guidance on vacuum dewatering, refer to ACI 302.1R.

15.7.2 *Reduction in slab thickness*—Removing water by the vacuum process can noticeably reduce a slab's thickness. Ordinarily, a 6 in. (150 mm) slab cast with concrete having a 6 in. (150 mm) slump will be reduced in thickness 1/8 to 1/4 in. (3 to 6 mm). If the slump is 3-1/2 in. (90 mm) or less, the thickness reduction usually is less than 1/8 in. (3 mm). When necessary, the thickness of the slab as placed should be increased to obtain the full design thickness after dewatering. Slumps of concrete used in adjacent areas should be approximately the same so that surface elevations will not vary.

15.8—Pumped concrete

Pumped concrete is conveyed by pressure through rigid pipes or flexible hose for discharge directly into the desired area. Pumping can be used for most types of concrete construction.

15.8.1 *Types of equipment*—The two primary pump types presently available are piston and squeeze-pressure.

A piston pump has a hopper with remixing blades for receiving the mixed concrete, an inlet valve, an outlet valve, and a piston in a cylinder. The outlet valve is located in the discharge line. When the piston starts its backward stroke, the inlet valve opens and the outlet valve closes, and concrete fills the cylinder. On the forward piston stroke, the inlet valve closes and the outlet valve opens, and the piston pushes the concrete from the cylinder into the pipeline or hose. Some pumps have two cylinders so that one can pump on the forward stroke while the other fills on the backward stroke.

A squeeze-pressure pump has a receiving hopper with remixing blades, a flexible hose, and rollers that rotate inside a metal drum kept under high vacuum pressure. The flexible hose connects the receiving hopper to the drum. It enters the drum at the bottom, and then circles upward around the inside surface to exit at the top. The hydraulically powered roller assembly within the drum rolls along the flexible hose to squeeze the concrete out at the top. The vacuum inside the drum helps restore the hose to normal shape after the roller flattens it, allowing the receiving hopper to deliver a steady supply of concrete to the hose.

Usually, rigid pipe or a combination of rigid pipe and heavy-duty flexible hose carries pumped concrete to the placement area. Rigid pipe (normally made of steel) is available in diameters of 3 to 8 in. (75 to 200 mm). Flexible hose can be made of rubber, spiral-wound flexible metal, or plastic, and is available in diameters of 3 to 5 in. (75 to 125 mm). Flexible hose with one or more sections is usually used at the discharge end of the pipeline to facilitate placement, especially in hard-to-reach locations such as areas within wall forms or between reinforcing mats. Because flexible hose offers greater resistance to pumping than rigid pipe, however, it is advisable to extend rigid pipe from the pump for as long as possible when long pump lines are needed.

The use of aluminum pipe to carry pumped concrete should be avoided. The concrete can abrade flakes of aluminum from the pipe wall, causing the formation of hydrogen gas that can lead to abnormal expansion of the concrete.

Line couplings should provide a full internal cross section, with no constrictions or crevices that can disrupt the smooth flow of concrete. It should be verified that the couplings are rated to resist the anticipated line pressures and are designed to allow replacement of any line section without movement of other line sections. Leaky couplings should be replaced immediately.

15.8.2 *Mixture requirements*—Although pumped mixtures contain the same ingredients as those placed by conventional methods, more emphasis on reducing variability is essential to obtaining good results.

The maximum size of angular coarse aggregate should be limited to 1/3 of the minimum inside diameter of the pipe or hose, and the maximum size of well-rounded aggregate to 40% of the inside diameter. Normalweight coarse aggregate and sand should meet ASTM C33 grading requirements, with the additional requirements that 15 to 30% of the sand particles pass the No. 50 (300 μm) sieve, and 5 to 10% pass the No. 100 (150 μm) sieve. Sand having a fineness modulus less than 2.40 or greater than 3.0 should not be used. To improve pumpability, it may be necessary to reduce the volume of normalweight coarse aggregate per unit volume of concrete by up to 10% compared with concrete placed by other methods. The amount of reduction depends on aggregate shape, cement, and SCM contents, whether a pumping admixture is used, and the capability of the pumping equipment.

Lightweight aggregates absorb more water than normalweight aggregates, and under pumping pressures, can absorb considerably more water. This can result in water loss from the concrete in the line, impairing its pumpability and placeability. Thoroughly presoaking the lightweight aggregate can help prevent this problem. The minimum moisture content after soaking should be equal to or greater than the average 24-hour absorption of the aggregate. Free water should be allowed to drain from presoaked stockpiled aggregates for 2 to 4 hours before mixing to permit uniform slump control. Lightweight aggregates should meet the grading limits of ASTM C330. In addition, 20 to 35% of the sand should pass the No. 50 (300 μm) sieve, and 10 to 20% should pass the No. 100 (150 μm) sieve.

Most lightweight concrete designed for pumping contains all natural sand for the fine aggregate. Gradation of this natural sand should conform to requirements for sand for normalweight concrete. Lightweight sand having a fineness modulus of less than 2.20 or greater than 2.80 should not be used.

Because pumped mixtures usually have high slumps along with aggregate gradings that increase water demand, they often require more cement than conventionally placed concrete.

Admixtures frequently used in pump mixtures include water-reducing admixtures, air-entraining admixtures, and finely divided mineral admixtures, or pumping aids. Properly used, these admixtures can improve lubrication, reduce segregation, improve workability, and decrease bleeding. The loss of air content that can occur during pumping should be considered when establishing the mixture requirements.

Trial mixtures intended for pumping should be prepared and tested in a laboratory in accordance with all applicable ASTM standards. For normalweight concrete, the method described in ACI 211.1 should be followed to select the volume of coarse aggregate per unit volume of concrete. For lightweight concrete, the volume of atmospherically soaked coarse aggregate should be selected following methods described in ACI 211.2 and 304.2R, and the volume of vacuum-saturated or thermally saturated aggregate should be selected following ACI 304.2R.

Because there is no standardized way to test the pumpability of a mixture in the laboratory, a test under actual field conditions is recommended to determine acceptability of a mixture for a particular job. When testing for pumpability, it is important to duplicate anticipated job conditions, including use of the same batching and mixing equipment,

Fig. 15.9—A placement boom supports and positions hose to deliver pumped concrete.

the same pump equipment and operator, the same pipe and hose layouts, and the same weather conditions, if possible. Previous use of a mixture on another job may furnish evidence of pumpability, but only if all conditions are essentially the same.

15.8.3 *Control of placement*—The pump should be located as close to the placing area as practical. To ensure a continuous supply of concrete, nothing should impede delivery of concrete to the pump. Continuous pumping is desirable because if the pump is stopped, movement of the concrete in the line may be difficult or impossible to start again.

Pump lines should be firmly supported to restrict excessive movement in response to pump-stroke pressure, and laid out with a minimum of bends. Use of alternate lines and flexible pipe or hose permits placing of concrete over a large area (Fig. 15.9), directly into the work, without rehandling of the concrete. For important concrete placements or large jobs, standby power and pumping equipment should be readily available.

When concrete is to be pumped downward 50 ft (15 m) or more, it may be desirable (depending on the pump manufacturer's recommendations) to provide an air-release valve at the middle of the top bend of the line to prevent vacuum or air buildup. When pumping upward, it is desirable to have a valve near the pump to prevent reverse flow of concrete during the fitting of cleanup equipment or maintenance of the pump.

The pump operator should maintain direct communication with the concrete placing crew. Good communication between the pump operator and the batch plant is also important. As a final check, the operator should start the pump and run it without concrete to be certain that all moving parts work properly. Mortar or a batch of regular concrete without coarse aggregate should be fed into the line ahead of the concrete to lubricate the pipes and reduce friction. This lubricating mixture should be wasted and not placed with the concrete. As soon as the pump receives concrete, the operator should run the pump slowly until the lines are completely full and the concrete is moving steadily.

The point of discharge should be monitored carefully to ensure that concrete segregation, displacement of reinforcement, and damage to forms do not occur. No water should be added to the concrete in the pump hopper.

During hot weather, shading the pipeline, covering it with damp material, or painting it white will reduce slump loss and concrete temperature rise.

When a delay in pumping occurs due to interruptions in concrete delivery, form repairs, or other factors, the operator should slow the pump, but maintain some movement of the concrete to avoid plugging. If, after a delay, concrete cannot be moved in the line, it will be necessary to clean out one line section, several sections, or even the entire line, and start over. When the form is nearly full, and enough concrete is in the line to complete the placement, the pump is stopped and a go-devil or rabbit (a ball of rolled-up burlap, paper, or specially fabricated material) is inserted and forced through the line to clean it out. Water or air under pressure can be used to push the go-devil. When water is used, the go-devil should be stopped several feet from the end of the line so the water will not spill into the placement area. If air is used, careful regulation of air pressure and installation of a trap at the end of the line are necessary to prevent the go-devil from being ejected as a dangerous projectile. An air-release valve installed in the line will help prevent pressure buildup. After removing all concrete from the lines, pumping crews should immediately clean all lines and equipment.

15.8.4 *Taking concrete samples*—When concrete is pumped, corresponding samples should be obtained at both truck and pump discharge points to measure any change in slump, air content, density, temperature, or other significant property of the mixture that can occur as a result of the pumping operation. Sample points should be selected for acceptance testing that represent the quality of the concrete as placed into the forms. This will usually require obtaining samples for acceptance testing at pump discharge points, especially when air-entrained concrete is pumped.

Whenever possible, flowing pump discharge samples should be obtained by intercepting or diverting the pump stream in a manner that best represents the way concrete is being placed into the forms. Stopping the pump flow during sampling, changing the angle of the boom or the discharge line, or allowing the concrete to freefall excessively can lead to nonrepresentative air content in the pump discharge sample.

Precautions should be taken to ensure that strength specimens cast at pump discharge points are not stored at locations where they will be subject to vibration or damage from nearby construction activities. When necessary, pump discharge samples should be obtained and then transported to test locations on stable ground before conducting tests and casting strength specimens. ACI 301 requires the contractor to furnish any labor necessary to assist the testing agency in obtaining and handling samples at the project site. For additional guidance on pumping concrete, refer to *Concrete Manual* (U. S. Bureau of Reclamation 1998), *Standard Practice for Concrete in Civil Work Structures* (U. S. Army Corps of Engineers 1994), and ACI 304R and 304.2R.

15.9—Shotcrete

Shotcrete is mortar or concrete pneumatically projected at high velocity onto a surface. Because shotcrete generally requires no exterior forms, mixtures should have low slumps to prevent sagging, particularly in vertical and overhead applications. In the past, shotcrete has been commonly called air-blown mortar, sprayed concrete, pneumatically applied mortar, or gunned concrete (also known by proprietary terms such as Gunite and Jetcrete).

Shotcreting can be an economical method for both new construction and repair work because it reduces forming costs and uses small, portable plants for mixing and placement. It is particularly suitable for structures having complex shapes, such as folded or curved roofs or walls, prestressed tanks, reservoir and canal linings, tunnel linings, and swimming pools. Repair applications for shotcrete include reservoir linings, dams, tunnels, waterfront structures, pipes, bridge and stadium superstructures, and masonry and concrete structures damaged by earthquake or fire.

For more information about shotcreting, refer to ACI 506R.

15.9.1 *Shotcreting processes*—There are two basic shotcreting processes: dry-mix and wet-mix.

In the dry-mix process, cement, sand, and possibly coarse aggregate are mixed dry and fed by compressed air through hose lines to a mixing nozzle. A separate line delivers water and admixtures (if used) to the nozzle, which provides final mixing. The mortar or concrete is then jetted at high velocity onto the surface to be shotcreted. Depending on the application technique used, the sand may be damp or dry before mixing, and the mixture may be dry until it reaches the nozzle or dampened before delivery to the nozzle.

In the wet-mix process, all ingredients are mixed, inserted into the delivery hose, and conveyed to the nozzle by compressed air or pump pressure. Additional compressed air injected at the nozzle propels the shotcrete at high velocity onto the surface (Fig. 15.10).

ACI 506R provides a comparison of the two processes (Table 15.1). Either one can produce shotcrete suitable for normal construction requirements. The contractor should determine which process to use based on job conditions and requirements of the contract documents.

15.9.2 *Mixture proportions*—For dry-mix fine-aggregate shotcrete, mixture proportions are typically one part cement to three to four and one-half parts damp sand, measured by volume. The sand should contain 3 to 6% moisture to ensure a homogeneous mixture and to prevent discomfort to the nozzle operator from buildup of static electricity.

Weight batching is preferred for proportioning of dry- or wet-mix coarse-aggregate shotcrete. Mixtures may contain 560 to 850 lb/yd^3 (335 to 505 kg/m^3) of portland cement, and generally contain more sand than conventional concrete. A typical coarse-aggregate shotcrete, before application, has proportions of cement, fine aggregate, and coarse aggregate ranging from 1:3:2 to 1:3:1, depending on job requirements (ACI 506R). That these mixture proportions, however, do not represent the shotcrete in place because rebound will cause some loss of solid ingredients. Establishing the in-

Fig. 15.10—Workers apply wet-mix shotcrete.

place quality of shotcrete requires testing, as discussed in a following section.

Shotcretes containing quick-setting admixtures or fast-setting modified portland cements are required for underground structural support. Such shotcrete should have high early strengths of 500 to 1000 psi (3.5 to 7 MPa) or higher at 8 hours. Both liquid and powdered admixtures have been used for high-early-strength shotcrete. In most cases, these admixtures are transmitted by air or water to the nozzle to produce fast initial set times of 1 minute or less.

Fast-setting, modified portland cement can produce shotcrete with higher early and later strengths than shotcrete with other admixtures.

The quality of shotcrete in structures depends largely on the skill of the application crew. All members of the crew should provide evidence of training and experience in performing satisfactory work. The nozzle operator should have ACI certification as a shotcrete nozzle operator.

The duties of nozzle operators include:
- Verifying that all surfaces to be shot are properly prepared, as described previously, and free of laitance or loose material;
- Verifying that the operating air pressure is uniform and provides proper nozzle velocity for good compaction;
- Regulating the water content so the mixture will be plastic enough to provide good compaction and a low percentage of rebound, but stiff enough not to sag. In the dry-mix process, the nozzle operator directly controls the mixing water; in the wet-mix process, the operator calls for changes in consistency as required;
- Achieving maximum compaction with minimum rebound by holding the nozzle at the proper distance and as nearly perpendicular to the surface as the type of work will permit; and
- Directing the crew when to start and stop the flow of material, and stopping the work when material is not arriving uniformly at the nozzle.

The nozzle operator usually has at least one assistant, who operates a blowpipe to help keep all rebound and other loose, porous material out of the new construction (except in work where the trapped rebound can readily be removed by the

Fig. 15.11—The nozzle operator builds up shotcrete gradually in layers to prevent sagging.

Table 15.1—Comparison of dry- and wet-mix shotcrete processes (ACI 506R)

Dry-mix process	Wet-mix process
Instantaneous control over mixing water and consistency of mixture at nozzle to meet variable field conditions.	Mixing water controlled at delivery equipment and be accurately measured.
Better suited for placing mixtures containing lightweight aggregates and refractory materials, and for shotcrete requiring high early strengths.	Better assurance that mixing water is thoroughly mixed with other ingredients.
Capable of being transported longer distances.	Less dusting and cement loss accompanies gunning operation.
Start and stop placement characteristics better, with minimal waste and greater placement flexibility.	Normally has lower rebound, resulting in less material waste.
Capable of higher strengths.	Capable of greater production.

nozzle operator). Another assistant may be required to handle the delivery hose.

Generally, the nozzle operator works from bottom to top, keeping the nozzle a distance of approximately 3 ft (1 m) from the work and the direction of the gunning as perpendicular to the surface as possible. Corners should be filled first. When placing shotcrete behind reinforcing bars, the operator should hold the nozzle at a slight angle and build up the concrete from both sides to better encase the bars.

To prevent sloughing or sagging of the shotcrete, the operator should build up the full thickness of shotcrete in layers (Fig. 15.11). In sloping, overhanging, or vertical work, it is necessary to wait between successive applications to permit the concrete to become strong enough to support the next layer. Construction joints should be sloped off to a thin, clean, regular edge, preferably at a 45-degree slope.

Inspectors should closely observe shotcreting operations to make sure the crew is using the proper procedures. All sand pockets, sloughed material, and rebound should be continually removed to avoid being covered by shotcrete. In most cases, rebound is best removed with a compressed-air jet. To detect inclusions of rebound or the presence of hollow spots, sound hardened surfaces with a hammer.

It is also important to be alert to weather conditions. It may be necessary to halt shotcrete placement temporarily if high winds separate sand and cement at the nozzle, if freezing of the shotcrete seems imminent, or if damaging rain occurs.

15.9.3 *Safety*—Shotcreting requires special attention to safety. The nozzle operator and assistants should wear safety glasses with side shields to protect their eyes from flying rebound. In particularly dusty situations, air-ventilated helmets may be required. It is often necessary for applicators to turn the nozzle away from the work during periods of unsatisfactory feed conditions, but they should be careful not to direct the nozzle at other crew members. Care should also be taken to prevent caustic accelerators from contacting skin or eyes.

15.9.4 *Forms and ground wires*—Where forms are required, they should be designed to permit the escape of air and rebound. Forms are needed only on one side of a wall, but they should be adequately braced to withstand the considerable force and vibration of shotcrete application. Use ground wires to establish the thickness, surface planes, and finish lines of the shotcrete. These alignment guides should be taut, secure, and true to line and plane.

15.9.5 *Surface finishing*—When finishing shotcrete, workers should bring the surface to an even plane with well-formed corners by working to ground wires or other thickness and alignment guides. If screeding is required, they should use a thin slicing edge to trim off high spots and expose low spots.

A thin finish or flash coat can be applied to remove rough areas after the ground wires have been removed, or the flash coat can be applied over the ground wires. To produce a float finish, the final surface can be lightly rubbed in a circular or spiral motion with a flat burlap or rubber pad. If a troweled finish is desired, the steel troweling should follow careful screeding to obtain satisfactory results with the least trowel pressure. Finishers should perform troweling within 1 hour of shotcrete placement.

15.9.6 *Curing and protection*—Curing and cold weather protection should follow approved concrete practices unless otherwise specified. Application of curing compounds is generally the most practical type of curing procedure, but may not be as effective as water curing. Refer to Chapter 10 for additional information on curing methods.

15.9.7 *Control testing*—Because there is no way to directly form shotcrete test specimens, the most reliable specimens come from cores—either from the structure itself or from special test panels. The nozzle operator performing the on-site work should produce these test panels, holding the nozzle at the same angle to be used on the structure. For fine-aggregate shotcrete, test panels are generally 18 x 18 x 3 in. (450 x 450 x 75 mm) and the cores approximately 3 in. (75 mm) in diameter, with a finished length somewhat less than 3 in. (75 mm). For coarse-aggregate shotcrete, panels should be larger and at least 6 in. (150 mm) thick, with a minimum core diameter of 3 in. (75 mm). When coring test panels for specimens, material typical of that in the structure should be obtained by avoiding areas near panel edges.

To ensure that the test panels properly reflect the quality of the shotcrete in the structure, cores from the structure should occasionally be taken as well. The cores should be obtained and tested in accordance with ASTM C42/C42M. Project specifications usually require that extracted cores be visually graded for uniformity and consolidation. ACI 506.2 provides details on visually grading shotcrete cores.

A measure of strength within the first 8 hours is desirable, particularly for underground support structures, but the shotcrete at this early age cannot usually be cored without damage to the specimen. Concrete pullout tests can be used to monitor early strength gain; however, the correlation between pullout and formed cylinder strength needs to be established for each type of mixture.

CHAPTER 16—SPECIAL TYPES OF CONCRETE

Inspectors may occasionally encounter projects that require the use of special types of concrete to satisfy specific design and economic concerns. These concretes have unique characteristics such as low or high densities, insulating properties, shrinkage compensation, low temperature rise, and high performance. Because these concretes have properties somewhat different from those of conventional portland-cement concrete, it may be necessary to modify procedures for mixture proportioning and testing, batching and mixing, and placing and consolidation. To ensure satisfactory production using these special types of concrete, inspectors should study the requirements of the contract documents that will serve as the inspector's primary guide.

16.1—Structural lightweight-aggregate concrete

Structural lightweight-aggregate concrete is sometimes used in buildings and bridges when obtaining the required strength level at a reduced density will result in greater economy. For most structures, designs require the same compressive strength levels as those used for normalweight concrete. Structural lightweight-aggregate concrete typically has a 28-day compressive strength exceeding 2500 psi (17 MPa) and an equilibrium density of less than 115 lb/ft^3 (1840 kg/m^3). Concretes having compressive strengths exceeding 5000 psi (35 MPa), however, can be economically produced with most manufactured lightweight aggregates available in the United States and Canada.

16.1.1 *Aggregates*—There are two general classes of structural-grade lightweight aggregates:
- Those produced by extruding, calcining, or sintering products such as slag cement, clay, diatomite, fly ash, shale, and slate; and
- Those produced by processing natural materials such as scoria, pumice, and tuff.

ACI 213R describes the manufacture of these aggregates.

To be acceptable for use in structural concrete, lightweight aggregates should meet the requirements of ASTM C330. This specification covers grading, density, deleterious substances, and other properties.

Most structural lightweight-aggregate concrete contains natural sand for the fine aggregate. Natural sand usually is more economical, and can improve concrete properties such as tensile strength, modulus of elasticity, creep, and drying shrinkage.

16.1.2 *Mixture proportioning and control*—Mixture proportioning methods for structural lightweight-aggregate concrete differ somewhat from those for normalweight concrete. Lightweight aggregates typically exhibit greater total water absorption and a higher rate of absorption than conventional aggregates. This absorption has little effect on compressive strength as long as enough water is supplied to saturate the aggregate. Thus, it is essential to know the moisture content of the aggregate and to make adjustments from batch to batch to provide constant cement and air contents, similar slumps, and a constant volume of aggregates. The yield of the concrete can change radically if the moisture content of the lightweight aggregate is not corrected.

Entrained air is desirable in all lightweight-aggregate concrete, regardless of exposure conditions. In addition to improving durability, entrained air enhances workability and decreases bleeding. Air-entrained concrete can blister when steel troweled, however. Recommended air contents are 4 to 8% with 3/4 in. (19 mm) maximum-size aggregate, and 5 to 9% with 3/8 in. (9.5 mm) aggregate. Pressure air meters should not be used for measuring air content of lightweight-aggregate concretes; the compression of air into the aggregate particles during the test often makes the results unreliable. The volumetric method (ASTM C173/C173M) provides the most reliable results. In some areas, however, low-pressure (3 psi [21 KPa]) air meters have been used with success.

ACI 211.2 details three methods of mixture proportioning for structural lightweight-aggregate concrete. Two of these—the weight method and the pycnometer method—are not widely used. With some lightweight aggregates, particularly those having a lower rate of absorption and lower total absorption, the same proportioning procedures as those for normalweight concrete can be used (refer to Chapter 6).

The most widely used method of proportioning structural lightweight-aggregate concrete is to make trial mixtures to determine the relationship between cement content and compressive strength at a fixed bulk volume of lightweight aggregate. This method is used instead of the traditional method of correlating compressive strength with the *w/cm* because lightweight aggregates can absorb water, making it difficult to calculate the *w/cm*.

For large or especially important structures, separate mixture proportioning studies should be made using trial mixtures with three different cement contents (for example, 500, 600, and 700 lb/yd^3 [300, 355, and 415 kg/m^3] of cement) and with slumps of less than 4 in. (100 mm) and air contents of 5 to 6%. To start, a trial batch should be made using one of the selected cement contents. Presaturated lightweight aggregate should be used so that the aggregate will not absorb water as it is introduced into the mixer, resulting in rapid slump loss. If the lightweight aggregate has not been presaturated, all of the aggregate (including the sand) should be premixed with approximately 2/3 of the estimated water for approximately 1 minute. The cement and most of the additional water can then be introduced (including water with the air-entraining admixture mixed in). The last of the water should be added carefully, and the total water adjusted to achieve the proper slump. Water held back from or added to the original estimated amount should be carefully weighed and recorded to arrive at the total water actually in the batch.

When mixing is complete, slump, density, and air content should be measured.

This first trial batch should be examined for workability, over- or under-sanding, harshness, and proper air content. It will probably be necessary to make a second trial mixture at the same cement content, adjusting quantities of sand, coarse aggregate, water, and air-entraining admixture as determined from the results of the first trial. Refer to ACI 211.2 for information on making these adjustments.

Proportions should be obtained for this second trial mixture by determining the specific gravity factor. The absolute volumes of cement, sand (SSD), water, and air content (per cubic yard of fresh concrete) should be summed based on the final proportions of the first trial mixture, then this sum is subtracted from 27 ft^3 to find the in-place volume of the coarse lightweight aggregate. Next, the specific gravity factor for the dry lightweight aggregate is computed using the following equation

$$\text{specific gravity factor} = \frac{\text{weight of coarse aggregate}}{\text{in-place volume} \times 62.4}$$

This factor can then be used to adjust proportions as if it were a true specific gravity. It cannot be used to determine the *w/cm* because the aggregate has absorbed part of the total water.

After obtaining a satisfactory set of proportions with one selected cement content, compressive-strength specimens should be cast for testing at 7 and 28 days and other ages as desired, and specimens should be prepared for plastic (freshly mixed) density and air-dry density. Mixtures with the other two selected cement contents should then be made to obtain strengths and densities for these contents. In the proportions for these mixtures, the volume of coarse lightweight aggregate should be maintained and water adjusted as needed to meet the target slump range. The absolute volume of sand should be varied inversely as cement content is changed, and the air-entraining admixture dosage adjusted as necessary for each cement content.

After obtaining the compressive strengths, a curve should be established that shows the relationship of strength and cement content. The proper cement content for the specified strength (plus necessary overdesign) can then be determined. This strength should be confirmed with additional trial mixtures.

Pozzolans can be used for partial replacement of portland cement, just as they are in normalweight concrete. Water-reducing admixtures, accelerators, and retarders can also be used in structural lightweight-aggregate concrete in essentially the same manner as in normalweight concrete.

16.1.3 *Testing*—In addition to laboratory testing for minimum compressive strength (ASTM C39/C39M) and maximum air-dry density (ASTM C567), specifications for structural lightweight-aggregate concrete often require tests for modulus of elasticity (ASTM C469) and splitting tensile strength (ASTM C496/C496M).

Field testing of structural lightweight concrete is generally limited to slump, fresh density, air content, temperature, and compressive strength. (ACI 318 states that splitting tensile-strength tests shall not be used as a basis for field acceptance of concrete.) All field testing and molding of specimens should be performed at the job site after all water has been added and the concrete has been thoroughly mixed.

16.1.4 *Batching and mixing*—The procedures for batching and mixing structural lightweight-aggregate concrete are the same as for normalweight concrete. The lightweight aggregate should be prewetted to as uniform a moisture content as possible or premixed with the mixing water before the addition of other ingredients. This will help to avoid slump control problems at the mixer as well as slump loss after discharge from the mixer resulting from water absorption of lightweight aggregate. Lightweight aggregates with low water absorption may not require extensive prewetting before batching and mixing. If the concrete will be exposed to severe cold before sufficient drying occurs, complete saturation of the aggregate during prewetting operations may have to be avoided to maintain good durability.

If the structural lightweight-aggregate concrete is to be placed by pump, it will often be necessary to completely saturate the aggregate to prevent severe slump loss during pumping. Under these conditions, the concrete should have a long drying period before exposure to freezing and thawing. ACI 213R and 304R should be carefully studied before monitoring the pumping of structural lightweight concrete.

16.1.5 *Placing, consolidation, and finishing*—Procedures for placing structural lightweight-aggregate concrete (except if by pumping, as noted previously) differ little from those for placing normalweight concrete. As with all concrete, segregation of coarse aggregate from the mortar should be avoided. It is particularly important to not overvibrate the concrete. The coarse-aggregate particles are the lightest solid ingredients in the mixture, and overvibration can cause the particles to rise, resulting in finishing problems from floating aggregate and nonuniform strength through the depth of the member. In addition to care in vibration, the use of slumps less than 4 in. (100 mm) greatly helps in avoiding segregation during concrete handling, consolidating, and finishing.

Properly proportioned structural lightweight-aggregate concrete can generally be finished earlier than is practical with normalweight concrete. Finishers should avoid overworking of the surface, however, which tends to bring excess lightweight coarse aggregate to the surface. The use of magnesium or aluminum screeds and floats will minimize surface tearing and pullouts. Grate tampers (jitterbugs), both fixed and roller type, may be useful in depressing coarse-aggregate particles and bringing mortar to the surface. Their use, however, should be closely monitored and limited to only one pass over the surface (refer to ACI 302.1R, Section 8.3.1). If grate tampers are used, it is necessary to keep the concrete slump low—not greater than 2 in. (50 mm). Vibrating grate tampers should never be used for structural lightweight-aggregate concrete.

16.1.6 *Curing and protection*—Curing and protection of structural lightweight concrete is no different from that of normalweight concrete. Refer to Chapter 10 for information on curing procedures.

16.2—Lightweight fill concrete

Lightweight fill concrete has oven-dry densities of 50 to 90 lb/ft^3 (800 to 1440 kg/m^3) and compressive strengths of less than 2500 psi (17 MPa). Applications for such concretes range from insulating fills to structural elements (Fig. 16.1). Detailed information regarding lightweight fill concrete is provided in ACI 523.3R.

16.2.1 *Aggregates*—All aggregates for fill concretes should meet the requirements of ASTM C33, C144, C330, or C332, whichever is applicable. Lightweight coarse aggregates of these types generally are used with similar lightweight fine aggregates, but may at times be used with natural sand. Cellular concrete with added sand or other fine aggregate also is used for fill concrete.

16.2.2 *Mixture proportioning and testing*—The general objective for proportioning lightweight fill concrete is to achieve the proper strength without exceeding the specified density. This is accomplished by preparing trial mixtures. The procedures for making trial mixtures with expanded or sintered shales, clays, or processed natural materials are the same as those for structural lightweight-aggregate concrete, described previously. The procedures for vermiculite, perlite, or foamed concretes are the same as for insulating concrete, described in this chapter. Laboratory and field testing procedures for lightweight fill concrete are also identical to those for structural lightweight aggregate concrete and insulating concrete.

Pozzolans, water-reducing admixtures, accelerators, and retarders can be used in lightweight fill concrete in essentially the same manner as in normalweight concrete.

16.3—Low-density concrete

Primarily used for insulating purposes, low-density concrete is made with or without aggregate, and has an oven-dry density of 50 lb/ft^3 (800 kg/m^3) or less. Field-placed, low-density concrete is commonly used in roof decks to provide thermal insulation and add stiffness. It can also be used to reduce heat transmission through floors and walls (Fig. 16.2).

The two generic types of low-density concrete are:
- *Aggregate type*—Made predominantly with low-density mineral aggregates, such as expanded perlite or vermiculite, or low-density synthetic aggregates; and
- *Cellular type*—Made by forming a cement matrix around air voids that are generated by preformed foams or special foaming agents, with or without the addition of mineral aggregates.

ACI 523.1R provides detailed information on low-density concrete. Information on aggregates, foams, and mixture proportions also is available from producers and trade associations.

16.3.1 *Aggregates*—There are two kinds of mineral aggregates for low-density insulating concrete:
- Group I consists of aggregates made by expanding products such as perlite and vermiculite. They generally produce concretes having air-dry densities of 15 to 50 lb/ft^3 (240 to 800 kg/m^3); and

Fig. 16.1—Flowable fill material is placed in a trench.

Fig. 16.2—Residential building that is constructed of low-density cellular concrete weighing between 300 and 1000 kg/m^3 (19 and 63 lb/ft^3). Material can float in water.

- Group II aggregates consist of the same materials used for structural lightweight-aggregate concrete. They generally produce concretes having air-dry densities of 45 to 90 lb/ft^3 (720 to 1440 kg/m^3).

The most commonly used natural aggregates are the expanded minerals—vermiculite and perlite. Vermiculite is a mica-like mineral. When expanded by heat, the particles are accordion-shaped. Its dry loose density is 6 to 10 lb/ft^3 (95 to 160 kg/m^3), and it usually is produced in graded sizes. Perlite is a naturally occurring glassy, siliceous rock. When expanded by heat, it produces light-colored spherical particles containing closed air cells. Perlite aggregate is also produced in graded sizes and has a dry loose density of 7-1/2 to 12 lb/ft^3 (120 to 190 kg/m^3).

Other aggregates used to produce insulating concrete are primarily manufactured materials such as ceramic and glass granules, hollow or low-density polystyrene beads, ground paper, and sawdust. Most of these are single-size particles.

Aggregates for low-density concrete should meet ASTM C332 requirements. The maximum aggregate size is usually less than 1/8 in. (3.2 mm) and seldom exceeds 3/8 in. (9.5 mm). Sand should meet the requirements of ASTM C33 (for concrete) or C144 (for mortar).

16.3.2 *Foams for cellular concrete*—Both preformed and mixer-generated foams are used in cellular low-density concrete. Preformed foam is generated by introducing controlled quantities of air, water, and foaming agent under pressure into a foaming nozzle. The foam is blended with a cement or cement-aggregate slurry, either in batched volumes or by continuous batching. The foam should have sufficient stability to maintain its structure until the concrete hardens.

Mixer-generated foams are produced by high-speed, high-shear mixing of water, foaming agent, cement, and aggregate (if required) with simultaneous air entrapment. Air bubbles are large initially, but become smaller as mixing proceeds.

Trial mixtures should be used to determine the quantity of preformed foam or foaming agent required. Up to 80% of the volume of the final concrete mixture may be air, depending on the desired concrete density.

16.3.3 *Mixture proportioning and control*—For most applications, proportions should be chosen for insulating concrete to provide a specified dry density, because thermal properties are primarily a function of density. If the concrete is to be conveyed by pumping, all laboratory mixtures should be trial-pumped under field conditions before construction begins. Pumping can affect water requirements, wet and dry densities, and mixture uniformity. It may be necessary to start with additional amounts of air or foam to make up for losses in air caused by mixing, pumping, and placing of insulating concrete.

16.3.3.1 *Aggregate type*—Mixtures containing lightweight aggregate often are specified in terms of cubic feet (bulk volume) of aggregate per bag of cement. A 1:6 mixture, for example, would contain one bag of portland cement and 6 ft^3 (0.17 m^3) of aggregate. A better method is to specify the total loose bulk volume of lightweight aggregate per cubic yard of concrete along with the weight of cement and the slump and air content required for the mixture. Required cement contents generally range from 330 to 630 lb/yd^3 (195 to 375 kg/m^3).

Insulating concrete made with lightweight aggregates typically includes an air-entraining admixture to act as a wetting agent, lower the specific gravity of the paste, and increase relative specific gravity of the coarse-aggregate particles. This reduces the mixing water content and substantially reduces the tendency of the aggregate to float. The use of an air-entraining admixture is particularly important in fluid, nearly self-leveling mixtures that are to be pumped through small (2 to 4 in. [50 to 100 mm] diameter) hose lines. It is often necessary to adjust the amount of air entrainment to produce concrete with the required dry density.

Water requirements of insulating concretes made with lightweight aggregates vary greatly with the absorption of the aggregates and the desired fluidity of the mixture. Vermiculite aggregate is highly absorptive, and typically requires 600 to 700 lb (355 to 415 kg) of water per cubic yard (meter) of concrete for fluid mixtures. Most perlites are less absorptive, with water requirements of 300 to 500 lb/yd^3 (180 to 295 kg/m^3).

16.3.3.2 *Cellular (foam type)*—Cement contents for cellular concrete range from 470 to 940 lb/yd^3 (280 to 560 kg/m^3). No aggregate is used when the desired dry density is less than 30 lb/ft^3 (480 kg/m^3). When densities greater than 30 lb/ft^3 (480 kg/m^3) are desired, fine sand usually is added, and the cement contents then range from 470 to 550 lb/yd^3 (280 to 325 kg/m^3). The water contents of cellular insulating concretes without aggregate are generally 300 to 500 lb/yd^3 (180 to 295 kg/m^3); with sand in the mixture, water contents are 200 to 375 lb/yd^3 (120 to 220 kg/m^3).

16.3.4 *Testing*—Laboratory tests of trial mixtures of low-density concrete are generally limited to compressive strength and plastic (freshly mixed) and dry densities. Compressive-strength and dry-density specimens (molded 3 x 6 in. [75 x 100 mm] cylinders) should be tested in accordance with ASTM C495. Plastic densities should be determined in a manner similar to that for other concretes (ASTM C138/C138M), but the concrete should be consolidated by tapping the sides of the container rather than by rodding. To permit construction control based on plastic density, the plastic density should be correlated with the dry density.

Once satisfactory mixture proportions have been established, other laboratory tests may be required. Because these concretes are used for insulation, the thermal resistivity is measured with a guarded hot plate (ASTM C177) or a calibrated hot conductometer (ASTM C518). Specific heat and thermal diffusivity are sometimes needed for design purposes.

If measurements of tensile strength, modulus of elasticity, Poisson's ratio, and drying shrinkage are required, the same techniques should be used as those for structural concrete. The testing equipment, however, should have sufficient sensitivity for the low values generally encountered. Drying shrinkage of insulating concrete is greater than that of structural concrete (as much as 0.5%).

Penetration resistance is sometimes used to define the ability of low-density concrete to sustain normal construction foot traffic. For acceptable resistance to foot traffic, the Proctor penetrometer reading should indicate an average bearing value of 200 psi (1.4 MPa) or greater.

A measure of nailing characteristics of low-density concrete may also be required. For satisfactory nailing, the concrete should be able to receive a specified type of nail without shattering and withstand a withdrawal force of 40 lb (0.18 kN).

Field control tests are typically limited to compressive strength and plastic density. Because of variations in the weights of the aggregates, cement, and water, density measurements accurate within ±1% are generally acceptable. Unless otherwise specified, an ordinary galvanized 10 quart pail (approximately 1/3 ft^3 [9.5 L]) or similar calibrated container and a scale should be used to determine density. The pail should be calibrated before using, and the scale's accuracy checked at least once a week during use.

16.3.5 *Batching and mixing*—To ensure uniform density at the point of placement, all materials should be added to the mixer at a constant rate, in their correct proportions and in the correct sequence. The required amount of water goes into the mixer first, followed by the cement, air-entraining admixture or foaming agent, aggregate, preformed foam, and other additives. Materials should be mixed so that the design plastic density is obtained at the point of placement. Any mixture changes that may result from the method of placement, such as mechanical or pneumatic pumping, should be allowed for.

When transit-mixing equipment is used for low-density concrete containing aggregate, the mixer should not be operated on the way to the job site.

16.3.6 *Placing and consolidation*—Most insulating concretes are pumped as extremely fluid mixtures. Generally, only screeding and minor floating are necessary for placing and finishing. When an aggregate is highly friable or when higher strengths are desired at a specified dry density, the water content should be kept lower and the concrete placed without pumping. Because the concrete is light, it is easy to push into place. Vibration is rarely needed, unless the concrete is placed in molded shapes or formed cavities requiring external vibration.

16.3.7 *Curing and protection*—The surface of freshly finished low-density concrete should not be allowed to dry appreciably for the first 3 days. Because of their high water contents, vermiculite and perlite concretes do not usually require membrane or water curing under mild weather conditions. In hot, dry, windy weather, curing membranes should be used.

Cellular concretes typically have higher cement and lower water contents than low-density aggregate concretes, so curing should be specified to prevent premature drying. If curing is not used, low strengths and excessive drying shrinkage can result.

16.4—Heavyweight concrete

Concretes having densities substantially above that of normalweight concrete (150 lb/ft^3 [2400 kg/m^3]) are primarily used as shielding to protect people and equipment from harmful radiation, such as x-rays, gamma rays, and neutrons. These concretes are made with heavy aggregates having densities as high as 350 lb/ft^3 (5610 kg/m^3). For neutron shielding, high-density concrete may also contain hydrous ores as aggregate and high cement contents to increase the amount of hydrated water. Such concretes may not be heavier than normalweight concretes, but they are more efficient in neutron shielding because they contain more hydrogen. The desired high density should be obtained if the concrete is to provide effective shielding.

16.4.1 *Aggregates*—Heavyweight aggregates for preparing high-density concrete include natural minerals such as barite, ferrophosphorus, goethite, hematite, ilmenite, limonite, and magnetite, plus man-made materials such as steel punchings and shot. Where a high fixed water content is desirable, hydrous iron ore, serpentine (slightly heavier than normal-weight aggregate), or bauxite may be used. Table 16.1 shows specific gravities and percents of fixed water for some of these materials. To produce concrete with a density of 230 to 240 lb/ft^3 (3680 to 3840 kg/m^3), the aggregate should have a specific gravity of at least 4.5. To produce concrete with a density of around 300 lb/ft^3 (4810 kg/m^3), the aggregate should have a specific gravity of at least 6.0.

Factors to consider when selecting heavyweight aggregates include physical properties, availability, and cost. The

Table 16.1—Specific gravities of aggregates used in heavyweight concrete

	Natural mineral		Synthetic	
Local sand and gravel	Calcareous (2.5 to 2.7)*	Crushed aggregates	Heavy slags (5.0)	
	Siliceous (2.5 to 2.7)		Ferrophosphorus (5.8 to 6.3)	
	Basaltic (2.7 to 3.1)		Ferrosilicon (6.5 to 7.0)	
Hydrous ore	Bauxite (1.8 to 2.3) [15 to 25%]†	Metallic iron products	Sheared bars (7.7 to 7.8)	
	Serpentine (2.4 to 2.6) [10 to 13%]		Steel punchings (7.7 to 7.8)	
	Goethite (3.4 to 3.8) [8 to 12%]		Iron shot (7.5 to 7.6)	
	Limonite (3.4 to 3.8) [8 to 12%]			
Heavy ore	Barite (4.0 to 4.4)	Boron additives	Boron frit (2.4 to 2.6)	
	Magnetite (4.2 to 4.8)		Ferroboron (5.0)	
	Ilmenite (4.2 to 4.8)		Borated diatomaceous earth (1.0)	
	Hematite (4.2 to 4.8)		Boron carbide (2.5 to 2.6)	
Boron additives	Calcium borates			
	Borocalcite (2.3 to 2.4)			
	Colemanite (2.3 to 2.4)			
	Gerstley borate (2.0)			

*Specific gravity shown in parentheses.
†Water of hydration indicated by brackets.

aggregates should be reasonably free of deleterious material, oil, and foreign coatings that affect either the bonding of the paste to the aggregate particle or cement hydration. They should also be nonreactive with alkalies. For good workability, maximum density, and economy, use aggregates roughly cubical in shape and free of flat or elongated particles. Natural aggregates and ferrophosphorus aggregate should be well-graded and within the limits shown in ACI 304.3R for coarse and fine aggregates.

16.4.2 *Mixture proportions and control*—Concrete with normal placing characteristics can be proportioned for densities as high as 350 lb/ft^3 (5610 kg/m^3). Although heavyweight aggregates have unique characteristics, they can be processed to meet standard requirements for grading, soundness, and cleanliness. The acceptability of a particular aggregate depends on its intended use.

Procedures for selecting materials and proportioning heavyweight concrete are similar to those for normalweight concrete, with the following additional considerations:

- In selecting an aggregate for specified density, the specific gravity of the fine aggregate should be comparable with that of the coarse aggregate to reduce settlement of the coarse aggregate through the mortar matrix;
- Ferrophosphorus and ferrosilicon (heavyweight slags) should be used only after tests of laboratory mixtures have shown the suitability of the materials. Hydrogen evolution in heavyweight concrete containing these aggregates can result in a reaction that produces over 25 times its volume of hydrogen before the reaction ceases. Antifoaming agents can help reduce entrapped air; and

- For radiation shielding, it should be determined which trace elements in the material may become reactive when subjected to radiation.

16.4.3 *Batching and mixing*—The techniques and equipment for producing heavyweight concrete are the same as those for normalweight concrete. Testing and quality-control measures, especially density determinations, however, assume even greater importance to ensure that weight specifications are met. Special attention should be given to the following:
- Measures taken to prevent contamination of heavyweight aggregate with normalweight aggregate in stockpiles and conveying equipment;
- Purging of all aggregate handling and batching equipment, premixers, and truck mixers before batching and mixing of heavyweight concrete;
- The accuracy and condition of conveying and scale equipment and concrete batching bins. Because of the greater density of the aggregate, the permissible volume batched in a bin is considerably less than the design capacity. For example, a 100 ton (90 tonne) aggregate bin designed for 75 yd^3 (55 m^3) of normalweight aggregate should be loaded with only 20 to 55 yd^3 (15 to 40 m^3) of heavyweight aggregate;
- The condition and loading of mixing equipment. Concretes with a density of 4800 to 9500 lb/yd^3 (2850 to 5640 kg/m^3) reduce the volume capacity of a truck mixer by 20 to 60%. Because of the high loads placed on mixing equipment, stopping and starting during loading of mixers should be avoided;
- Accurate weighing of aggregate to maintain the specified *w/cm*;
- Avoidance of overmixing. Some heavyweight aggregates are subject to breakage; and
- Frequent checking of fresh density.

For additional details on the proportioning, batching, and mixing of high-density concrete, refer to ACI 211.1.

16.4.4 *Placing, consolidation, and finishing*—Methods used for normalweight concrete generally apply, but care should be taken not to overload conveying equipment and to use forms designed to withstand the high density of the concrete. High-density concretes can be pumped, but the maximum feasible height or distance is less than for normalweight concrete.

To avoid segregation of dense coarse aggregates, concrete slumps should be kept low and overvibration avoided. Coarse aggregate is sometimes preplaced (refer to Chapter 15 for a discussion of preplaced-aggregate concrete).

Puddling is another way to obtain a uniform coarse-aggregate distribution. This involves placing mortar in layers of controlled thickness, and then placing a measured quantity of coarse aggregate over the mortar and vibrating it in.

16.4.5 *Curing and protection*—Requirements for curing and protection are the same as for normalweight concrete.

16.5—Mass concrete for dams

Mass concrete is primarily used in large structures, particularly dams, and in deep, thick structural members, such as heavy mats and walls for nuclear containment or heavy, long-span spandrel beams. Because the concretes for these two types of structures differ widely, they are discussed in this chapter separately.

16.5.1 *Mixture proportioning*—To save cement and reduce temperature rise, concrete in dams has a low cement content (often a pozzolan replaces part of the portland cement), large maximum-size aggregate (generally 3 to 6 in. [75 to 150 mm]), and a low percentage of fine aggregate. *Concrete Manual* (U. S. Bureau of Reclamation 1988) and ACI 207.1R and 304R provide detailed information on concrete for dams.

When proportioning mass-concrete mixtures, materials should be selected that will provide economy, low temperature rise, and adequate workability, strength, durability, and permeability. To achieve this objective, Types II or IV portland cement should be used along with pozzolans or portland-pozzolan cements. Water-reducing and air-entraining admixtures are primarily used to reduce cement content.

In gravity dams, stresses are low and develop slowly, permitting the use of low-strength concrete with a low cement content. Specifications often call for the design compressive strength to be achieved at 90 days or 1 year. Stresses also develop slowly in arch dams, but these structures may require somewhat higher strengths at those ages.

A wide range of coarse-aggregate grading (maximum size of 6 in. [150 mm]) can be used. Grading limits are usually approximately 20 to 35% retained on each of the 3, 1-1/2, 3/4, and 3/16 in. (75, 37.5, 19.0, and 4.75 mm) screens. Sand-grading limits are somewhat more restrictive than those provided by ASTM C33.

The recommendations of ACI 211.1 should be followed to select trial mixture proportions for mass concrete for dams. Slumps are usually limited to 1-1/2 to 2 in. (37.5 to 50 mm) to prevent segregation. The ratio of fine aggregate to total aggregate by absolute volume may be as low as 21% with natural aggregates and 25 to 27% with crushed aggregates.

16.5.2 *Testing*—Design compressive strength should be confirmed by testing cylinders of suitable diameter. Strength specimens should contain the full-sized aggregate, with the diameter not less than three times the maximum aggregate size. These specimens should be tested at the specified design-strength age.

Cylinders for job control should be tested at earlier ages than acceptance cylinders if they are to be useful in monitoring concrete uniformity. All aggregate larger than 1-1/2 in. (37.5 mm) should be wet-screened or handpicked from the fresh concrete sample before molding 6 x 12 in. (150 x 300 mm) cylinders. Slumps and air contents should be measured on similarly wet-screened or handpicked concrete specimens.

Because cylinders used for job control vary in diameter from those used for confirming design strength, correlation tests of the relative strengths of these specimens should be made in the laboratory well ahead of construction (U. S. Bureau of Reclamation 1988; ACI 207.1R).

16.5.3 *Temperature control*—Temperature control of mass concrete is essential for two reasons: to reduce cracking from large differential thermal strains caused by differences between high internal temperatures and lower ambient temperatures, and to reduce strength loss at later ages.

Mass concrete is usually placed at temperatures of 60 °F (16 °C) or less. Using ice to replace part of the mixing water and shading of damp (not wet) aggregate can reduce concrete temperatures to near 50 °F (10 °C) in all but the hottest weather. Coarse aggregate can be cooled by passing frigid air or nitrogen gas through the bins or by passing the coarse aggregate through ice water. Aggregate can also be cooled by a vacuum process, but this method can cause slump-control problems if the moisture content of the aggregate is not uniform. For other recommendations on cooling concrete, refer to ACI 305R.

Temperature rise can be minimized after concrete placement by using steel forms for quick heat transfer, spraying them with cold water if necessary, and by water curing of horizontal construction joints with controlled evaporative spraying. If these cooling measures are ineffective, embedded pipes circulating refrigerated water can be used. Pipe cooling may also be required to ensure that contraction joints remain open if these joints should be grouted. For other recommendations, refer to ACI 207.1R.

16.5.4 *Special equipment and procedures*—Because of the unusually low amount of mortar in mass concrete for dams, workability is more sensitive to variations in batching. Fortunately, it is usually economically feasible on large mass-concrete jobs to specify the most effective methods and equipment for batching, including:
- Finish screening of coarse aggregate;
- Automatic weighing and cutoff of ingredients;
- Interlocks to prevent recharging when material remains in the hopper;
- A device for instant reading of sand moisture content;
- Recording of weighing and mixing operations; and
- Bins and dispensers for pozzolans, ice, and admixtures (air-entraining, water-reducing, and set-controlling).

Mixers for mass concrete for dams are typically stationary, central-plant units with tilting discharge and capacities of at least 4 yd^3 (3 m^3). Specifications for mixing time relate to mixer capacity—ranging from a minimum of 1 minute for the first cubic yard (0.75 m^3) plus 15 seconds for each additional yard (0.75 m^3) of mixer capacity to 1-1/2 minutes for the first 2 yd^3 (1.5 m^3) plus 30 seconds for each additional yard (0.75 m^3). Specifications usually require mixer performance tests based on criteria given in ASTM C94/C94M, Table 1. During mixing of the concrete, the inspector and mixer operator should be stationed where they can see the batch in the mixer because this is the last opportunity to make adjustments, if necessary, to obtain the desired consistency and slump.

Before placing of mass concrete for dams, it is important for construction crews to clean horizontal construction joints, preferably by wet sandblasting or high-pressure water jet. Surface retarders are not effective because protruding coarse aggregate at horizontal joints in mass concrete is unnecessary. Specifications may call for brooming of a thin layer of sand-cement mortar on horizontal construction joints, but the value of this practice is questionable (ACI 207.1R).

Fig. 16.3—Thick mat foundation is an example of structural mass concrete. In hot weather, it is important to keep mixture ingredients cool to prevent cracking caused by high internal temperatures.

Before fresh concrete is placed, joint surfaces should be damp or approaching dryness, with no surface moisture.

Mass concrete with a maximum aggregate size of 6 in. (150 mm) is best placed in layers no deeper than 18 to 20 in. (450 to 500 mm), because shallower layers allow better consolidation. The layers should be carried forward and added in the block by successive rows of bucket dumps, so there will be a setback of approximately 5 ft (1.5 m) between successive layers.

Adequate vibration is key to the successful placement of lean, low-slump mass concrete. Ensuring full vibration at the interface between two placements requires particular attention. Vibrators operated in a vertical position should penetrate several inches into lower layers, and vibration should continue until large air bubbles have ceased to escape from the concrete. Overvibration of low-slump mass concrete is unlikely. Refer to ACI 309R for detailed recommendations for vibration of mass concrete.

To permit more rapid construction of dams, roller-compacted concrete (RCC) is sometimes used. RCC is a zero-slump concrete that is transported, placed, and compacted using earth and rockfill construction equipment. Instead of vertical construction with virtually monolithic blocks, the placement of RCC involves thin lifts placed over a large area. ACI 207.5R contains information on the use, design, and properties of RCC for mass-concrete placements.

16.6—Structural mass concrete

Structural mass concrete, such as for heavy mat foundations (Fig. 16.3) for tall buildings or power plants, large diameter

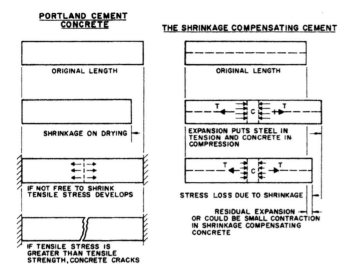

Fig. 16.4—Tensile stresses resulting from shrinkage can cause cracks in portland-cement concrete (left). Initial expansion compensates for drying and hydration shrinkage in shrinkage-compensating concrete (right).

piers or columns, nuclear containment walls, and long-span spandrel beams, is typically at least 3 ft (1 m) thick. ACI has not prepared guides or recommended practices for structural mass concrete, but inspectors can find useful information in ACI 349 and 359.

Structural mass concrete differs from that in dams in several respects:
- Specifications generally require higher compressive strengths, ranging from 3000 to 5000 psi (21 to 35 MPa) or greater. Strength is often to be achieved at 28 days, but ages up to 90 days are sometimes used;
- The maximum aggregate size usually is 1-1/2 in. (37.5 mm) or less; and
- Slumps generally are 3 to 4 in. (75 to 100 mm), but have been placed at higher slumps with the use of high-range water-reducing admixtures. Air contents, when specified, are 4 to 6%.

Limiting temperature rise is equally important in structural mass-concrete elements as it is in mass concrete for dams. The methods used to cool concrete ingredients in dams, however, are often too costly for structural mass concrete. Maximum placing temperatures of 70 °F (21 °C) are usually specified, but lower values are preferable when practical. In hot weather, chipped ice can be used to replace mixing water.

Trial mixtures of structural mass concrete should be proportioned in accordance with ACI 211.1. For nuclear construction, all concrete ingredients should be tested with particular care before use in the structure. Aggregates should be sampled and tested for grading and examined petrographically for quality and to identify deleterious materials. The chert content of fine aggregates should be minimized. Air-entraining admixtures, water-reducing admixtures, and fly ash should be tested for compliance with ASTM C260, C494/C494M, and C618, respectively.

Structural mass concrete can be placed by conventional methods, including pumping, and consolidated like other structural concrete. As with mass concrete for dams, construction crews should avoid cold joints and thoroughly clean planned construction joints before resuming concreting.

16.7—Shrinkage-compensating concrete

Shrinkage-compensating concrete is used in various types of construction to minimize cracking caused by drying shrinkage of hardened concrete. To compensate for this shrinkage, these concretes contain cements that expand after setting (Fig. 16.4).

The characteristics of shrinkage-compensating concrete are similar in most respects to those of other types of portland-cement concrete. Mixture proportions and procedures for placing and curing, however, should allow the expansion for shrinkage compensation to occur in the amount and at the time required. In addition, the reinforcement should provide the proper restraint to expansion. ACI 223 presents detailed information on shrinkage-compensating concrete.

16.7.1 *Materials*—ASTM C845 defines three types of expansive cement: Types K, M, and S. Although each type is produced by a different process, all form the same compound, ettringite, at levels typically in excess of that normally formed in other hydraulic cements. The formation of this compound is the source of the expansive force in the hardening concrete. Approximately 90% of expansive cement consists of the constituents of conventional portland cement, with added sources of aluminate and calcium sulfate. Thus, the oxide analysis as shown on mill test reports will not differ substantially from ASTM C150 portland cements.

Two basic factors are essential to the development of expansion: an appropriate amount of soluble sulfates and the availability of sufficient water for hydration during curing. Ettringite begins to form almost immediately when water is introduced, and its formation is accelerated by mixing. Cement ground more finely than normal also accelerates formation of ettringite. For ettringite to be effective, however, most of it should form after the concrete attains some strength; otherwise, the expansive force will dissipate as it deforms the plastic concrete. Extended mixing is detrimental to expansion, so if long delivery times of ready-mixed shrinkage-compensating concrete are expected, this should be considered in proportioning and trial mixing. SCMs with expansive cements should not be used without performing mixture-proportioning studies. Some SCMs can reduce the benefits of expansive cements.

Admixtures in shrinkage-compensating concrete should be used with care, making sure they will not be detrimental to expansion. In all cases, test admixtures in trial mixtures with job materials and proportions under simulated job conditions. During warm periods, shrinkage-compensating concrete can experience rapid slump loss, and a mid- or high-range water-reducing admixture may be desirable. Air-entraining admixtures are as effective in increasing the durability of shrinkage-compensating concrete as with portland-cement concrete.

16.7.2 *Mixture proportioning and control*—Concrete made with shrinkage-compensating cement has a high water demand. The best data for cement content and required *w/cm* for a specified compressive strength may be available from the producer of the expansive cement. If past performance data are not available, ACI 223 offers a guide to required *w/cm* for both air-entrained and non-air-entrained shrinkage-compensating concrete (Table 16.2).

When determining the required *w/cm* and corresponding cement content, consider the effect on restrained expansion. Expansion increases at higher cement contents, and decreases at lower contents. A lower limit of 515 lb (305 kg) of cement per cubic yard (meter) of concrete is recommended to achieve the required expansion with minimum reinforcement. A qualified, accredited laboratory should determine the degree of concrete expansion by means of restrained 3 x 3 x 10 in. (75 x 75 x 250 mm) prism specimens (ASTM C878/C878M).

Trial mixtures should be prepared following ACI 211.1 and ACI 211.2 procedures to develop satisfactory aggregate proportions, cement content, and water requirements. These mixtures should be made at the approximate concrete temperatures anticipated in the field. When the delay between mixing and placing of shrinkage-compensating concrete will not exceed 15 minutes, such as when mixing takes places at the job site or precast plant, the total mixing water required for a specified slump will not be much greater than that for a Type I or II portland-cement concrete.

When the water is added at the batch plant and delivery will require 30 to 40 minutes of travel time or the expected concrete temperature will exceed 75 °F (24 °C), slump loss will occur in some shrinkage-compensating concretes. To ensure that the specified slump is obtained at the job site, it is necessary to provide a higher slump at the batch plant while remaining within maximum allowable water limits. The importance of taking this slump loss into account in the mixture proportioning for shrinkage-compensating concrete cannot be over-emphasized. It becomes even more important during hot weather, when concrete temperatures are relatively high and reactions are accelerated. Under such conditions, both Procedures A and B for trial batch tests have been used successfully.

16.7.2.1 *Procedure A*

1. Prepare the batch according to ASTM C192/C192M, but add 10% more water than normally used for Type I cement;

2. Perform initial mixing in accordance with ASTM C192/C192M (3 minutes of mixing followed by 3 minutes of rest and 2 minutes remixing);

3. Determine slump, and record as initial slump;

4. Continue mixing for 15 minutes;

5. Determine slump, and record as placement slump. Experience has shown this slump correlates with that expected for a 30- to 40-minute delivery time. If this slump does not meet the required placement specification limits, discard and repeat the procedure with an appropriate water adjustment; and

6. Cast compressive-strength and expansion specimens and determine plastic properties—density, air content, and temperature.

16.7.2.2 *Procedure B*

1. Prepare the batch according to ASTM C192/C192M for the specified slump;

Table 16.2—Trial mixture guide for shrinkage-compensating concrete (from ACI 223)

Compressive strength at 28 days, psi (MPa)	Absolute *w/cm* by weight	
	Non-air-entrained concrete	Air-entrained concrete
6000 (40)	0.42 to 0.45	—
5000 (35)	0.51 to 0.53	0.42 to 0.44
4000 (28)	0.60 to 0.63	0.50 to 0.53
3000 (21)	0.71 to 0.75	0.62 to 0.65

2. Mix in accordance with ASTM C192/C192M, and confirm slump;

3. Stop the mixer, and cover the batch with wet burlap for 20 minutes;

4. Determine slump;

5. Remix for 2 minutes, adding water to produce the specified placement slump. The total water (initial plus the remixing water) is that required at the batch plant to give the proper job-site slump after a 30- to 40-minute delivery time; and

6. Cast compressive-strength and expansion specimens and determine plastic properties—density, air content, and temperature.

16.7.3 *Production, placing, and finishing*—Control of slump is of prime importance in the production of shrinkage-compensating concrete. For best results, slumps at the time of concrete placement should be within the maximum range specified by ACI 211.1 for the work involved when concrete temperatures do not exceed 75 °F (24 °C). At higher concrete temperatures, the following maximum slumps are recommended:

Type of construction	Slump, in. (mm)
Reinforced foundation walls and footings	5 (125)
Plain footings, caissons, and substructure walls	4 (100)
Slabs, beams, and reinforced walls	6 (150)
Building columns	6 (150)
Pavements	4 (100)
Heavy mass construction	4 (100)

Under adverse hot-weather conditions, ready mixed portland-cement concrete experiences significant slump loss during normal delivery times. Under the same conditions, some shrinkage-compensating concretes experience even greater slump loss. Slump-loss controls effective for portland-cement concrete in hot weather are generally effective for shrinkage-compensating concrete. These control measures should be strictly enforced when expansive cements are used because of the greater possible slump loss. Recommended controls include cooling concrete, reducing the speed of the truck mixer drum to a minimum during travel and waiting time, and efficient truck scheduling to minimize the time period between mixing and delivery. When job locations require extended travel times, dry-batched truck delivery with job-site mixing is effective. Refer to ACI 305R for a more complete discussion of hot weather concreting.

No special techniques or equipment are required to place shrinkage-compensating concrete. Placement patterns for slabs differ from those used for portland-cement concrete, however, as shown in Fig. 16.5.

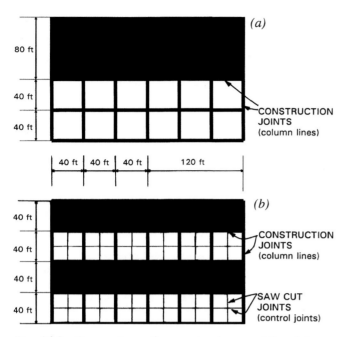

Fig. 16.5—Comparison of concrete placement for slab-on-ground using: (a) shrinkage-compensating concrete; and (b) portland-cement concrete. The shrinkage-compensating concrete can be cast in larger sections without saw-cuts and open space around the columns.

To obtain the full benefits of expansion, construction crews should take the following precautions:
- Soak the base material;
- Take measures to avoid plastic shrinkage cracking and uneven moisture loss;
- Set reinforcement in the proper position to provide the required restraint;
- Avoid placing delays when using truck mixers; and
- Limit the temperature of fresh shrinkage-compensating concrete to 90 °F (32 °C). At concrete temperatures of 85 to 90 °F (29 to 32 °C), limit the time between mixing and placing and finishing to 1 hour.

Finishing qualities of shrinkage-compensating concrete are similar to those of air-entrained concrete. In general, finishing should start earlier than with comparable portland-cement concrete because of faster setting.

16.7.4 *Curing and protection*—Shrinkage-compensating concrete, like portland-cement concrete, requires continuous curing at moderate temperatures for several days after final finishing to prevent early drying shrinkage and to develop strength, durability, and other desired properties. Inadequate curing can also reduce the amount of initial expansion, which is needed to offset later drying shrinkage. Such drying shrinkage will occur when wet curing stops. To ensure enough water for ettringite formation and expansion, curing methods should be used that add moisture to the concrete, such as ponding, continuous sprinkling, or wet coverings. For best results, water curing of shrinkage-compensating concrete should be continued for at least 7 days.

During the initial curing period, shrinkage-compensating concrete should be protected from extreme heat or cold. The former can cause thermal and plastic shrinkage cracking; the latter can reduce expansion. ACI 305R and 306R describe recommended methods.

16.8—High-performance concrete

High-performance concrete (HPC) is a general term for concrete that meets specific design concerns, such as ease of placement and compaction without segregation, enhanced long-term mechanical properties, high early strengths, extraordinary toughness, volume stability, and long life in severe environments. The term implies a concrete with properties and uniformity that cannot be routinely obtained using normal mixing, placing, and curing practices.

Because HPC is more dependent than normal concrete on material selection, proportioning, mixing, placing, and curing, careful inspection is required. Inspectors should thoroughly understand the specific concrete performance requirements and the roles that they and others are to perform in the quality-assurance process.

Concrete mixtures designed as HPC usually have higher compressive strengths. Because high-strength concrete requires a higher degree of quality control, the inspector and testing technicians need to pay special attention to quality issues. Guidance on performing quality control and testing of high-strength concrete is provided in ACI 363.2R.

CHAPTER 17—PRECAST AND PRESTRESSED CONCRETE

Precast concrete refers to members cast at some location other than their final position in the structure, usually at a precast plant. When the members have cured to sufficient strength for handling, they are removed from the forms and installed in the structure (Fig. 17.1).

Precast concrete can be structural or architectural. Structural precast concrete includes beams, girders, joists, purlins, lintels, columns, posts, piers, piles and pile caps, slab and deck members, and load-bearing wall panels. This concrete may be conventionally reinforced or prestressed (Fig. 17.2). Architectural precast concrete includes elements with textures or decorative features, as specified by the designer. Typical architectural precast concrete includes wall panels, window-wall panels, mullions, and column covers. This concrete may be unreinforced, conventionally reinforced, or prestressed.

Prestressed concrete contains tensile reinforcement— either pretensioning, post-tensioning, or a combination of the two—to induce internal stresses that counteract tensile stresses induced by imposed loads. Prestressed concrete can be precast or cast in place. When precast, tension is applied either by pretensioning or post-tensioning. When cast in place in the structure, the tendons are stressed by post-tensioning.

In pretensioning, the reinforcing tendons are installed and stressed as specified, and then the concrete is placed and consolidated around them to ensure adequate bond to the tendons. After the concrete develops the necessary minimum strength, releasing the tensile anchors of the tendons produces the required compression in the concrete, with the bond between the steel and concrete providing the initial tension in the steel.

Post-tensioning requires tensioning the tendons after the concrete has hardened. Concrete is placed around voids or ducts through which steel tendons are inserted, either before or after concrete placement. After the concrete develops the necessary minimum strength, the tendons are stretched to the required tension and anchored to the concrete at the ends to retain tension in the steel and thus develop compression in the concrete. The tendons can remain unbonded, with grease or wax inserted between tendons and ducts, or they can be bonded by grout inserted into the ducts. Some tendons are encapsulated in heat-sealed plastic sheathing over part or all of their length to prevent bonding where desired.

To ensure satisfactory production of these types of concrete, inspectors should study the requirements of the contract documents that will serve as the inspector's guide.

Fig. 17.1—Assembly of hollow precast members for a prestressed bridge superstructure. Keys cast in box-girder walls align each member with the previously installed member.

17.1—Precast concrete

Precast reinforced concrete has much in common with cast-in-place concrete. Thus, much of the information outlined in other chapters of this manual applies to precast construction, including procedures for forming, setting of reinforcement, special finishes, curing, and stripping of forms. Some procedures for cast-in-place work can be used as standard practice for fabricating precast components. Consequently, inspections of precast concrete fabrication and erection (Precast/Prestressed Concrete Institute 1999a) should follow recommendations developed specifically for the type of work being evaluated. In all cases, however, the specifications and dimensional tolerances given in the contract documents should be adhered to. If the documents do not specify dimensional tolerances, MNL 116 (Precast/Prestressed Concrete Institute 1999b) should be used as a guide to acceptable and reasonable practices.

17.1.1 *Scope of inspection*—Inspection duties to be performed in precasting plants include:
- Identifying, examining, and accepting materials;
- Inspecting and recording of tensioning (if the product is prestressed);
- Inspecting beds and forms prior to concreting;
- Checking dimensions of members; positions of inserts and tendons; reinforcing steel, type, grade, size, and placement; openings; and blockouts;
- Inspecting batching, mixing, conveying, placing, consolidation, finishing, and curing of concrete;
- Preparing or witnessing the preparation of concrete specimens for testing; and
- Performing or witnessing specified tests for slump, air content, and concrete strength.

The number of people needed to perform these inspection services varies with the size of the plant; the size, volume, and type of work being produced; and the inspectors' abilities. For small plants, one or two inspectors may be sufficient to perform all of these services. Larger plants may need to assign specific duties to several inspectors to ensure that all necessary inspections can be performed satisfactorily without limiting production volume. Clearly defined functions should be assigned to each inspector, and inspectors

Fig. 17.2—Installation of precast hollowcore roof decking.

should be given authority to require a uniform standard of quality in all phases of production.

17.1.2 *Inspecting for quality control and quality assurance*—All precasting plants should be PCI certified and have ACI or PCI Level I and II certified plant inspection personnel who regularly monitor all aspects of production to ensure compliance with prescribed methods. Inspectors should be responsible for quality only, not production techniques. They should never share proprietary methods of fabrication with other plants.

Inspection during production in the precasting plant may consist of both quality-control inspections conducted by the precast fabricator and quality-acceptance inspections performed by an independent agency on behalf of the owner. Plant certification by PCI requires plant management to maintain a quality-control group independent of the production staff. Inspectors employed by the plant should be responsible to the chief engineer or management rather than to the production superintendent. Not all precast plants, however, will have their own in-house quality-control staff and may rely instead on independent consultants. Some projects may require separate inspections of the work independent from the

> **REFERENCES FROM THE PRECAST/ PRESTRESSED CONCRETE INSTITUTE**
>
> Additional guidance is provided in the following manuals published by the Precast/Prestressed Concrete Institute (PCI):
> - *Architectural Precast Concrete* (MNL 122) (1989);
> - *Manual for Quality Control for Plants and Production of Structural Precast Concrete Products* (MNL 116) (1999b);
> - *Manual for Quality Control for Plants and Production of Architectural Precast Concrete Products* (MNL 117) (PCI 1996); and
> - *Erectors' Manual* (MNL 127) (PCI 1999a).
>
> These references detail quality-control requirements for precast and prestressed concrete components and should be reviewed by inspectors involved with such work. ACI 301, 423.3R, and 523.2R also contain detailed information on precast and prestressed concrete.

plant quality-control staff. In such cases, the outside agency, usually defined as a quality-assurance agent for the owner, should work closely with the plant staff to minimize the impact of inspection on schedules and production.

Because inspectors are responsible for verifying that precast members are fabricated in accordance with contract documents (and any referenced documents), they should have a complete and up-to-date set of contract documents covering the members being fabricated. Occasional discrepancies may occur between the contract specifications and typical plant practices. Each instance should be reported to the plant quality-control manager and to the engineer-of-record for clarification.

17.1.3 *Record keeping and test reports*—To establish evidence of proper manufacture and quality of precast concrete members, complete records of materials testing, tensioning, concrete proportioning, placing and curing, and condition of members should be kept. Each precast concrete member should be identified by bed and date cast, and assigned an identification number, which is referenced in design calculations, shop drawings, tensioning records, concreting records, cylinder strength tests, and erection plans. Complete and accurate records should also be kept documenting the materials used in fabrication and verification of their compliance with contract specifications.

For materials that are not plant-tested, certified test reports should be obtained from manufacturers. These reports should show that the materials comply with applicable contract document provisions. Examples of materials for which manufacturers should provide test reports include:
- Strand, wire, bars, or other tendon materials;
- Reinforcing steel;
- Cement and SCMs;
- Aggregates;
- Admixtures; and
- Curing materials.

Each report should be identified with reels, packs, heats, bins, cars, or other specific lots.

For general information about maintaining records and test reports, refer to Chapter 20.

17.1.4 *Formwork*—Conditions affecting finished surfaces, dimensional tolerances, and other details can usually be controlled more closely in members cast in the plant rather than in the field. For members with standard geometric shapes (planar wall panels, beams, columns, and piles), forms often are reused extensively. Forms for special or nonstandard members have more limited use. All forms, however, should be constructed to produce members within specified tolerances (Fig. 17.3). When the surfaces of casting beds, edge forms, beam sides, and other formwork can no longer produce the required level of quality, the forms should be removed and replaced.

The following steps should be taken to ensure that forming of precast members produces the quality level required by the contract documents:
- Regularly inspect bulkheads and similar equipment that influence the accuracy of dimensions and alignment;
- See that casting beds and forms are supported on unyielding foundations;
- Check form alignment and grade at each casting to ensure accuracy;
- Make sure that form joints are smooth and sufficiently tight to prevent leakage of cement paste and that holes and slots in the forms are plugged so the finished member will have an acceptable appearance;
- Inspect beds and forms after each use to ensure they are clean. Do not allow form-release coatings to build up on the formwork or dust to accumulate on the coating before concrete is placed;
- See that provisions are made to securely hold forms for internal voids, such as tendon ducts and hollow cores, in place during concrete placement and consolidation to ensure correct positioning per the specifications. The forms should also be strong enough to remain stable during concrete placing and handling. Improper location or shape of voids can change the structural properties of the member and result in reduced strength or improper performance; and
- Make sure removable templates remain in place long enough after concrete placement so that removing them will not displace the void. Holes left by template removal can be filled and the concrete compacted into the surrounding mass.

17.1.5 *Embedded items*—Tolerances for embedded items depend on their intended use. Close tolerances often are required for anchor bolts and bearing plates. Bearing shoes, in particular, should be level, aligned properly, and anchored in the exact location shown on the contract documents. Make sure all embedded items are firmly positioned so they will not become displaced during concrete placement. Plant personnel should insert anchor bolts, studded plates, or other embeds into the plastic concrete only using techniques expressly permitted by the contract documents. Built-in fixtures should not affect the position of principal reinforcement or the placement of concrete. Embeds for final erection or attachment to the structure may require welding inspection during their manufacture or some other certification of quality, as with other materials used in precast members.

Precast plants should avoid the use of wood inserts, which tend to become displaced and can swell, causing the concrete to crack. Plants also should not use aluminum in reinforced or prestressed members because of possible galvanic corrosion and embrittlement of steel caused by the generation of free hydrogen.

17.1.6 *Bar and wire reinforcement*—Bar and wire reinforcement should be fabricated as shown in the contract documents and positioned in the member within specified tolerances. Installers should adequately secure reinforcement to beds and forms by using chairs or blocking or by tying to tendons so that the steel will stay in position during concrete casting and vibration. Ends of tie wires used to fasten bars should be bent into the member to provide maximum concrete cover, which will help prevent corrosion of the wires and rust staining of the concrete. Bars also can be fabricated into cages by tying, or as specified in contract documents.

Particular attention should be paid to proper placement of bars extending out of the member that are intended to provide structural connections to cast-in-place concrete. Improper extension can result in a weak joint. The extensions should normally be within ±1/2 in. (13 mm) of plan dimensions. Workers should remove any cement paste adhering to the extended steel. Prestressed strands and anchorages should be examined for any sign of damage. It should be verified that damage strands and related hardware are replaced or repaired before concreting.

17.1.7 *Curing*—Common methods of curing precast members include steam and radiant heat. Steam provides accelerated curing of precast concrete to attain high early strength and, thus, faster production cycles. With any type of accelerated curing, however, achieving high levels of early strength compromises strength at later ages. Thus, specifications usually establish a preset period before temperature can be increased. Refer to Chapter 5 for a discussion of the effects of higher early temperatures on the properties of concrete.

The enclosure for steam curing should be designed to retain the steam to minimize moisture and heat losses and permit free circulation of the steam around the top and sides of the member. Steam jets should not be allowed to impinge directly on the concrete. Recording thermometers showing the time-temperature relationship should be installed no more than 200 ft (60 m) apart.

Radiant heat is applied to casting beds by pipes circulating steam, hot oil, or hot water, or to forms by electric blankets or heating elements. Pipes, blankets, or heating elements should not be in direct contact with fresh concrete.

During radiant-heat curing, no part of the member should experience moisture loss. Covering the member with a plastic sheet and an insulating cover can effectively retain moisture. Moisture can also be applied by moist burlap or cotton matting, or by flooding the exposed surface. Covering materials should be selected that will not stain the concrete.

Temperature limits and the use of recording thermometers are similar to those recommended for steam curing. Application of the heat cycle may be accelerated to offset climatic conditions and to obtain the desired concrete temperatures. The curing procedure, however, should always be carefully controlled.

17.1.8 *Handling, storage, and transportation*—The location of pickup points for handling of members and details of pickup devices are important parts of the design of precast

Fig. 17.3—Form for precast girder.

Fig. 17.4—Cranes lift a precast prestressed girder onto a flatbed.

concrete members, and should be in accordance with shop drawings. Misplaced or incorrect lifting inserts can result in cracked or damaged members, excessive deflections, and rust stains. In extreme cases, they can cause catastrophic structural failure. Usually, the precast plant's engineer determines the location of pickup points for handling of precast members and details of pickup devices. On occasion, the engineer will specify requirements for lifting-insert materials, coverage, and location. To avoid subjecting members to excessive dynamic loads or impact during lifting, they should be handled only by approved pickup devices at designated locations (Fig. 17.4).

Precast members should be stored on a prepared site with suitable foundations to prevent differential settlement or twisting of members (Fig. 17.5). This is especially important for members stored by stacking. To separate and support stacked members, plant personnel should place battens across the full width of each bearing point (Fig. 17.6). These battens should be arranged in vertical planes at a distance no greater than the depth of the member from designated pickup points. When used with stemmed members or planks, battens should not be continuous over more than one stack of precast units. It is important to use batten materials that will not stain the precast members, such as wood slats or precast concrete battens.

Fig. 17.5—Storing precast elements on stable foundations prevents differential settlement and twisting of members.

Fig. 17.6—Stacking hollow precast members on flatbed for transportation to job site.

All precast members to be transported by truck, railroad car, or barge should be supported as described previously, except that battens can be continuous over more than one stack of units. The stacks should also be braced to keep them vertical, with adequate padding materials between the concrete members and tie chains or cables to prevent abrading or chipping of the concrete (Fig. 17.7).

Any sign of lateral deflection or vibration of long, slender members should be watched for during transport. Use of lateral trusses or rigid bracing between members can usually correct this problem. In extreme cases, strong backs, stiffening trusses, or frames may be required.

17.1.9 *Erection*—At the erection site, the same storage and handling provisions should be followed as described in the previous paragraphs. Adequate bearings are particularly important. If the designed bearing surfaces do not make full contact with bearing supports, the structure can develop serious structural weakness or distortion.

It is essential for precast members to be connected in strict accordance with the contract documents. Welded connections to provide continuity should be made only by certified welders using equipment, procedures, and electrodes applicable to the base metals. Misplacement of matching embeds in the supporting structure or inconsistent dimensional tolerances

(a)

(b)

Fig. 17.7—Transporting precast concrete elements by truck mailer.

occasionally prevent the use of specified connection details. The engineer-of-record should approve alternate details before incorporation into the work. Welded connections should not be substituted without the engineer's approval.

The number of connections to be inspected in the field should be confirmed before erection begins. When precast components do not offer a redundancy of connections, each connection should be carefully inspected to verify compliance with contract documents. The failure of one connection can cause substantial movement, or possibly even structural collapse. Precast components that offer redundant connections might require a less-intensive inspection program. The engineer should determine the necessary inspection requirements.

Unless otherwise specified, bolted connections for ASTM A325 or A490 high-strength bolts should conform to the AISC "Code of Standard Practice for Steel Buildings and Bridges" and the AISC "Specification for Structural Joints Using ASTM A325 or A490 Bolts," which are both included in the AISC *Steel Construction Manual* (American Institute of Steel Construction 2006). Inspection of welded connections should be done by a certified welding inspector and should conform to criteria specified in the AWS D1.1.

17.1.10 *Repairs*—Precast elements often suffer chips, spalls, cracks, or stains as a result of lifting, storing, transporting, or erection. All completed members should be inspected at appropriate times during the construction schedule, and all distress noted. Deficiencies should be classified in the work as structural or cosmetic so they can be repaired accordingly. The precast plant will generally have established procedures for both types of repairs. The plant engineer should evaluate and approve these repair methods before the work begins. All completed repairs should be reviewed as part of the inspection program.

17.2—Precast prestressed concrete

Quality-control considerations for the fabrication of precast prestressed concrete are similar to those for nonprestressed concrete. Ensuring product integrity, however, requires additional attention to proper placement and tensioning of tendons. Incorrect tendon placement and tensioning can result in excessive or inadequate camber and cracking. It can also pose a serious safety hazard to plant personnel (refer to the rules for safe post-tensioning at the end of this chapter).

On the other hand, a properly designed and fabricated precast prestressed concrete element is resistant to cracking and, therefore, can be much easier to handle and transport than nonprestressed components.

17.2.1 *Concrete materials*—Concrete in prestressed members is usually higher in strength and lower in slump than concrete in conventionally reinforced precast members. Specifications often call for compressive strengths ranging from 4000 to 12,000 psi (28 to 85 MPa), and it is critical to achieve the required strengths within the specified time frame. Obtaining high-strength, low-slump concrete that is still workable often requires using a high-range water-reducing admixture in the concrete mixture. The specific contract requirements for concrete used in precast prestressed members should be identified, and it should be verified through conventional quality-control testing that those requirements are met.

High-strength prestressing steels under tension are particularly sensitive to corrosion, especially when exposed to chloride ions. Therefore, no admixtures or materials containing chloride ions should be used.

17.2.2 *Tendons for pretensioning*—Tendons used in precast members with bonded pretensioning are stress-relieved wires or strands. Almost all are seven-wire strands meeting ASTM A416/A416M requirements. Some small precast prestressed members use small-diameter two- and three-wire stress-relieved strands.

To compute elongations resulting from tensioning, it is necessary to know the stress-strain properties of all tendons. The stress-strain relationship (modulus of elasticity) varies among the several types of tendons and between tendons of the same type from different mills. Accompanying each shipment of tendons from the same manufacturer should be a certificate showing that the tendons have been manufactured and tested in accordance with applicable ASTM specifications. In the absence of an ASTM specification, or when requested by the prestressed concrete manufacturer, the strand manufacturer should furnish a certified test report for every 20 tons (18 tonnes) or part thereof to show compliance with applicable specifications. If the strand manufacturer furnishes a typical stress-strain curve in place of a specific stress-strain curve, the manufacturer should certify that it is representative of the material shipped.

In addition to verifying proper certification, the condition of the prestressing steel should be inspected. It should be free of deleterious materials that could prevent bond of the steel to the concrete, such as grease, oil, wax, dirt, paint, and loose rust.

17.2.3 *Tendon handling and storage*—Although stress-relieved strands are made of extremely high-strength wire, they are more susceptible to damage than ordinary reinforcing bars or structural steel. Thus, handling and storage of these materials require special care.

Prestressing steel is brittle in nature, making it highly susceptible to failure at nicks, notches, kinks, and other similar local damage. Even small nicks in a strand (sometimes called stress raisers) will increase local stress and thus reduce the ultimate strength of the wire. This makes the wire especially prone to failure under fatigue loading.

Cold-working of the prestressing strand after the final heat treatment produces more than half of the strand's ultimate strength. This process changes the internal structure of the metal from granular to fibrous. Excessive heat can instantly change the fibrous structure back to the low-strength, granular structure. Therefore, it is important for welders to avoid working near prestressing wires, strands, or high-strength bars. Heat from a welding torch, arc-welding currents, or hot weld metal can reduce the tendon's strength by more than 50%. The damage may not be visible to the naked eye, and may only become apparent when the tendon fails during tensioning.

Slight rusting of prestressing steel can be desirable because it increases the bond of the tendon to the concrete. As ASTM A416/A416M states: "Slight rusting, provided it is not sufficient to cause pits visible to the naked eye, shall not be cause for rejection." Pits visible to the naked eye act as stress risers, much like the nicks from mechanical damage. Heavy rusting is not acceptable because it reduces the area of steel available to carry load. Any active corrosion should be removed or the affected tendon replaced before installation. Active corrosion can lead to brittle tendon failure.

17.2.4 *Attachments for tendons*—Strand grips for pretensioning should be capable of anchoring the strand positively without slippage after seating, with grip lengths and serration configurations designed to prevent strand failure within the grips at stresses less than 90% of ultimate strength. It should be verified that the proper wedges are placed in their respective anchor bodies. Using jaws for 3/8 in. (No. 9) strands with 1/2 in. (No. 13) anchor casings, for example, can cause premature failure of strands well below ultimate load. The manufacturer should proof-test steel cases for strand vises to at least 90% of the ultimate strength of the strand.

17.2.5 *Deflection devices*—Pins used to deflect or otherwise change the path of a tendon from a straight line should not cause a critical increase in strand stress due to pressure. The following are factors that affect the amount of pressure the deflecting pin places on individual wires:
- The larger the pin diameter, the lower the pressure on the strand;
- The smaller the angle through which the strand is bent, the lower the pressure;
- If the pin has a semicircular groove just slightly larger than the strand, several wires will bear on the surfaces of the groove at one time, greatly reducing the pressure on individual wires; and
- If several strands in one vertical row are deflected at one point so that each strand bears on the strand above it and the top strand bears on a pin, the pressure on the top strand will be high.

Successful use of deflection devices depends on the pin detail, the number of strands, and the angle of the bend. The design engineer should provide shop drawings showing pin requirements.

17.2.6 *Tensioning of tendons*—Producing acceptable prestressed members requires accurate stressing of tendons. Moderate variations in stress levels of tendons will usually not affect the ultimate capacity of a member, but they can affect camber, cracking load, and other properties.

In all methods of tensioning, the stress induced in the tendons is directly determined by observing the application of the required force using a calibrated pressure gauge/hydraulic jack system, dynamometer, or load cell and confirmed by measurement of tendon elongation. The gauging system indicates that the proper force has been applied, and the measure of elongation provides a check on the gauging system. The inspector should verify proper calibration of the gauging system employed. If discrepancies in forces determined by elongation measurements or by gauge readings exceed 5% of engineering established values, the entire operation should be checked for possible errors before proceeding further. The inspector should record the force and elongation measured for each strand and submit results to the engineer for review and approval. MNL 116 (Precast/Prestressed Concrete Institute 1999b) gives methods of measuring stress in prestressing strand, descriptions of gauging systems, and information on controlling jacking forces.

17.2.7 *Wire failure in tendons*—Failure of some wires in a pretensioning strand or a post-tensioning tendon is usually acceptable, provided that the total area of wire failure is not more than 2% of the total area of tendons in any member and the engineer determines that the failure is not symptomatic of more extensive distress. Depending on the type of structure involved, however, these acceptance criteria may not be appropriate. The engineer should always be consulted to determine acceptability when strand wires are found damaged or broken.

17.2.8 *Detensioning*—Stress transfer of pretensioned members should not be performed until concrete strength, as indicated by test cylinders, has reached the specified transfer strength. For certain dry-mix, machine-cast products, cylinders cannot be made that are representative of the units in the bed. In these instances, concrete strength should be verified by approved test methods recommended by manufacturers of machine-cast products.

For heat-cured concrete, detensioning should be performed following the curing period, while the concrete is still warm and moist. Allowing the concrete to dry and cool before detensioning can result in dimensional changes that lead to cracking or undesirable stresses in the concrete. This is especially true if hold-downs are used to deflect the strands.

In all detensioning operations, prestressing forces should be kept nearly symmetrical about the vertical axis of the member and applied in a manner that will minimize sudden loading. Maximum eccentricity about the vertical axis should be limited to one strand. For unusual and asymmetrical shapes, shop drawings should show detensioning procedures.

Detensioning should be performed in a manner and sequence that will minimize longitudinal movement. It is still necessary, however, to verify that forms, ties, inserts, hold-downs, blockouts, or other devices that could restrict longitudinal movement of the members along the bed are removed or loosened.

17.2.8.1 *Draped strands*—For prestressed members having draped strands, any longitudinal movement along the beds before removal of hold-down devices can cause serious cracking of the concrete, destruction of the hold-down devices, or both. Releasing hold-downs and removing bolts before releasing the stresses at the anchorages can prevent such movement. Release of hold-downs without release of anchorage stress, however, can result in dangerous concentrated vertical loads that can crack the top of the member. Hold-down forces should always be computed and compared with the weight of the member if hold-downs are to be released before release of anchorage stress. The plant engineer will usually specify the sequence for releasing hold-down devices and anchorages.

17.2.8.2 *Multiple strands*—In multiple-strand detensioning, hydraulic jacks release the strands simultaneously. The total force is taken from the header by the jack, then gradually released. With this method, some sliding of the members on the beds is inevitable. The amount of sliding is proportional to the exposed lengths of stressed strands between members and between the last member and the fixed end. Holding these lengths to the practical minimum will minimize sliding.

17.2.8.3 *Single strands*—In single-strand detensioning, heat cutting using a low-oxygen flame releases the strands.

To reduce sliding of members, strands are cut simultaneously at both ends of the bed. Strands should not be cut quickly; however, heating of the metal should allow for gradual release of stress. The responsible engineer should approve the sequence used to cut the strands.

17.2.8.4 *Post-tensioned tendons*—In some instances, post-tensioned tendons provide all or part of the prestressing force in a precast prestressed member (Fig. 17.8). The requirements for these tendons, including placing, tensioning, and grouting, are essentially the same as for post-tensioned tendons in cast-in-place concrete, discussed in a following section of this chapter.

17.3—Cast-in-place prestressed concrete

Inspection of cast-in-place prestressed concrete requires paying close attention to the following operations:
- Accurate placement and secure tying of the tendons to be post-tensioned or the tendon ducts;
- Placement of concrete without damage to coverings around the tendons;
- Proper tensioning and anchoring of tendons after the concrete has reached its specified strength; and
- Prompt grouting of bonded tendons soon after stressing.

Information on cast-in-place prestressed concrete requirements and tolerances can be found in ACI 318, 301, and 117.

17.3.1 *Concrete materials*—Requirements for concrete used in cast-in-place prestressed members are similar to those for precast prestressed members. Workability of the mixture, however, is more critical because of the difficulties in field placement of concrete in the congested and confined regions adjacent to and beneath tendon anchorage assemblies. It is important to be familiar with the contract documents and any supplementary requirements for mixture workability, placement requirements in the anchorage region, and safety requirements. Impact on embedded tendons or tendon ducts or contamination with aggressive chemicals should be prevented. High-strength prestressing steels are particularly sensitive to corrosion and embrittlement, especially when exposed to chloride ions and hydrogen. It should be verified that all admixtures and curing agents meet requirements for allowed chloride levels.

17.3.2 *Post-tensioning tendons*—Typical post-tensioning tendons are:
- Seven-wire, stress-relieved strands (ASTM A416/A416M);
- Stress-relieved wires (ASTM A421/A421M); or
- High-strength, specially processed alloy steel bars (ASTM A722/A722M).

Post-tensioning systems should be installed in accordance with manufacturers' directions and proven procedures. Manufacturer recommendations should be observed for end-block details and special reinforcement in anchorage zones.

Handling and storage requirements for post-tensioning tendons to be grouted are the same as those discussed previously for pretensioning tendons. Unbonded (or ungrouted) tendons are factory-coated with a lubricating corrosion inhibitor, such as wax or grease. Sheathing (typically plastic) protects this coating and prevents bond

Fig. 17.8—Hydraulic jacks are used to post-tension tendons of precast beams.

with the concrete to allow post-tensioning of the tendon. If small tears or holes occur in the sheathing during shipment or placement, they should be repaired before concrete is placed. The use of slings or padded carriers to handle tendons can help prevent sheathing damage and tendon kinking. Storage facilities should protect the tendons from exposure to corrosion or damage.

Ducts for grouted post-tensioning tendons typically are rigid, galvanized tubing through which the tendons are inserted and stressed after concrete has hardened. Flexible ducts with preassembled prestressing steel can also be used. All joints in the ducts should be sealed to prevent intrusion of cement paste and premature bonding of tendons.

Tendons or tendon ducts should be placed at the specified locations and profiles. Deviations, such as kinks or wobbles, can place unintended forces on the concrete, resulting in excessive frictional losses in prestressing force and possible concrete spalling. Horizontal alignment should vary no more than 1/2 in. (13 mm) in 10 ft (3 m) from specified tolerances. Locations requiring maximum effective depth should be within 1/8 in. (3 mm) of intended profile for members 8 in. (200 mm) thick or less, and within 1/4 in./ft (6 mm/300 mm) of depth for deeper members.

Supports such as tie bars, chairs, and bolsters can help maintain tendon alignment during concrete placement. These support accessories should be tied securely to the tendon to prevent displacement. Maximum support spacing should not exceed:
- 4 ft (1.2 m) for 0.5 in. (No. 13) strand tendons or multiple-wire tendons;
- 4-1/2 ft (1.4 m) for 0.6 in. (No. 15) strand tendons or flexible ducts for grouted tendons; and
- 6 ft (1.5 m) for semi-rigid ducts 2-1/2 in. (65 mm) or greater in diameter.

For grouting, the inside diameter of ducts should be at least 1/4 in. (6 mm) larger than the nominal diameter of single wire, bar, or strand tendons; for multiple wire or strand tendons, the inside cross-sectional area of the duct should be at least twice the net area of the prestressing steel. Similar tolerances should be used if the duct voids are to be filled with corrosion inhibitor after stressing operations.

> **RULES FOR SAFE TENSIONING**
> Because large tensioning forces are necessary in prestressing operations, this type of construction can be extremely hazardous. Tendons under tension represent stored energy that, if suddenly released, could cause serious injury or death to personnel in the vicinity. The following basic rules for tensioning should be included in the safety requirements of all operations:
> - Before tensioning of any bed, give visible and audible signals to warn all nonessential personnel to leave the area adjacent to the bed;
> - Have provisions in place to prevent jacks from flying longitudinally or laterally in case of tendon failure;
> - Never allow personnel to stand behind either anchorage end or to walk on the bed or concrete member directly in line with the tendon being tensioned;
> - Make elongation measurements by jigs or templates from the side or from behind shields. Do not make elongation measurements while standing over tendons;
> - Provide shields at both ends of the bed to stop flying tendons;
> - Provide eye protection for personnel engaged in wedging and anchoring operations to shield against flying pieces of steel; and
> - Avoid welding of prestressing wires, strands, or high-strength bars.

17.3.3 *Anchorages and tensioning*—Anchorages for all post-tensioning tendons should align with the tendon axis at the point of attachment. Concrete surfaces the anchorage devices bear on should be perpendicular to this line. The contract documents should specify the minimum retained tendon elongation after anchor set. The elongations attained in the field are affected by field procedures and friction losses. If the specified force or measured elongations differ from engineering established values by more than 7%, all aspects of the operation should be checked before proceeding. The inspector should record the force and elongation measurements for each tendon and submit results to the engineer for review and approval. When making measurements, the tendons should be marked in a consistent manner and in a slack-free state. Tendons to be stressed from both ends should be marked at each end before stressing. Stressing from two ends may be simultaneous or sequential, provided that the forces at each end are reasonably equal and that any anchor set caused by sequential stressing is properly excluded from the total elongation measurement.

17.3.4 *Grouting procedures*—Proper grouting of tendon ducts after stressing operations is essential. Improperly grouted tendons do not adequately bond to the rest of the member and, therefore, do not fully contribute to the strength of the member. Spaces not filled with grout also permit the intrusion of water or chemicals, ultimately leading to serious corrosion of the tendon.

Tendons should be grouted as soon as possible after inspection and acceptance of stressing. Tendons that cannot be grouted within 5 days of stressing should be protected from corrosion until grouted. Any water standing in ducts subject to freezing temperatures should be removed because it can freeze and split the member.

The grout can be a mixture of cement and water only, or it can include fine sand, fly ash, pozzolan, grouting aids, or an admixture to increase flowability. The use of admixtures containing chlorides should not be allowed. Trial mixtures should be used to determine optimum grout proportions for a particular job. Grout water content should be limited to the minimum volume that will result in a pumpable mixture. The w/cm should not exceed 0.45.

The grout-pump operator should always pump material toward open vents to force out all entrapped air. Grout vents should be high enough above the tendon to allow for sedimentation of the solids in the grout. Even though grout is discharged through the vent before sealing, free water from the sedimentation rises to the uppermost part of the duct/vent system (typically approximately 5% of the vertical height). The pump operator should apply grout continuously under moderate pressure at one point in the duct until the open vent or vents discharge a steady stream of grout. With the entire duct filled and with discharge vents closed, the pressure should be raised to a minimum of 50 psi (0.35 MPa) and held for at least 1 minute, after which the injection point is plugged to prevent any loss of grout.

Because grouting is so important and there are so many chances for error, an experienced, qualified inspector should be present during the entire grouting procedure.

17.3.5 *Postconstruction inspection*—Following installation or field construction of a prestressed member, it may be necessary to verify force levels of unbonded tendons and identify any time-dependent losses. With certain member configurations and construction systems, this may be relatively easy. With other types, in-service tendon tests may not be possible without local concrete excavation. Usually, lift-off tests using a load cell or dynamometer are used for in-service tendon testing. If severe corrosion in the tendon system is suspected, additional visual inspection may be warranted after exposure and testing. In-service testing of bonded (grouted) tendons is very limited unless the tendons were instrumented previously. For more information on postconstruction inspection, refer to *Post-Tensioning Manual* (Post-Tensioning Institute 1990).

CHAPTER 18—GROUT, MORTAR, AND STUCCO

Some construction materials not specifically used for concreting are used for closely related operations (Fig. 18.1). These include various types of grout and mortar for structural applications and stucco for surface finishes. This chapter briefly describes these materials and methods of placement and testing. As with concrete construction, inspectors should ensure that all materials and construction procedures conform strictly to the contract documents. It is the responsibility of the architect/engineer to ensure that the contract documents require the use of materials suitable for the loads to be carried and environmental conditions to be encountered, including weather, chemical attack, and extreme temperatures.

18.1—Pressure grouting
Pressure grouting has a variety of applications, including:
- Consolidation grouting of large dam foundations and other rock foundations;

- Grouting of contraction joints in concrete dams;
- Contact grouting behind tunnel liners and similar items;
- Grouting of rock anchors and soil nails;
- Grouting of voids under slabs and subsurface voids, such as mine voids and gobs; and
- Repair of cracks in pavements, bridges, and buildings.

For all of these applications, the grout should penetrate into fissures or openings under the applied pump pressure. In addition, the pressures used should not cause distress to adjacent slabs or structures.

The portland-cement grout typically used for these applications is a neat cement grout or a cement-sand mixture, but some grouts may also contain admixtures or pozzolans. Several chemical materials also are used for pressure grouting, particularly epoxy resins and polymers.

Typically, neat cement grout is used for foundations. When fissures are very fine, it may be necessary to use a special air-separated cement. The grout should be well mixed and preferably used within 1 hour after mixing. Refer to ACI 304.1R for information on grout mixers and pumps. Grouting of dam foundations and other rock foundations is done through drill holes under a wide range of pressures, depending on conditions. Higher pressures are usually applied at deep locations, and lower pressures at shallower ones.

Fig. 18.1—Self-leveling underlayments and toppings are poured or pumped into the floor. The grout forms a smooth, level surface ready to take any type of floor covering.

18.2—Grouting under base plates and machine bases

The requirements for grouting under base plates for structural members and for support of machinery are similar. In both cases, it is important for the hardened grout to remain in permanent contact with the underside of the base and have sufficient strength to resist stresses applied by the member or the machinery. Materials commonly used for grouting machinery and base plates include:
- Damp-pack mortar;
- Aluminum-powder and other gas-forming grouts;
- Catalyzed metallic grouts;
- Cementitious systems, including expansive-cement grouts; and
- Polymer grouts, 100% solids (most commonly epoxy resins).

To perform effectively, the grout should establish close contact with the entire underside of the plate at placement and maintain that contact while hardening. Relief holes in the plates are often required to allow release of trapped air.

To ensure satisfactory production using these materials, inspectors should study the requirements of the contract documents that will serve as the inspector's primary guide.

18.2.1 *Damp-pack mortar*—Packing with damp mortar is an efficient method for setting heavy machinery on a concrete base and for fastening anchor bolts in concrete (Fig. 18.2). Damp-pack mortar is usually one part cement to three parts well-graded sand, by weight. First, the dry cement and sand are thoroughly mixed, then water is added and all materials are mixed again. Water content should be such that a mass of mortar tightly squeezed by hand will moisten but not significantly soil the hand, and will not crumble easily.

Fig. 18.2—Damp-pack (dry pack) mortar is typically one part cement to three parts sand and enough water to produce a ball when compacted by hand.

Before setting the machine or base plate, workers should properly prepare the base concrete by:
- Roughening and chipping as necessary to provide a strong, clean surface and to expose the sand;
- Removing all dust, preferably by suction, and then scrubbing the base with fiber-bristled brushes and water to remove all loose material and coatings; and
- Thoroughly saturating the base concrete for 24 hours, removing all free water from the surface immediately before placing the damp-pack mortar.

The damp-pack mortar should be mixed ahead of time and allowed to age for at least 30 minutes before application. This significantly reduces the potential for shrinkage. Before packing the mortar under the machine or base plate, workers should block one side of the open space. They can then begin placing all of the mortar from the other side, using wood blocks to ram in small amounts at a time. Careful use of a hammer to pound the blocks will ensure complete consolidation of the grout without warping of the base plate. After filling the space between the plate and base, workers can remove

Table 18.1—Dosage of aluminum-powder blend*

Type of grout	Placing temperature	
	70 °C (21 °C)	40 °F (4 °C)
Sand-cement grout	5.5 (160)	8.5 (240)
Neat cement grout	4.5 (130)	7.0 (200)

*oz/bag (g/bag) of cement, based on 1:50 blend of aluminum powder with either cement or dry sand.

the backup block and ram the face of the mortar from that side. Mortar can be packed around bolts by hammering down on a section of pipe that fits the annular space between the bolt and the walls of the hole.

18.2.2 *Gas-forming grouts*

18.2.2.1 *Aluminum-powder grout (job-mixed)*—When added to grout, aluminum powder reacts chemically with the soluble alkali constituents of the cement and generates hydrogen gas. The resulting expansion before the grout sets is intended to compensate for shrinkage, causing the grout to harden in contact with the plate it is to support. Gas expansion from aluminum powder or other sources does not compensate for hydration shrinkage that occurs as the grout gains strength after initial set or for later drying shrinkage.

The ground aluminum powder can be of any variety that produces the desired expansion, but it should not contain polishing agents such as stearates, palmitates, or fatty acids. Because soluble alkalies in the cement react with the aluminum powder, the alkali content of the cement has a major effect on the expansion obtained. A few cements with extremely low alkali contents produce so little reaction that they are not suitable for use in this type of grout.

Laboratory tests have demonstrated that a gas-forming grout suitable for use under machine bases can be produced by adding extremely small quantities of aluminum powder, equal to 50 or 60 millionths of the weight of cement used (about a teaspoonful per bag of cement). The aluminum powder should first be blended in the proportions of one part powder to 50 parts by weight of cement or dry sand passing a No. 100 (150 μm) sieve. This blend is then added by sprinkling over the batch. Governing the required dosage of the blended materials is the amount and chemical composition (alkali content) of the cement, the placing temperatures, and whether the aluminum admixture is used in a neat cement grout or a sand-cement grout.

To determine the required amount of powder and the effectiveness of the powder and portland-cement combination, tests should be performed with the materials at job temperatures, adjusting the dosage of powder as necessary to obtain effective expansion. Table 18.1 gives suggested dosages for preliminary trial mixtures.

Because aluminum powder has a tendency to float on water, the blend should be thoroughly mixed with the cement and sand before water is added. After the addition of all ingredients, the batch should be mixed for 3 minutes. The action of the aluminum weakens approximately 45 minutes after mixing, so it should be verified that the batches are small enough to allow immediate placement of the freshly prepared mortar.

Forms should be provided to confine the grout on all surfaces (including the top), because unconfined expansion can drastically reduce grout strength. The forms should allow for slight expansion, however, because completely confined grout can exert pressures of up to 100 psi (0.7 MPa).

Concrete Manual (U. S. Bureau of Reclamation 1988) gives more detailed information on using grout containing aluminum powder.

18.2.2.2 *Other gas-forming grouts*—Finely ground activated carbon or fluid coke added to grout produces gases (hydrogen or nitrogen) by reacting chemically in the presence of water. As with aluminum powder, these reactions take place only while the grout is plastic.

Proprietary premixed grout fluidifiers available from a number of manufacturers can also produce expansion-causing gases. These products usually achieve more uniform, reliable results than job-mixed aluminum-powder grouts.

18.2.3 *Catalyzed metallic grouts*—These grouts contain iron aggregates and an oxidation catalyst that causes the metallic particles to enlarge, thus compensating for drying shrinkage and settlement. The grouts should be mixed, placed, finished, and cured according to manufacturer recommendations. They are intended to be fluid, and should not be used as stiff mortars.

For satisfactory service, catalyzed metallic grouts require rigid confinement during the first 7 to 14 days of hardening. After hardening, extended exposure of unconfined areas of grout to wetting and drying in air can lead to self-destructive expansion. Therefore, unconfined shoulders should be cut flush with the edges of the plate after initial set and the exposed surface dressed with a cement-sand mortar or heavily coated with curing compound. Manufacturer recommendations should be adhered to for grouts that will be exposed to weather or wet conditions.

18.2.4 *Cementitious systems*—Cementitious grouts contain expansive cement, plaster of Paris, or other expanding components to trigger expansion or shrinkage compensation. The cement may be a blend of portland and expansive cements. These grouts usually contain natural fine aggregates or a combination of natural aggregates and similarly graded iron, with the latter used to increase fatigue resistance. Many nonmetallic cement-based grouts use trisulfate hydrate, which produces calcium sulfoaluminate crystals responsible for volume changes in the hardened phase.

Cementitious grout systems can be mixed on the job, or premixed blends containing fine-graded silica aggregate are available. With job-mixed cementitious grouts, sufficient water must be present to initiate the formation of ettringite, which causes the expansive force (ACI 223). As with shrinkage-compensating concrete (discussed in Chapter 16), it is important for cementitious grouts to be adequately restrained. As normally used, however, the grout is subject to drying shrinkage because the restraint is not resilient.

Because each grout type has limitations, initial acceptance testing of the grout system is advisable to ensure that the material will perform at the consistency and range of temperatures specified. Compressive-strength data should accompany information on the age at test, flow of the grout as sampled, initial grout temperature, and curing conditions. Depending on job requirements, grout consistency can readily be altered by changing the water content. Retempering should not be allowed. In general, cementitious grouts will perform as proven by test at any consistency thicker (lower water content) than that used in the test. At thinner consistencies (higher water content), bleeding is likely to occur or increase while expansion and strength decrease.

18.2.5 *Nonshrink grouts*

18.2.5.1 *Preblended*—Specially formulated proprietary grouts preblended to compensate for shrinkage while providing the desired control of volume change have the advantages of being ready to use and offering a fairly wide range of placing consistencies. These grouts are often used when requirements cannot be met by ordinary cement-sand grouts, such as when high fluidity is required and bleeding is undesirable, or when nonshrink or expansive action is required.

Preblended grouts can be grouped by their expansion-producing (shrinkage-compensating) characteristics, such as gas-forming, catalyzed-metallic, cementitious, or polymer. Special preblended grouts also are available for post-tensioning applications and rock or concrete anchorage systems. Because the steel used for these installations should be protected against corrosion, these grouts usually contain fine-graded silica aggregate low in chlorides and sulfides.

18.2.5.2 *Polymer systems*—With epoxy-resin grouts, the cementing agent is a resin and a hardener (polyamide curing agent) that, when mixed, forms a high-strength, nonshrink, thermosetting plastic (provided that the epoxy resin is 100% solids with no diluents). These grouts should be mixed, placed, and cured according to the manufacturer's instructions. Hydrophobic hybrid polyurethanes also are available for grouting.

18.2.5.3 *Placement*—The same rules of good practice necessary for conventional grouting apply to field placement of nonshrink grouts. All surfaces in contact with the grout should be free of dirt, oil, laitance, and other foreign substances, and the grout should be properly cured to prevent moisture loss during early stages of hydration. Additional measures may be required to ensure that preblended nonshrink grouts perform as intended and accomplish the desired results under all conditions.

Grouting of foundations and large base plates requires a fluid grout with stable flow characteristics. To determine stable grout fluidity, flow tests should be conducted with a flow cone (ASTM C939) 30 minutes after mixing. To maintain plate contact during the plastic stage, specifications usually require batter boards or sideforms to extend above the bottom of the plate to provide a slight head for uniform bond. Strapping or vibration should not be used to eliminate air voids and pockets unless testing positively demonstrates that vibration does not induce bleeding and settlement. The use of chains, once a popular consolidation method, should also be discouraged. The links could drag in air bubbles that rise to the surface of the grout, and thus reduce the contact area.

Nonshrink grouts are not intended for use in self-stressing concrete because they do not provide adequate expansion for stressing of reinforcement. In addition, too much expansion, such as with gas-producing grouts, can be undesirable for load-bearing applications because excessive volume changes usually reduce strength. Expansion not exceeding 0.2% can safely compensate for settlement shrinkage. Most specifications prohibit grout shrinkage in either the plastic or hardened stage.

18.2.5.4 *Field testing*—Quality-control testing of the grout during placement is necessary to monitor performance and provide test data in case of questionable service. At a minimum, the as-mixed grout sampled at the mixer should be tested to determine consistency, specific water content, expansion, bleeding, and compressive strength.

Standard tests, such as the ASTM C109/C109M cube strength test (modified as appropriate for various grout consistencies) and the ASTM C939 flow test, will disclose critical performance limits. Plastic consistencies of 110 to 125% at five drops, as measured by the flow table, are suitable for grouting small plates. Plastic consistencies of 125 to 145% at five drops and fluid consistencies of 25 to more than 30 seconds measured with the flow cone (ASTM C939) are appropriate for grouting of medium- to large-sized plates and bases, structural columns, and anchor bolts. An arbitrary increase of fluidity above the specified consistency at the job site will probably cause bleeding, thus preventing contact with the base plate and proper load distribution to the grout.

With any type of grout, the adequacy of contact between the grout and a simulated or actual grouted plate (by sounding the plate at various ages after the grout hardens) should be checked to ensure that:

- The interface is not weakened by the collection of gas bubbles at the contact surface; and
- All bleedwater has been absorbed or displaced so that physical contact exists between grout and plate over at least 90% of the plate area.

18.3—Mortar and stucco

Mortar and stucco often are used for bedding concrete slabs or for plastering over other surfaces (Fig. 18.3). Typically, the materials are applied pneumatically.

If specifications do not provide mortar proportions, a ratio of cementitious material (cement or cement and lime) to sand of approximately 1:3 to 1:5 by dry volume should be used, with allowance for bulking. Leaner mixtures are more porous; richer mixtures require more mixing water and thus have greater drying shrinkage and a tendency to crack. It should be verified that the sand is clean and free from excessive fines and that the mortar is well-mixed, preferably by machine.

Stucco, or portland-cement plaster, is a special form of mortar coating for walls and soffits, usually built up in three coats. The final surface coat typically is a factory-prepared material consisting of portland cement, color, and plasticizer, and it often receives a decorative texture during application (Fig. 18.4). ACI 524R provides detailed recommendations and specifications for portland-cement plaster.

When inspecting stucco work, particular attention should be paid to:

- Preparation of the backing;

Fig. 18.3—Mortar and stucco are often used for bedding concrete slabs or for plastering over surfaces. To restore workability, mortar may be retempered.

Fig. 18.4—Applying stucco plaster onto a masonry wall.

- The method of applying each coat, whether by trowel, darby, or dashing;
- The thickness of various coats;
- Uniform proportioning and mixing of colored cements; and
- Thorough curing of each coat.

Admixtures to enhance impermeability and bonding of stucco should be used in strict accordance with manufacturer recommendations. It should be verified that the admixtures are compatible with other admixtures that may be present in the prepared material.

Adequate curing is especially important for thin applications of stucco because they tend to crack and loosen if allowed to dry at an early age. Preferably, burlap should be hung against coated surfaces and kept wet by spraying for 3 to 7 days (depending on weather conditions). The burlap should be allowed to dry slowly in place.

CHAPTER 19—TESTING OF CONCRETE

Concrete test methods are detailed in national standards published by ASTM International, AASHTO, the U.S. Army Corps of Engineers, and the U.S. Bureau of Reclamation. The most commonly used test methods for the largest variety of construction projects are those of ASTM. This chapter presents an overview of ASTM methods for testing concrete in the field or field laboratory. (ASTM methods for testing aggregate are described in Chapter 3.) For complete details, refer to the *Annual Book of ASTM Standards*, Volume 04.02. This publication also contains the *Manual of Aggregate and Concrete Testing* prepared by ASTM Committee C-9. Although not an ASTM standard, the manual provides useful commentary and interpretation of the various test methods.

To ensure strict compliance, it is important to precisely follow the test methods and procedures required by the contract documents. If the contract documents do not describe or specify the test methods directly to be used, then the appropriate ASTM or other test methods should be applied. Some of the test methods require the technician to be certified. ASTM C1077 and E329 define duties and responsibilities and establish the minimum requirements for testing agencies and personnel, including laboratory accreditation, technician certification, and equipment requirements for testing and inspection of concrete and concrete materials.

19.1—Sampling

One of the most important aspects of testing is obtaining a representative sample for measuring a specific property. ASTM C172 explains methods of sampling from various types of concrete production equipment. The contract documents may define where to obtain samples, and would thus govern options given in ASTM C172. Samples should be taken at random. Selective sampling should be avoided, which may not represent the actual construction. Refer to Chapter 2 for examples of statistical sampling concepts.

For production quality control, concrete samples should be taken as the concrete is delivered from the final mixer. Sampling may be from a stationary mixer at a mixing plant or from a truck mixer as it prepares to discharge concrete at the job.

The goal of job-site sampling should be to obtain concrete samples representative of concrete properties at the point of placement. Some placement methods, such as pumping, may significantly change slump and air content of the concrete. Thus, it is important to test concrete properties at mixer discharge and again following discharge from the delivery system at the point of placement. ACI 301 requires that concrete possess the specified characteristics in the freshly mixed state at the point of final placing.

ASTM C172 describes sampling methods for the following concrete equipment:

Revolving-drum truck mixers or agitators—Two or more portions should be taken throughout the discharge of the middle part of the batch and composited for compliance testing of the concrete. The discharge gate should be open, and the samples obtained either by passing a receptacle through the discharge stream or by diverting the stream into the sample container. When the stream is too fast to sample, the entire load should be discharged and sampled in accordance with the applicable method.

Stationary mixers—A receptacle should be through the discharge stream at two or more regularly spaced intervals during discharge of the middle portion of the batch.

Paving mixers—After the mixer has discharged, portions from at least five points should be collected to make a representative sample. Contamination with subgrade material should be avoided.

Open-top truck mixers or agitators, receiving hoppers, and buckets—Any of the preceding methods can be used, as applicable.

Pump systems—The discharge stream is diverted into the container at two or more regularly spaced intervals during discharge. Diversion should not be such as to create excessive changes to boom angle, pump rate, or hose configuration because these can affect test values.

The amount of concrete sampled should be greater than required for the specimens or tests, and not less than 1 ft^3 (28 L) for acceptance tests. All portions of the composite sample should be obtained within 15 minutes after the start of sampling. The composite samples should then be mixed with a shovel only until uniform. Tests for air content and slump should begin within 5 minutes after compositing, and molding of strength-test specimens should begin within 15 minutes after compositing.

19.2—Tests of freshly mixed concrete

19.2.1 *Consistency*—The consistency of concrete is a measure of its workability, which may be defined by its slump characteristics, flow characteristics, or other consistency indicators. Following is a summary of the ASTM C143 slump test, which is illustrated in Fig. 7.10:

1. Place a damp, clean cone on a damp, nonabsorbent flat level surface;
2. Fill the cone with fresh concrete in three layers of equal volume, with the top layer heaped above the cone. Rod each layer 25 times. Hold the cone firmly in place during filling and rodding;
3. After the last layer has been rodded, strike off the concrete level with the top of the cone, and remove fallen concrete from around the base of the mold; and
4. Lift the cone in a smooth, vertical motion without twisting, and measure slump at the center of the displaced top to the nearest 1/4 in. (5 mm).

19.2.2 *Air content*—ASTM provides three methods for determining total air content of freshly mixed concrete: the pressure method, the volumetric method, and the gravimetric method.

19.2.2.1 *Pressure method (ASTM C231)*—Two types of meters (A and B) are used to determine total air content by the pressure method. The Type A meter is based on the correlation of reduction of water level with reduction of air volume in the concrete sample at a predetermined air pressure. The Type B meter (Fig. 19.1) operates on the principle of equalizing a known volume of air at a known pressure in a sealed chamber to obtain the value of the unknown volume of air in the concrete sample.

The general requirements of air-content testing using the air-pressure method are:

1. Calibrate the air meter in accordance with manufacturer's instructions;

Fig. 19.1—The pressure method (ASTM C231) provides a determination of total volume of air content in fresh concrete. Type B meter is shown above.

2. Fill the bowl with fresh concrete in three equal layers, rodding each layer 25 times. Tap the bowl with a mallet 10 to 15 times after each layer has been rodded;
3. Remove excess concrete with a sawing motion of the strike-off bar, and assemble the meter;
4. Add the necessary water and pressurize; and
5. Read the result from a gauge or standpipe, and use the aggregate correction factor to obtain the true air reading. ASTM C231 gives a procedure for determining the aggregate correction factor.

19.2.2.2 *Volumetric method (roll-o-meter) (ASTM C173/C173M)*

1. Verify volumes of bowl and cup and the accuracy of neck graduations;
2. Fill the meter (Fig. 19.2) in two layers, and rod each layer 25 times. Tap the bowl 10 to 15 times with the mallet after rodding of each layer;
3. Use the strike-off bar with a sawing motion to level and smooth concrete even with the rim of the bowl;
4. Assemble the meter and, using the funnel, add a water/alcohol mixture. Adjust water level to the zero mark. Use a sufficient amount of alcohol to ensure that, after agitation and mixing, foam in the neck does not exceed the height of 2 air-content percent divisions. Measure alcohol addition using a container with at least 4 oz [100 mL] graduations;
5. Tighten the cap on the meter, and invert and agitate for a minimum of 45 seconds. With the neck elevated, rock and roll the meter for a minimum of 60 seconds;
6. Set the meter upright, loosen the cap, and allow air to rise until a stabilized liquid level can be read. For high-air-content concrete where the liquid level is below the neck window, remove the cap and add water with the 1% cup until a liquid level is readable;

Fig. 19.2—The volumetric method (ASTM C273) measures the air content of normalweight and lightweight concrete (the meter shown above is known as a roll-o-meter).

7. Repeat the rolling operation until the stabilized water level reading is repeated within 0.25% after two successive rollings; and

8. Read the total air content as the final, stabilized liquid level plus the number of 1% cups of water added during the test (if any). Correct air content, when more than 2.5 pints (1.2 L) of alcohol are used in the test, per the table provided in ASTM C173/C173M.

19.2.2.3 *Gravimetric method (ASTM C138/C138M)*— The ASTM C138/C138M gravimetric method of determining total air content is based on the difference between the measured plastic density and the calculated air-free density of the concrete. The difference in density divided by the air-free density is then multiplied by 100 to calculate the percent air content. Knowledge of exact batch quantities and specific gravities of mixture materials is needed to calculate the air-free density of concrete.

19.2.3 *Density of freshly mixed concrete (ASTM C138/ C138M)*—The density test of freshly mixed concrete can be used to determine yield, cement factors, and gravimetric air content (Fig. 19.3). Density determinations are based on the mass of concrete required to exactly fill a measure of known, calibrated volume, as follows:

1. Fill the measure in three equal layers, rodding each layer 25 times for a 1/2 ft^3 (14 L) size or less and 50 times for a 1 ft^3 (28 L) measure;

2. Tap the measure 10 to 15 times after each rodding, and then strike off the top and finish smooth with a flat strike-off plate 2 in. (50 mm) larger than the diameter of the measure;

Fig. 19.3—The density of fresh concrete (ASTM C138/ C138M) can be used to determine the yield factor based on weighing the concrete in a unit measure of known volume.

3. Clean the exterior and weigh;

4. The density of the concrete is the net mass (total mass less the mass of measure) divided by the volume of the measure. The volume of the measure is calculated by determining the net mass of water that it can hold, as described in ASTM C29/C29M; and

5. Depending on the size of coarse aggregate in the concrete, the air-meter bucket used for the pressure-method determination of air content may be suitable for use in conducting density determinations.

19.2.4 *Temperature*—The temperature of freshly mixed concrete samples should be measured per ASTM C1064/ C1064M (Fig. 19.4). Temperature is usually taken with an immersion thermometer having a range of approximately 30 to 120 °F (0 to 50 °C). Higher-than-specified temperatures can lead to lower later-age strengths, and lower-than-specified temperatures can lead to lower early-age strengths or early-age freezing damage. If monitoring of internal concrete temperatures is specified, it is usually conducted by casting thermocouples or temperature sensors within the concrete during placement.

19.3—Strength tests

The standard methods used for determining concrete strength during construction involves making and curing structural concrete compressive- or flexural-strength specimens in the field. A minimum of two strength specimens should be cast for testing at each test age specified for acceptance (usually 28 days). To provide an indication of expected 28-day strength, additional specimens should be cast for testing at 7 days or earlier. Chapter 2 discusses the evaluation and acceptance of concrete strength. All specimens should be cast at or near the place of initial curing to avoid the possible detrimental effects of moving freshly made test specimens.

19.3.1 *Compressive strength: ASTM C31/C31M, C192/ C192M, and C39/C39M*—ASTM C31/C31M (field) (Fig. 19.5) and ASTM C192/C192M (laboratory) cover the requirements for molds and for making and curing of cylindrical specimens.

19.3.1.1 *Molds*—Molds may be reusable or single-use, and be made of steel, cast iron, plastic, coated cardboard, or other materials that are nonabsorbent and nonreactive with cement. They should be watertight and sufficiently strong and tough to resist tearing, crushing, or deforming during use. They also should meet specified absorption, elongation, and dimensional tolerances (ASTM C470/C470M). Mold height should be equal to twice the diameter.

19.3.1.2 *Making specimens*—To make standard (6 x 12 in. [150 x 300 mm]) cylindrical specimens according to ASTM C31, sample the concrete in accordance with ASTM C172, as described previously in this chapter. When casting standard specimens in the laboratory per ASTM C192/C192M, representative portions of the batch mixture should be selected for use. Molding of the specimens should begin within 15 minutes after the sample is obtained, as follows:

1. Fill the mold uniformly in three approximately equal layers with a scoop or trowel (or two layers, if vibrated);

2. Consolidate each layer as follows: for slumps less than 1 in. (25 mm), vibrate; for slumps of 1 in. (25 mm) or greater, rod or vibrate. Rod each layer 25 times, or insert the vibrator two times per layer. Tap the sides after each rodding or vibration to close voids, and strike off the top with a rod, trowel, or float; and

3. Cover the molds with glass, metal plate, plastic, or wet burlap to avoid loss of moisture. Avoid contact of coated cardboard cylinder molds with wet burlap or other moist surfaces.

Requirements for casting other sizes of cylindrical specimens can be found in appropriate tables of ASTM C31/C31M and C192/C192M.

19.3.1.3 *Curing specimens*

1. Store acceptance test cylinders in the field up to 48 hours at 60 to 80 °F (16 to 27 °C). Protect specimens during transport from damage, moisture loss, and freezing. Store laboratory test cylinders at laboratory temperatures (68 to 86 °F [20 to 30 °C]) for 24 ± 8 hours;

2. After the initial storage period, remove cylinders from the molds and place in moist-cure storage at 73 ± 3 °F (23 ± 2 °C) until the moment of testing. Moist storage may be by immersion in saturated limewater, placement in curing cabinets, or storage in curing rooms (Fig. 19.6) conforming to ASTM C511 requirements;

3. Field cure cylinders to be used for determining when to remove forms, when the structure can be put into service, or the adequacy of curing by storing the cylinders in or on the structure or as near as possible. They should receive, insofar as is practical, the same protection and curing as the structure. Remove specimens for determining when a structure may be put into service from the molds at the time of form removal.

19.3.2 *Capping cylindrical concrete specimens for compressive strength tests: ASTM C617*—Freshly molded concrete cylinders may be capped with high-strength gypsum-plaster or neat cement, but this usually is not

Fig. 19.4—Monitoring the concrete temperature (ASTM C1064/1064M) during placement informs the inspector if the concrete is above or below allowable tolerances. Excessively high temperatures can cause thermal cracking, whereas colder temperatures may delay setting and strength gain.

Fig. 19.5—Cast concrete cylinder (ASTM C31) at or near place of initial curing to avoid disturbance while moving freshly made specimens.

expedient. Capping hardened concrete with molten sulfur mortar (5000 psi [35 MPa] or more) is presently the most practical method (Fig. 19.7). The cap should be plane to a tolerance of 0.002 in. (0.05 mm), perpendicular to the cylinder axis within 0.5 degrees, and sound without hollow spots. ASTM C617 contains strength and thickness tolerance limits for the capping material.

19.3.3 *Use of unbonded caps for compressive strength tests: ASTM C1231/C1231M*—Unbonded elastomeric pads restrained by metal rings may be used in place of capping compound when testing hardened concrete cylinders. Neoprene caps are now the most commonly used test cap. ASTM C1231/C1231M, however, does not allow unbonded caps for acceptance testing of concrete with a compressive strength below 1500 psi (10 MPa) or above 12,000 psi (85 MPa). Reuse of unbonded pads is generally limited to 100 tests unless higher reuse values are qualified by comparison testing of cylinders with companion cylinders whose ends

Fig. 19.6—Storage of compressive strength test cylinders in a laboratory curing room.

Fig. 19.7—Capping test cylinders (ASTM C617) with sulfur compound.

have been properly ground or capped with a qualified sulfur-capping material. Testing of specimens above 7000 psi (50 MPa) should also be qualified by comparison testing.

19.3.4 *Testing concrete cylinders: ASTM C39/C39M*—Compression tests of concrete cylinders should be conducted on a power-operated, calibrated testing machine capable of providing a constant loading rate of 20 to 50 psi/s (0.15 to

Fig. 19.8—Compression testing of a concrete cylinder (ASTM C39) by a calibrated test machine.

0.35 MPa/s) and conforming to ASTM E4 requirements for testing machines. The bearing surfaces should be flat and clean, and the cylinder should be centered in the test heads (Fig. 19.8). Testing concretes with compressive strengths exceeding 6000 psi (45 MPa) requires special attention to the end preparation of the test specimens and to the rigidity of the testing machine (ACI 363.2R.) After failure, the type of fracture should be noted and the compressive strength calculated to the nearest 10 psi (0.1 MPa) by dividing the maximum load by the cross-sectional area of the specimen.

19.3.5 *Flexural strength of concrete: ASTM C31/C31M, C192/C192M, C78, and C293*—The flexural strength of concrete is most commonly determined by testing rectangular beams, 6 x 6 x 20 in. (150 x 150 x 500 mm), for concrete made with maximum-size coarse aggregate up to 2 in. (50 mm).

19.3.6 *Molding flexural specimens*

1. Use molds that are nonabsorbent, nonreactive, smooth, free of blemishes, and watertight;

2. Fill the mold in two equal layers if rodded, and one layer if vibrated. Vibrate if slump is 1 in. (25 mm) or less, rod or vibrate for a 1 to 3 in. (25 to 75 mm) slump, and rod when slump exceeds 3 in. (75 mm);

3. Rod each layer uniformly—once per 2 in.^{2} (14 cm^2)—and tap the sides of the mold to fill rodding voids. When rodding each layer, penetrate into the layer below. After tapping each layer, spade the concrete along the sides and ends of the mold with a trowel or other suitable tool. If consolidating by vibrator, insert the vibrator at intervals not exceeding 6 in. (150 mm). Tap the mold 10 to 15 times with a mallet to close any holes left by the vibrator;

4. Strike off the surface with a wood or magnesium float; and

5. Immediately after finishing, cover the surface to prevent evaporation.

19.3.7 *Curing flexural specimens*

1. Initially store beams cast in the field or in the laboratory in the manner described previously for cylinders;

2. After initial storage and transportation, remove beams from the molds and place in moist-cure storage at 73 ± 3 °F (23 ± 2 °C) (similar to cylinders);

3. Immerse beams in saturated limewater a minimum of 20 hours before testing, and maintain specimens in a wet condition until the moment of testing; and

4. Field cure flexural specimens by storing them under conditions representative of the concrete in the structure. After 48 hours, remove beams from the molds. Store specimens representing slab-on-ground pavements on the ground, in the "as-molded" position. Bank damp earth or sand against the sides, leaving only the top surface exposed. Store specimens representing structures as near as possible to the structure they represent, in a manner similar to concrete within the structure.

19.3.8 *Testing beams for flexural strength*—Flexural strength may be determined using a simple beam with center-point loading (ASTM C293) or with third-point loading (ASTM C78) (Fig. 19.9). Center-point loading usually gives higher indicated strengths, but produces more inherent scatter in the data. Third-point loading is recommended for most applications. Strict conformance to ASTM procedures is critical. The following are general requirements for beam testing:

1. The testing machines should be motorized and capable of applying the load at a constant rate. ASTM C78 permits conditional use of positive-displacement hand pumps;

2. Turn the test specimen on its side with respect to its position as molded, and center it on the bearing blocks;

4. Apply the load at a rate that constantly increases the extreme fiber stress 125 to 175 psi/min (0.86 to 1.21 MPa/min) until rupture occurs; and

5. Measure average width and depth (three measurements across each dimension at fracture face), and calculate modulus of rupture per ASTM C78 or C293, as appropriate.

19.3.9 *Splitting tensile strength of cylindrical concrete specimens: ASTM C496/C496M*—Splitting tensile strength is primarily used to evaluate the shear resistance provided by low-density concrete. This test is intended for laboratory use, although some specifications require that it be performed on field-cast specimens. ACI 318 states that splitting-tensile-strength tests shall not be used as a basis for field acceptance of concrete.

1. Cast and cure test specimens in accordance with ASTM C31/C31M or C192/C192M, as described previously. Moist cure low-density concrete specimens for 7 days, and then store them at 73 ± 3 °F (23 ± 2 °C) and a relative humidity of 50 ± 5% for 21 days;

2. Mark the cylinders with diametral lines on each end using a suitable device that will ensure the lines are in the same axial plane, and take an average of three diameter measurements;

3. Center the specimen in the testing machine with a 1/8 x 1 x 12 in. (3.2 x 25 x 300 mm) plywood strip on the top and bottom. Do not reuse plywood strips. Use a supplemental bearing bar or plate to apply the load if the diameter of the upper bearing block is less than the length of the specimen;

Fig. 19.9—Casting (ASTM C31) and testing (ASTM C78) of beam for flexural strength of concrete.

4. Apply the load without shock at a constant rate of 100 to 200 psi/min (0.7 to 1.4 MPa/min) splitting tensile stress until failure. For a 6 x 12 in. (150 x 300 mm) cylinder, splitting tensile stress will increase at a rate of 100 psi/min (0.7 MPa/min) if the load increases at a rate of 11,300 lb/min (50 kN/min); and

5. Calculate splitting tensile strength, to the nearest 5 psi (0.05 MPa)

$$T = 2P/\pi l d$$

where T = splitting tensile strength, psi (MPa); P = maximum load, lb (N); l = length, in. (mm); and d = diameter, in. (mm).

19.3.10 *Compressive strength of lightweight insulating concrete: ASTM C495*—ASTM C495 covers procedures for compressive strength testing of lightweight insulating concrete having an oven-dry weight not exceeding 50 lb/ft^3 (800 kg/m^3). This method is restricted to the use of 3 x 6 in. (75 x 150 mm) cylinders.

1. Sampling is in accordance with the standard procedures described previously in this chapter. When sampling from pump equipment, however, pass a bucket through the discharge stream of the concrete pump hose at the point of placement;

2. Mold cylinders by placing the concrete in two equal layers. Tap the sides of the mold lightly after each layer is placed until the surface has subsided. Do not rod the concrete;

3. Cure specimens at 60 to 80 °F (16 to 27 °C) for the first 24 hours after molding. After 24 ± 2 hours, remove specimens from the molds and moist cure at 73 ± 3 °F (23 ± 2 °C) for 6 days. At 7 days of age, store specimens at 60 to 80 °F (16 to 27 °C) and a relative humidity of 50 ± 30% for 18 days. At 25 days of age, place specimens in a drying oven at 140 ± 5 °F (60 ± 3 °C) for 3 days. Cool specimens to room temperature before testing; and

4. Determine compressive strength per ASTM C39/C39M, as described previously.

19.4—Accelerated curing of test specimens

Contract documents for some projects may require or permit accelerated strength testing of standard concrete cylinders. ASTM C684 defines standard methods of accelerated curing to provide strength data earlier than conventional 7- and 28-day compressive strength tests. To properly evaluate the relationship between accelerated cured cylinders and standard cured cylinders, a correlation should be developed for each concrete mixture, using the same materials.

ASTM C684 presents four methods for accelerated curing of test cylinders:
- Procedure A: Warm water method;
- Procedure B: Boiling water method;
- Procedure C: Autogenous method; and
- Procedure D: High temperature and pressure method.

Procedure A is the simplest with respect to the apparatus required. Immediately after casting, the cylinders are cured in a warm-water bath at 95 ± 5 °F (35 ± 3 °C) for 23-1/2 hours ± 30 minutes and are then demolded. The cylinders are tested at an age of 24 hours ± 15 minutes.

Procedure B adds a short accelerated curing period to a longer conventional curing period. Initial curing of 23 hours ± 15 minutes at 70 ± 10 °F (21 ± 6 °C) is followed by immersion in boiling water for 3-1/2 hours ± 5 minutes. The cylinders are then demolded and cooled for at least 1 hour at room temperature. They are tested at 28-1/2 hours ± 15 minutes.

Procedure C uses heat generated by cement hydration to accelerate strength development. After casting, the cylinders are stored in a thermally insulated container for 48 hours ± 15 minutes. They are then removed from the container, demolded, and allowed to stand for 30 minutes at room temperature. They are tested at 49 hours ± 15 minutes.

Procedure D involves simultaneous application of elevated temperature and pressure to the concrete using special containers and apparatus. All curing takes place while the cylinders are in their molds. Specimens are cured at a pressure of 1500 ± 25 psi (10.3 ± 0.2 MPa) for a period of 5 hours ± 5 minutes. Heating at 300 ± 5 °F (150 ± 3 °C) is maintained for the first 3 hours. At the end of the curing period, pressure is released, and the cylinders are extruded from the molds and tested.

For all four methods, the concrete cylinders should be sampled, molded, and tested in accordance with the ASTM C31/C31M and C39/C39M procedures described previously in this chapter, except that cardboard molds should not be used.

19.5—Uniformity tests of mixers

19.5.1 *Truck mixers*—ASTM C94/C94M stipulates that concrete mixed completely in a truck mixer, 70 to 100 revolutions at the manufacturer's specified mixing speed, should meet the uniformity requirements in Table 19.1. A uniformity test compares two samples of freshly mixed concrete, one from near the front of the mixer drum and one from near the back. Concrete uniformity is satisfactory when at least five of the six tests shown in Table 19.1 are within the specified limits. For a quick check of probable degree of uniformity, slump tests should be made of individual samples taken after discharge of approximately 15 and 85% of the load. These two samples should be obtained within an elapsed time of not more than 15 minutes. If the slumps differ more than the amount specified in Table 19.1, the mixer should not be used unless the condition is corrected to permit the requirements of Table 19.1 to be met, such as by changing to a longer mixing time, a smaller load, or a more efficient charging sequence.

19.5.2 *Stationary mixers*—ASTM C94/C94M specifies a mixing time based on mixer capacity, requiring 1 minute for the first cubic yard plus 15 seconds for each additional cubic yard or fraction thereof. These minimum time requirements will produce uniformity, but can be unnecessarily time-consuming for large-capacity mixers. Therefore, ASTM C94/C94M permits a reduction in this mixing time, provided that tests show that uniformity can be achieved in less time.

Samples for testing are taken after discharge of 15 and 85% of the batch. Alternatively, the mixer can be stopped after the designated mixing time without discharging, and samples removed from locations near the front and back of the drum. If the mixer drum must be entered for this sampling, the electric power fuses should be disconnected first and the equipment should be tagged out in accordance with safety regulations.

19.5.3 *Washout test for coarse-aggregate content*—One measure of uniformity of a concrete batch is the percentage of coarse aggregate contained in two different portions of the batch (Tables 19.1 and 19.2). This determination should be made using the following procedure:

1. Determine the mass of a sample of plastic concrete in accordance with ASTM C138/C138M. To partially combine this determination with tests for air-free density of concrete and mortar, use the base of an air meter as the volume measure for the density determination. Then test the sample for air content in accordance with ASTM C231;

2. After the air test, sieve and wash the concrete sample over a No. 4 (4.75 mm) sieve;

3. Remove the aggregate retained on the sieve, and determine the mass while it is immersed and suspended in water;

4. Determine the saturated surface-dry (SSD) mass of the sample by towel drying and weighing the sample. Alternately, formulas given in ASTM C127 for bulk specific gravity (SSD) can be used with the known specific gravity of the aggregate to calculate the SSD mass of the sample; and

5. Calculate the percent of coarse aggregate by dividing the SSD mass of the coarse aggregate by the mass of the concrete before washing and multiplying by 100.

19.5.4 *Air-free density of concrete test*—The method for determining air-free density of concrete is as follows:

1. Perform tests for density and air content as described previously;

2. Obtain the volume of air-free concrete by subtracting the volume of air measured for the sample from the calibrated volume of the container used in the density determination; and

3. The air-free density of concrete M_c, in pounds per cubic foot (kilograms per cubic meter), is calculated as the mass of the concrete sample divided by the volume of air-free concrete as follows

$$M_c = \frac{b}{V - A}$$

where

b = mass of concrete sample, lb (kg);
V = volume of the (sample) container, ft³ (m³); and
A = volume of air, computed by multiplying the container volume V by the percent air divided by 100.

19.5.5 *Air-free density of mortar test*—The method for determining air-free density of mortar is as follows:

1. Perform tests for density, air content, and percent of coarse aggregate as described previously;

2. Determine the mass of the mortar in the sample by subtracting the SSD mass of the coarse aggregate from the mass of the concrete sample;

3. Obtain the volume of the air-free mortar by subtracting the volume of the coarse aggregate and the volume of air measured for the sample from the calibrated volume of the container used in the density determination; and

4. The air-free density of mortar, M_m, in pounds per cubic foot (kilograms per cubic meter), is its mass divided by its volume, calculated as follows

$$M_m = \frac{b-c}{V - \left(A + \frac{c}{G \times 62.4}\right)} \quad \text{(in.-lb)}$$

$$M_m = \frac{b-c}{V - \left(A + \frac{c}{G \times 1000}\right)} \quad \text{(SI)}$$

where
b = mass of the concrete sample, lb (kg);
c = SSD mass of the aggregate retained on No. 4 (4.75 mm) sieve, lb (kg);
V = volume of the (sample) container, ft³ (m³);
A = volume of air, computed by multiplying the volume V by percent of air divided by 100;
G = SSD bulk specific gravity of the coarse aggregate; and
62.4 = density of water, lb/ft³ (or 1000 kg/m³).

The following example illustrates the method of determining the density of air-free mortar and coarse-aggregate content.

Given: Concrete sample containing 1-1/2 in. maximum nominal-size aggregate, with an air content of 5.0%.
G = 2.65;
b = 35 lb;
V = 0.25 ft³;
S = 12.0 lb (immersed-suspended mass of the aggregate retained on No. 4 sieve); and
C = SSD mass of aggregate.

Calculation

$$C = \frac{S \times G}{G-1} = \frac{12 \times 2.65}{2.65 - 1} = 19.27 \text{ lb}$$

W = coarse aggregate content, percent by weight

$$W = \frac{19.27}{35} \times 100 = 55.0\%$$

Table 19.1—Requirements for uniformity of concrete

Test	Maximum permissible difference in samples taken from two locations in concrete batch
Density of fresh concrete calculated to air-free basis	1.0 lb/ft³ (16 kg/m³)
Air content, volume percent of concrete	1.0% numerical difference
Slump — If average slump 4 in. (100 mm) or less	1.0 in. (25 mm)
Slump — If average slump 4 to 6 in. (100 to 150 mm)	1.5 in. (40 mm)
Coarse aggregate content, portion by weight of each sample retained on No. 4 (4.75 mm) sieve	6.0% numerical difference
Density of air-free mortar	1.6% difference
Average compressive strength at 7 days for each sample	7.5%* difference

*Tentative approval of the mixer may be granted pending results of the 7-day compressive strength tests.

Table 19.2—Test for uniformity of concrete

Test	Front	Back	Variation	Specification requirements expressed as maximum permissible difference between samples, ASTM C94/C94M
Coarse aggregate content, portion of each sample percent retained on No. 4 sieve	55	52.5	2.5	6.0
Density of air-free mortar	130 lb/ft³	132 lb/ft³	1.5%	1.6%

Note: Percent variation in density of mortar may be based on average of comparative samples as follows:

Average density of air-free mortar = (130 + 132)/2 = 131 lb/ft³

Variation in density expressed as percent of average = [(132 – 130)/131] × 100 = 1.5% < 1.6% (OK)

$$A = \frac{0.25 \text{ ft}^3 \times 5.0}{100} = 0.0125 \text{ ft}^3$$

This gives the values necessary to calculate M_m, the density of air-free mortar, using the formula given previously

$$M_m = \frac{35 - 19.27}{0.25 - \left(0.0125 + \frac{19.27}{2.65 \times 62.4}\right)} = 130 \text{ lb/ft}^3$$

To check compliance of a stationary mixer or truck, determinations should be made on each of two samples, one from near each end of the batch. It should be assumed that the previous calculations represent the first sample from a batch, and that the second sample is found to have 52.5% coarse aggregate and 132.2 lb/ft³ density of air-free mortar. Table 19.2 compares these results to see if the uniformity is within ASTM C94/C94M limits. Data and the calculated mass and volume for both samples are listed in Table 19.3.

19.6—Density of structural lightweight concrete

ASTM C567 provides procedures for determining oven-dry and equilibrium densities of structural lightweight

Table 19.3—Example of computation for density of air-free mortar

Data	Sample from front of mixer or first portion of batch as discharged from mixer		Sample from back of mixer or last portion of batch as discharged from mixer	
	Mass, lb	Volume, ft³	Mass, lb	Volume, ft³
Mass and volume of sample in air meter	35.00	0.2500	35.16	0.2500
Air content by air meter	—	(5%)	—	(4.75%)
Volume of air	—	0.0125	—	0.0119
Mass and air-free volume of sample	35.00	0.2375	35.16	0.2381
Submerged mass of sample retained on No. 4 screen	12.00	—	11.50	—
Computed SSD mass and solid volume of plus No. 4 material	19.27	0.1167	18.47	0.1119
Mass and volume representing mortar in sample	15.73	0.1208	16.69	0.1262
Computed density of air-free mortar, lb/ft³	130.22	—	132.25	—

concrete. The measured or calculated equilibrium density of the concrete determines whether the specified density requirements have been met.

1. Oven-dry density is calculated as:

$$O_c = (M_f + M_c + 1.2M_{ct})/V$$

where

O_c = calculated oven-dry density, lb/ft³ (kg/m³);
M_f = mass of dry fine aggregate in batch, lb (kg);
M_c = mass of dry coarse aggregate in batch, lb (kg);
M_{ct} = mass of cementitious material in batch, lb (kg);
1.2 = factor to approximate mass of cement plus chemically combined water; and
V = volume of concrete produced by the batch, ft³ (m³).

2. The oven-dry density should be measured as follows:
   a) After 24 to 32 hours, remove cylinders from the molds, measure the mass of the cylinders while they are immersed and suspended in water, and then determine the SSD mass of the cylinders;
   b) Place cylinders in a drying oven for 72 hours or until constant mass is reached. Allow cylinders to cool, and determine the dry mass; and
   c) Calculate measured oven-dry density as follows

$$O_m = (D \times 62.4)/(F - G) \quad \text{(in.-lb)}$$

$$O_m = (D \times 1000)/(F - G) \quad \text{(SI)}$$

where

O_m = measured oven-dry density, lb/ft³ (kg/m³);
D = mass of oven-dry cylinder, lb (kg);
F = SSD mass of cylinder, lb (kg); and
G = suspended-immersed mass of cylinder, lb (kg).

ASTM C567 states that extensive tests have determined that the equilibrium density of structural lightweight concrete will be approximately 3 lb/ft³ (50 kg/m³) greater than the oven-dry density. Therefore, equilibrium density can be calculated by adding 3 lb/ft³ (50 kg/m³) to either the calculated oven-dry density or the measured oven-dry density described previously.

3. Equilibrium density should be measured as follows:
   a) Six days after molding, remove cylinders from molds and immerse in water at 73.5 ± 3 °F (23 ± 2 °C) for 24 hours.
   b) At 7 days, determine the SSD mass and the suspended-immersed mass of the cylinders, as described previously;
   c) Air dry the cylinders in a room or enclosure at 73.5 ± 3 °F (23 ± 2 °C) and 50 ± 5% relative humidity until the mass of the specimens does not change by more than 0.5% in successive determinations 28 days apart. (ASTM indicates that approximately 40 to 180 days may be required to reach the equilibrium density.); and
   d) Calculate measured equilibrium density as

$$E_m = (A \times 62.4)/(B - C) \quad \text{(in.-lb)}$$

$$E_m = (A \times 1000)/(B - C) \quad \text{(SI)}$$

where

E_m = measured equilibrium density, lb/ft³ (kg/m³);
A = mass of dry cylinder, lb (kg);
B = SSD mass of cylinder, lb (kg); and
C = suspended-immersed mass of cylinder, lb (kg).

ASTM C138/C138M should be used to determine the density of freshly mixed concrete for control purposes or to check compliance with concrete placement specifications.

19.7—Tests of completed structures

19.7.1 *Cores from hardened concrete*—The compressive strength of concrete is sometimes determined from cores extracted from a structure. This is usually done when the compressive strength of a given concrete is not known or the results of concrete cylinder tests indicate that concrete strength is questionable. The architect/engineer should specify where and how many cores are to be obtained. Usually, three cores are averaged to comprise a test. ACI 318 and ASTM C42/C42M cover core test requirements.

Cores should be extracted using a core drill, exercising care not to cause overheating. Obvious defects, such as rock pockets and joints, should be avoided. Cutting the reinforcing steel, which can compromise the strength of both the core and the cored member, should also be avoided. Magnetic detection devices, penetrating radar, and radiography can be used to nondestructively determine the location of reinforcing steel, as care should be exercised to miss reinforcing steel, especially post-tensioned tendons.

The length of cores used for compressive-strength determinations should be as near as possible to twice the core diameter, and the core diameter should be at least twice the nominal maximum size of the aggregate in the concrete, and preferably at least three times the nominal maximum aggregate size. Cores having lengths less than the diameter after capping cannot be used for compression testing. Those with length-diameter ratios less than 1.8 to 1.0 may be used with a correction factor, as specified in ASTM C42/C42M. If necessary, the ends should be sawed to produce an even surface, and the core should be grinded or capped to enable proper testing.

Moisture gradients at the time of testing will affect core strength. ACI 318 states that cores should be prepared for transportation and storage by wiping drilling water from their surfaces and placing them in watertight plastic bags or containers immediately after drilling. Cores should be tested no earlier than 48 hours and no later than 7 days after coring unless approved by the licensed design professional.

19.7.2 *Load tests*—If a load test is required to determine the structural integrity of a reinforced concrete slab or beam, the member being tested should be isolated so that the influence of adjacent structural members, components, or whole structures can be accounted for during the load test and when evaluating the results. Safety precautions should be taken. All participants in the tests and all passersby need to be safe during setup and performance of the tests. Refer to ACI 318 and 437R for information on how to perform load tests.

19.7.3 *Nondestructive tests*—Several methods are available for nondestructive testing or evaluation of concrete in place, including:
- Test Method for Pulse Velocity through Concrete (ASTM C597);
- Test Method for Rebound Number in Concrete (ASTM C805);
- Test Method For Penetration Resistance of Hardened Concrete (ASTM C803/C803M);
- Test Method for Measuring the P-Wave Speed and the Thickness of Concrete Plates Using the Impact-Echo Method (ASTM C1383);
- Test Method for Determining the Thickness of Bound Pavement Layers Using Short-Pulse Radar (ASTM D4748);
- X-ray; and
- Test Method for Detecting Delaminations in Bridge Decks Using Infrared Thermography (ASTM D4788).

Each of these methods, properly used, can quickly provide information about concrete quality without invasive or destructive consequences. All have limitations, however, and highly skilled operators are essential because data analysis is subject to interpretation. Refer to ACI 228.1R for guidance on in-place methods to estimate concrete strength. Any estimates derived should be evaluated with sound engineering judgment and caution.

When it is desirable to infer relative compressive strength from data produced by the following methods, measurements should be carefully calibrated to laboratory tests and simulations, and correlated to the actual conditions encountered on each job, including concrete strength, aggregate type, and concrete surface conditions. The accuracy of the strength correlation is highly variable among the different methods. Each ASTM should have a precision statement that discusses the accuracy of the test method.

Pulse-velocity tests (ASTM C597) measure the velocity with which an impinged sound wave is propagated through concrete. They can be used on structures to detect progressive deterioration, concealed cracking, honeycombing, or other defects. Pulse velocity can also be used to locate concrete of considerably lower quality than that specified. Figure 2.11 of ACI 228.1R provides a typical relationship between pulse velocity and compressive strength of a given concrete mixture.

If it is necessary to interpret pulse velocities in terms of compressive strength, six cores should be obtained from each area of the concrete in question. These cores should cover the range of pulse velocities found, and they should be taken at locations where the pulse velocity has actually been measured. After proper preparation, the cores should be tested for compressive strength, and this strength correlated with pulse velocity. The correlation will probably not be linear, and it may be necessary to identify and reject outlying data points. Only experienced personnel should obtain and interpret pulse-velocity data.

Impact-hammer testing (ASTM C805) provides a relative measure of the surface uniformity and can detect areas of variability that should be further investigated. The impact hammer (sometimes called a Schmidt or Swiss hammer) is a spring-loaded steel hammer that, when released, strikes a steel plunger in contact with the concrete surface. The rebound distance of the hammer from the plunger after impact is recorded on a built-in linear scale. The instrument carries a chart that attempts to correlate the rebound number to approximate compressive strength. This correlation should be used only to compare the numbers and determine areas of variation, because results are predominately influenced at a shallow depth from the surface. Determinations made with this instrument should be used only for comparing concretes of the same composition, age, and moisture content. Generally, rebound readings of concrete suspected to be of poor quality are compared with readings conducted on similar concrete of known quality. Judgment should be used in the interpretation because rebound is affected by many factors, such as moisture content of the concrete, type of surface finish, type of aggregate, age of the concrete, and proximity of reinforcing steel to the point of impact.

Penetration-probe testing (ASTM C803/C803M) is a method for nondestructive estimation of approximate compressive strength of concrete in the field when properly correlated in the laboratory with the same concrete mixture. The test is based on the principle that penetration of the probe (originally called a Windsor probe) is inversely proportional to the compressive strength of the concrete being tested. Typically, three probes are driven into the concrete by a powder-actuated device that delivers a specific energy-producing force. Spring-loaded devices also are used to drive the probes. The compressive strength is determined experimentally, usually by correlating with cores from the same area or with laboratory comparisons of the same mixture. It is subject to many of the same limitations as the impact hammer. In addition, the test method is not totally nondestructive because the probes damage the concrete surface.

19.8—Shipping and handling samples

When shipping samples to a laboratory, they should be identified with complete information including project name or number, sample number, material identification (type and grade), source, date of sampling, quantity represented, location where the sample was obtained, tests required, and shipper's identification.

When shipping concrete specimens by common carrier, they should be packed in suitable cushioning materials and protected from damage, moisture loss, and freezing temperatures. When specimens are to be moved by vehicle, they should be secured so they do not sustain damage from bouncing or falling.

Concrete specimens that are 1 or 2 days old should be handled with particular care because they can easily sustain serious damage. Handling freshly made specimens the first day should be avoided.

CHAPTER 20—RECORDS AND REPORTS

Contract documents, building codes and standards, and regulatory agencies require written records and reports of inspections and tests. These documents provide a permanent record of the construction, including verification that construction materials and standards of workmanship are in accordance with the contract documents. They are often used to settle disputes or as a basis for future modifications to the project.

Concrete inspection and testing services may entail sampling and testing of ingredients in the concrete mixture, checking of production equipment and procedures, preplacement inspection of forms and reinforcing, observation of placement and curing, field testing of plastic concrete, and laboratory testing of hardened specimens. This chapter lists general and specific data and observations to include in inspection and test records and reports, as applicable. Many factors, however, affect the exact details to include, such as the type of test or inspection requested by the client, contract requirements, legal requirements of the jurisdiction in the project's locale, and the specific sampling and test methods used. Building codes and standards and the quality-management program of the inspection and testing agency may dictate additional requirements. For additional guidance, refer to ASTM C1077 and E329.

The records and reports shown in this chapter (Fig. 20.1 to 20.12) are examples only, and should be modified to suit project conditions and contract requirements.

20.1—General information
- Report title, identifying the nature of the test or inspection;
- Date of issue;
- Name and address of inspection or testing agency;
- Project name and identification number;
- Report identification number;
- Number of each report page and total number of pages;
- Client name;
- Name of technician or inspector performing the work and the inspector's certification number, when applicable;
- Date of test or inspection;
- Signature of licensed professional engineer or designee accepting responsibility for the report; and
- Identification of any results obtained from subcontracted laboratories or inspection agencies used to perform tests.

20.2—Specific information
- Identification and description of the activity observed or item inspected or tested;
- Identification of the inspection or test method used, or a description of any nonstandard test method or procedure used;
- Sampling location and procedure (test method may provide options);
- Any deviations from, additions to, or exclusions from the inspection or test method;
- Other information relevant to a specific observation, inspection, or test, such as environmental conditions and equipment used;
- Identification of the standard or specification to which the results are to be compared;
- All information required by the test method(s), such as results of the inspection or test, observations, measurements, examinations, and derived results;
- Supporting data such as tables, graphs, sketches, and photos, as appropriate;
- Interpretation statement relating the results to the identified standard or requirements;
- Statement of measurement uncertainty, if applicable; and
- Corrections or additions to reports that clearly reference the report being amended.

20.3—Maintaining records

Testing and inspection agencies should record all data and observations accurately, clearly, and objectively, in a format meaningful to the reader and in accordance with the requirements of the inspection or test method. The client, test specifications, or code may specify the format. Included in this chapter are examples of report forms for various purposes. Copies of these forms can be downloaded from the ACI website (http://www.concrete.org/general/SP2_Chap2_Forms.doc). If a particular format is not specified, the report should conform to the requirements of the applicable test standard referenced. In addition to formal report forms, testing agencies will need to develop a series of internal use forms to organize and control the processing of test samples. Examples of these types of forms would include sample receipt logs, test schedule logs, temperature storage logs, and equipment calibration records.

ASTM C1077 requires reports and related records to be retained for at least 3 years, unless otherwise required by law or other governing specifications. Documents such as work sheets, computer printouts, calibration data, graphs, and photographs should be retained as part of the inspection or test records.

20.4—Quality-control charts

Some records and reports are used to control the construction and ensure timely action in taking corrective steps to avoid substandard quality. Typical records of this type are included in

ACI 214R. Refer to the quality-control chart in ACI 214R and Chapter 2. Early-strength data are plotted to provide a graphical representation of quality trends. These charts are also used to establish that the specified strength criteria have been achieved or to indicate when corrective action is required.

20.4.1 *Concrete delivery ticket*—The concrete supplier is required by ASTM C94/C94M to submit a delivery ticket with each load of concrete. This ticket serves as a certification of proportions and quantities delivered. The minimum information should include:

1. Name of ready-mix supplier and identification of batch plant;
2. Series number of ticket;
3. Date and truck number;
4. Name of contractor;
5. Specific designation of the job (name and location);
6. Specific class or designation of the concrete in conformance with that required by the contract documents;
7. Amount of concrete delivered (yards);
8. Time the mixer was loaded or time of first mixing of cement and aggregates;
9. Time of arrival at site;
10. Time discharge started;
11. Time delivery was completed; and
12. Amount of water added by receiver of concrete and the receiver's initials.

Additional information that may be included on the delivery ticket includes:

1. Reading of revolution counter at the first addition of water;
2. Signature of ready mixed concrete representative;
3. Type and brand of cement;
4. Amount of cement and pozzolans;
5. Total water content by the producer;
6. Maximum size of aggregate;
7. Weight of fine aggregate and coarse aggregate;
8. Type, name, and amount of each admixture;
9. Free water on the aggregate; and
10. Indication that all materials are as previously certified or approved.

The delivery ticket may be supplemented by any additional information required by the project specification. For example, a copy of the printout for each batch (from an automatic printing unit incorporated in the batching system) containing actual weights of all materials may be submitted by the producer when required by the purchaser.

CONCRETE BATCH PLANT INSPECTION

[Name and Address of Laboratory]

CLIENT:_____   REPORT NO:_____ PAGE_____OF_____

PROJECT:_____   PROJECT NO:_____

LOCATION:_____   INSPECTION DATE:_____

Plant: **Class/Design:**

S = Satisfactory   **U = Unsatisfactory**   **N/A = Not Applicable**

1. _____ Do stockpiles appear to be homogeneous?
2. _____ Are stockpiles located to prevent contamination; arranged to assure that each aggregate as removed from its stockpile is distinct and not intermingled with others?
3. _____ Are separate storage bins or compartments for each size and type of aggregate properly constructed and charged to prevent mixing of different sizes or types?
4. _____ Is cement the type and brand required for use?
5. _____ Is the fly ash the brand required for use?
6. _____ Are aggregates the required type for use?
7. _____ Are admixtures the required type for use?
8. _____ Are admixture dispensers clearly marked?
9. _____ Are scales calibrated?   Date of last calibration_____
10. _____ Cement mass within tolerance of $\pm 1\%$?
11. _____ Fly ash mass within tolerance of $\pm 1\%$?
12. _____ Coarse aggregate mass within tolerance of $\pm 2\%$?
13. _____ Fine aggregate mass within tolerance of $\pm 2\%$?
14. _____ Water mass within tolerance of $\pm 1\%$?
15. _____ Admixture volumes within tolerance of $\pm 3\%$
16. _____ Scales return to zero?
17. _____ Dispensers zeroed?
18. _____ No water in drum of truck before batching?
19. _____ Truck-mix concrete is completely mixed at a minimum of 70-100 revolutions?
20. _____ Truck counters zeroed?
21. _____ Truck water gauges operative?
22. _____ Chutes and rear fins clean?

Truck No.(s) inspected:_____

REMARKS:

_____
Inspector – Level/Certification No.

_____   _____
Reviewer – Level/Certification No.   Date

Fig. 20.1—Sample concrete batch plant inspection report.

MANUAL OF CONCRETE INSPECTION (ACI 311.1R-07)

CONCRETE CURING RECORD

[Name and Address of Laboratory]

CLIENT:	REPORT NO:	PAGE	OF
PROJECT:	PROJECT NO:		
POUR LOCATION:	DATE POURED:		
	CONCRETE CLASS:		

Concrete Placement Card No.

TYPE OF CONCRETE

Normal	Mod. Mass	Mass

Building	
Member	
Elevation	

TEMPERATURE MONITORING DEVICES USED

DATES/TIMES

Placement completed	
Heat discontinued	
Forms removed	
Curing completed	

CODE
Type of Curing:
1 – curing compound
2 – polyethylene sheets
3 – ponding water
4 – damp burlap or fabric
5 – damp sand or earth
6 – water spray or ponding
7 – curing paper
8 – insulation (tarps)
9 – forms
10 – not applicable

DAY / DATE	Time of day	Outside air temp. (°F)	Temp. within enclosure (°F)		Concrete surface temperature (°F)		Type of curing See Code			REMARKS
			High	Low	High	Low	Top surface	Sides	Constr. joints	
1										
2										
3										
4										
5										
6										
7										

If curing compound is used, record the following:

Manufacturer:_____ Surface preparation:_____
Brand name:_____ Req'd surface finish:_____
Air temp. @ application:_____ Rate of application:_____
Locations sprayed (if not entire placement): Number of coats:_____
_____ Application date:_____

If curing/protection requirements are not met, identify areas not meeting requirement and corrective action implemented.

_____ _____ _____ _____
Inspector – Level/Certification No. Date Reviewer – Level/Certification No. Date

Fig. 20.2—Sample concrete curing record.

CONCRETE PLANT – DAILY INSPECTION

[Name and Address of Laboratory]

CLIENT:_____ REPORT NO:_____ PAGE_____ OF _____
PROJECT:_____ PROJECT NO:_____
LOCATION:_____ DATE:_____

WEATHER:_____ AIR TEMP.: AM_____F PM_____F
AGGREGATE HEATED/COOLED: YES_____ NO_____ HOT WATER/ICE: YES_____ NO_____
TOTAL CUBIC YARDS:_____ MIXING METHOD: CENTRAL_____ TRUCK_____
DESIGN MIX CLASS:_____ DESIGN SLUMP:_____ DESIGN AIR:_____

ONE YARD DESIGN PROPORTIONS

CEMENT	FLYASH		FINE AGG. lb			COARSE AGG. lb			H_2O	ADMIX oz		AEA	FIBER
lb	lb	lb	SSD	M%	Adj lb	SSD	M%	Adj lb	Adj. Gal	#1	#2	oz	lb

BATCH WEIGHTS / MATERIALS/SOURCES

TICKET NO.				Brand/Producer/Location	Size/Type	S.G.
CEMENT						
FLY ASH						
FINE AGG.						
COARSE AGG.						
AEA						
ADMIX #1						
ADMIX #2						
FIBER						
MIXING H_2O						

MIXING WATER CALCULATIONS

1) Water in F. Agg.=_____ 2) Design Mix Water =_____ 3) Adjusted Mix Water=_____
  + Water in C. Agg.=_____   - Total Water in Agg.=_____   - Plant Added Water=_____
  = Total Water in Agg.=_____ = Adjusted Mix Water=_____   = Allowable Field Water=_____

MOISTURE CALCULATIONS

F. Agg: Wet Mass =_____   C. Agg: Wet Mass =_____
  Dry Mass =_____   Dry Mass =_____   % Moisture Formula
  Difference =_____   Difference =_____
  % Total Moisture =_____   % Total Moisture =_____ $\dfrac{\text{wet mass} - \text{dry mass}}{\text{dry mass}} \times 100$
  % Absorption =_____   % Absorption =_____
  % Free Moisture =_____   % Free Moisture =_____

TIME:_____ 1/2 Day _____Full Day _____OT (hours)/ Client Signature:_____/_____
                                                                          Company

Proportions Batched (did) (did not) meet required tolerances.

_____ _____ _____
Inspector – Level/Certification No. Reviewer – Level/Certification No. Date

Fig. 20.3—Sample concrete plant daily inspection report.

MANUAL OF CONCRETE INSPECTION (ACI 311.1R-07)

FINE AGGREGATE ANALYSIS

[Name and Address of Laboratory]

CLIENT:_____ REPORT NO.:_____PAGE____OF____
PROJECT:_____ PROJECT NO:_____
MATERIAL SOURCE:_____ DATE SAMPLED:_____
SAMPLE LOCATION:_____ DATE RECEIVED:_____

SIEVE ANALYSIS – ASTM C136-xx

SIEVE SIZE	TOTAL MASS RETAINED	PERCENT RETAINED	CUMULATIVE PERCENT RETAINED	CUMULATIVE PERCENT PASSING	ASTM C33 SPEC. % PASSING	DOT SPEC. % PASSING
3/8"					100	100
#4					95 to 100	
#8					80 to 100	
#16					50 to 85	45 to 95
#30					25 to 60	
#50					10 to 30	8 to 30
#100					2 to 10	1 to 10
PAN		FM:				

MOISTURE ASTM C566-xx	WASH LOSS ASTM C117-xx	SP. GRAVITY & ABSORPTION ASTM C128-xx
Wet mass =	Initial dry mass =	Sample mass (SSD) =
Dry mass =	Dry mass after wash =	Mass of water Disp. =
Difference =	Difference =	Sp. gravity =
% Total moisture =	% - 200 =	Sample mass (Dry) =
% Absorption =		Difference =
% Free moisture =	Spec. req't:	% Absorption =

ORGANIC IMP. ASTM C40-xx	SAND EQUIVALENT - AASHTO T 176-xx	
Color Plate No. =	Sand reading:	x 100 = _____
Spec. req't:	Clay reading:	Spec. req't:

REMARKS:

_____   _____   _____
Technician – Level/Certification No.   Reviewer – Level/Certification No.   Date

Fig. 20.4—Sample fine aggregate analysis report.

FORMS/REINFORCING CHECKLIST

[Name and Address of Laboratory]

CLIENT:_____ REPORT NO:_____ PAGE____ OF____
PROJECT:_____ PROJECT NO:_____
LOCATION:_____ DATE:_____

REINFORCING STEEL INSPECTION

Plans Used: Contract Drawing_____ OR Shop Drawing:_____ Stamped Approved:_____
Date:_____
Drawing Numbers:_____

Type of Bar Markings: Numbers_____ Line:_____ Grade Specified:_____ Correct:____YES ____NO

Manufacturer's certifications supplied/checked:	____YES ____NO	
Rebar size per the approved drawing:	____YES ____NO	
Horizontal spacing within tolerance:	____YES ____NO	
Vertical spacing within tolerance:	____YES ____NO	
Proper bnds:	____YES ____NO	
Did you count every vertical bar:	____YES ____NO	--Any Missing___YES___NO; Corrected__Y__N
Did you count every horizontal bar:	____YES ____NO	--Any Missing___YES___NO; Corrected__Y__N
Overlaps corrected:	____YES ____NO	
Overlap specified:	_____Inches _____Bar diameters	
Supporting system (chairs or blocks) correct:	____YES ____NO	
Splices of proper length and location:	____YES ____NO	
Epoxy coated:	____YES ____NO	
Repair of epoxy coating:	____YES ____NO	
Rebar straight (not bent or kinked):	____YES ____NO	
Clean and tied:	____YES ____NO	
Clearance from forms/ground as specified:	____YES ____NO	
Was rebar disturbed after placement:	____YES ____NO	
Dowels placed properly:	____YES ____NO	
Welding – as required/permitted:	____YES ____NO	

FORM INSPECTION

Subbase satisfactory:	____YES ____NO	Blockouts in:	____YES ____NO
List form:	Width_____ Depth_____	Loose soil removed:	____YES ____NO
Within tolerance?	____YES ____NO	Water and debris removed:	____YES ____NO
Clean:	____YES ____NO	Rebar properly supported:	____YES ____NO
Oiled:	____YES ____NO	Joints placed properly:	____YES ____NO
Tight:	____YES ____NO	Embedments in and correct:	____YES ____NO

REMARKS:_____
_____

_____
Inspector – Level/Certification No.

_____ _____
Reviewer – Level/Certification No.   Date

Fig. 20.5—Sample reinforcing steel/form.

MANUAL OF CONCRETE INSPECTION (ACI 311.1R-07)

REINFORCING STEEL/FORM INSPECTION

[Name and Address of Laboratory]

CLIENT:_____ REPORT NO:_____ Page_____of_____

PROJECT:_____ PROJECT NO:_____

LOCATION:_____ INSPECTION DATE:_____

Drawing#:_____ Rev/Approval Date:_____

Placement location per the drawing:_____

Type of placement: Walls_____ Footings_____ Slabs_____ Other_____

Corrective action: YES_____ NO_____

Type of action needed:_____

Satisfactory for concreting: YES_____ NO_____ (Explain in remarks) NA: _____

Remarks:_____

Work satisfactory (Except as noted above): _____YES _____NO _____NA (see checklist on back)

Any NCR's written ___YES ___NO ___NA NCR#:____ Office faxed ____YES ____NO ____NA

TIME:_____ 1/2 Day _____ Full Day _____ OT (Hours): Authorized by:_____

_____ _____ _____
Inspector's Name (print)   Inspector's Signature   Level/Certification No.

Client Representative's Signature:_____ FOR:_____
                                                                    COMPANY

_____ _____
Reviewer – Level/Certification No.   Date

Fig. 20.6—Sample reinforcing steel/form inspection report.

REPORT OF CONCRETE TESTS

[Name and Address of Laboratory]

CLIENT:_____ REPORT NO:_____ PAGE_____ OF _____
PROJECT:_____ PROJECT NO:_____
LOCATION:_____ CLASS/DESIGN:_____
CONTRACTOR:_____ SUPPLIER:_____

MATERIALS/PROPORTIONS   SAMPLING DATA

Cement, lb	_____	Date sampled	_____
Fly ash, lb	_____	Time sampled	_____
Coarse aggregate, lb	_____	Yards batched	_____
Fine aggregate, lb	_____	Time batched	_____
Admixture, oz	_____	Truck/Ticket No.	_____
Admixture, oz	_____	Yards placed	_____
Other	_____	Cylinders molded	_____
Total water, lb	_____	Water added at site	_____

Location of placement_____

PROPERTIES OF FRESH CONCRETE

	Field measurement	Specification requirement	Notes:
Air temp., °F	_____	_____	1. Cylinders molded and cured according to ASTM C31-xx, prepared per ASTM C617-xx or C1231-xx, and tested according to ASTM C39-xx. Nominal cylinder 6" x 12" within ASTM tolerances.
Concrete temp., °F	_____	_____	
Slump, in.	_____	_____	
Air content by vol., %	_____	_____	
Density, lb/ft^3	_____	_____	

2. Field tests performed according to ASTM 172-xx (sampling), ASTM C1064-xx (concrete temp.), ASTM C143-xx (slump), ASTM C173-xx or ASTM C231-xx (air content) and ASTM C138-xx (unit weight).

COMPRESSIVE STRENGTH TEST RESULTS

Cylinder No.	Test date	Age, days	Area, in.^{2}	Maximum load, lb	Compressive strength, psi	Type of fracture
1A						
1B						
1C						
1D						

Remarks:

Field Technician:_____ Level/Certification No:_____
Lab Technician:_____ Level/Certification No:_____

_____ _____
Reviewer – Level/Certification No.   Date

Fig. 20.7—Sample report of concrete tests (combined format).

TENSIONING REPORT

[Name and Address of Laboratory]

CLIENT:_____ REPORT NO:_____ PAGE____ OF____
PROJECT:_____ PROJECT NO:_____
LOCATION:_____ DATE POURED:_____
_____ DATE STRESSED:_____

Tendon No.	Actual elong.	Calc. elong.	Gage psi	Actual stress	Req'd stress	7% Tol. limits		Remarks
						Sat.	Unsat.	

Hydraulic System No.:_____
Date calib:_____

_____
Inspector – Level/certification No.

_____ _____
Reviewer – Level/Certification No.   Date

Fig. 20.8—Sample tensioning report.

COMPRESSIVE STRENGTH OF CONCRETE - ASTM C39

[Name and Address of Laboratory]

CLIENT:_____   REPORT NO.:_____
PROJECT:_____   PROJECT NO:_____
PROJECT LOCATION:_____
PLACEMENT LOCATION:_____
PLACED BY:_____   CONCRETE SUPPLIER:_____
SPECIMENS CAST BY:_____   TOTAL CUBIC YARDS PLACED:_____
CONCRETE CLASS:_____ f'_c:_____PSI
DATE CAST:_____   DATE RECEIVED:_____

TIME	TRUCK NO.	TICKET NO.	SLUMP (in.)	AIR CONT. (%)	CONC. TEMP (°F)	TEMP. (°F)	DENSITY (lb/ft^3)

I.D.	TEST DATE	AGE (DAYS)	DIA. (in.)	AREA (in.^{2})	TOTAL LOAD (lb)	COMP. STR. (PSI)	* TYPE BREAK	**CYL. MASS (lb)

* If other than visual cone   **Optional

REMARKS: Test specimens (DID) (DID NOT) attain the required f_c'._____
_____

Test Results (DID)   (DID NOT) comply with project specification_____

_____OF_____ WAS NOTIFIED OF THE TEST RESULTS
    SUPERVISOR/ENGINEER   NAME OF COMPANY

CYL. STORAGE LOCATION:_____   MIN-MAX TEMP:_____

APPLICABLE TEST PROCEDURE (U.N.O)
Sampling: ASTM C172-xx
Slump: ASTM C143-xx
Air content: ASTM C231-xx or ASTM C173-xx   Field Technician – Level/Certification No.
Temperature: ASTM C1064-xx
Density: ASTM C138-xx
Casting specimens: ASTM C31-xx   Lab Technician – Level/Certification No.
Unbonded caps: ASTM C1231-xx
Compressive strength: ASTM C39-xx

                                                             Reviewer – Level/Certification No.   Date

Fig. 20.9—Sample report of concrete lab and field tests.

CONCRETE FIELD TESTING DATA SHEET

[Name and Address of Laboratory]

CLIENT:									REPORT NO:		PAGE OF
PROJECT:									JOB NO:		
POUR LOCATION:									DATE:		
									CONCRETE CLASS:		
CU. YDS.	TIME	TRUCK NO.	TICKET NO.	SLUMP (in.)	AIR CONT. (%)	CON. TEMP. (°F)	DENSITY (lb/ft^3)	TEST SET NO'S	NOTES		

REMARKS:

APPLICABLE TEST PROCEDURES (U.N.O)

Sampling: ASTM C172-xx
Slump :   ASTM C143-xx
Air Content :   ASTM C231-xx or ASTM C173-xx
Temperature :   ASTM C1064-xx
Density :   ASTM C138-xx
Casting Specimens: ASTM C31-xx

TOTAL CU. YDS. PLACED:

SPEC. REQ'TS:
SLUMP: TEMP:
AIR: DENSITY:

_____
TECHNICIAN - Level/Certification No.

_____ _____
REVIEWER – Level/Certification No.   Date

Fig. 20.10—Sample concrete field testing data sheet.

FLEXURAL STRENGTH OF CONCRETE
ASTM C78

[Name and Address of Laboratory]

CLIENT:_____ REPORT NO.:_____ PAGE_____ OF_____
PROJECT:_____ PROJECT NO.:_____
LOCATION/PIECE ID'S:_____
_____
CONCRETE CLASS:_____ SPECIFIED STRENGTH:_____

DATE SAMPLED:_____
DATE RECEIVED:_____

TIME	TRUCK NO.	TICKET NO.	SLUMP (in.)	AIR CONT. (%)	CONC. TEMP (°F)	AMB. TEMP. (F)	DENSITY (lb/ft^3)

SAMPLE NUMBER	TEST DATE	AGE (days)	WIDTH (in.)	DEPTH (in.)	MAXIMUM LOADS (lb)	MODULUS OF RUPTURE (PSI)

NOTE: Lab specimens were immersed until time of test
Span length of all samples = 18 inches
Leather shims were used

APPLICABLE TEST PROCEDURE (U.N.O)
Sampling   ASTM C172-xx
Slump: ASTM C143-xx
Air Content: ASTM C231-xx or ASTM C173-xx
Temperature: ASTM C1064-xx
Density: ASTM C138-xx
Casting Specimens: ASTM C31-xx
Flexural Strength: ASTM C78-xx

_____
Field Technician – Level/Certification No.

_____   _____ _____
Lab Technician – Level/Certification No.   Reviewer – Level/Certification No.   Date

Fig. 20.11—Sample flexural strength of concrete sheet.

COMPRESSIVE STRENGTH TESTS OF CONCRETE CORES
ASTM C42

[Name and Address of Laboratory]

CLIENT: _____
PROJECT: _____
STRUCTURE OBTAINED FROM: _____
CORES OBTAINED AND IDENTIFIED BY: _____
DATE PLACED: _____
DATE CORED: _____
DATE TESTED: _____

REPORT NO.: _____ PAGE ____ OF ____
PROJECT NO.: _____
CONCRETE CLASS: _____ SPECIFIED STRENGTH: _____
MAX. NOMINAL AGG. SIZE: _____
TECHNICIAN: _____ Level/Certification No. _____
REVIEWER: _____ Level/Certification No.: _____

TEST RESULTS

SPEC. ID	LOCATION IN STRUCTURE	SPEC. LGTH. (IN.) BEFORE CAP	SPEC. LGTH. (IN.) AFTER CAP	SPEC. DIA. (IN.)	CROSS-SEC. AREA (IN.²)	DIRECTION OF LOAD TO HORIZ. PLANE	L/D	TOTAL LOAD (LB)	CORR. PSI*	TYPE FRAC.	DRILLED INCHES

NOTES: (1) Moisture condition at time of test: _____
(2) Test device I.D. Nos.: _____

REMARKS: * PSI corrected for L/D, as per ASTM C42

Fig. 20.12—Sample report of compressive strength of cores.

CHAPTER 21—INSPECTION CHECKLIST

For quick reference, this chapter lists the various items and activities that might be covered by inspection. For a particular job, inspectors should prepare a similar checklist that applies to the given specifications and job conditions. Detailed information regarding these subjects is contained in the foregoing chapters, as noted. Also refer to Appendix II of ACI 311.4R.

21.1—Materials (Chapters 3 and 4)
1. General (applies to all materials);
- Identification;
- Acceptability requirements;
- Quantities (used, on hand);
- Uniformity;
- Storage and handling;
- Sampling procedures;
- Schedule of testing;
- Cement and SCMs;
- Material certifications;
- Signs of contamination or exposure to moisture; and
- Temperature.

2. Aggregates
- Acceptability tests (gradation, organic impurities, deleterious substances, soundness, resistance to abrasion); and
- Control tests (moisture content, absorption, specific gravity, density, voids).

3. Admixtures
- Material certifications; and
- Stored without contamination or deterioration.

4. Reinforcing steel
- Material certifications;
- Surface condition (no damage or excessive rusting);
- Cleanliness (free of oil or nonadherent mortar); and
- Results of physical tests.

21.2—Proportioning of concrete mixtures (Chapter 6)
- Aggregate grading and proportioning;
- Statistical analysis of mixture performance;
- Mixture computations;
- Batch quantities;
- Yield computations;
- Air content; and
- Water-cementitious material ratio (w/cm).

21.3—Batching and mixing (Chapter 7)
1. Batching
- Allowable tolerances for cement, SCMs, aggregates, water, and admixtures;
- Calibration of scales and measuring equipment;
- Bins (adequate size, separation of compartments, free discharge of materials);
- Allowable water adjusted for free moisture in aggregates;
- Admixtures discharged separately; and
- Yield of concrete.

2. Mixing
- Condition of mixer and mixing blades;
- Charging of mixer;
- Mixing capacity of drum not exceeded;
- Duration of mixing;
- No batches delayed in mixer;
- Number of revolutions of drum;
- Water temperature; and
- Mixer uniformity tests.

21.4—Before concreting (Chapter 8)
1. Preliminary study
- Plans, specifications, and building codes;
- Division of duties among parties;
- Permissible tolerances of measurement;
- Provisions for records and reports; and
- Contractor's plant, calibrations, equipment, organization, and methods.

2. Excavations and foundations
- Location;
- Dimensions and slope;
- Compaction and drainage;
- Stability; and
- Preparation of surfaces.

3. Forms
- Specified type and dimensions;
- Location and alignment;
- Provision for settlement;
- Stability (adequate bracing, shores, ties, and spacers);
- Provision for inspection openings;
- Preparation of surfaces (cleanliness, coating for release);
- Tight joints; and
- Cleanliness.

4. Reinforcement in place
- Grade;
- Size (diameter, length);
- Bending details;
- Location (horizontal and vertical spacing, concrete cover);
- Splicing, welding, and anchoring; and
- Adequate support (wiring, chairs, spacers).

5. Embedded fixtures
- Location;
- Firmly fixed in position; and
- No displacement of reinforcement.

21.5—Concreting operations (Chapter 9)
1. Working conditions
- Adequate tools, personnel, and equipment for all operations;
- Preparations completed;
- Adherence to specified placing sequence;
- Lighting for night work;
- Provisions for curing; and
- Protection against sun, rain, and hot or cold weather.

2. Control of consistency
- Observation of concrete being placed;
- Schedule of testing;
- Adjustments of water or admixtures in mixture;
- Monitoring of air content; and
- Temperature check.

3. Conveying
- No segregation or loss of materials;
- Prevention of contamination; and
- No excessive stiffening or drying out.

4. Placing
- Preparation of contact surfaces;
- Vertical drop to avoid segregation;
- No displacement of forms or reinforcement;
- Little or no flow after depositing;
- Depth of layers;
- Prevention of rock pockets;
- Removal of temporary ties and spacers; and
- Disposal of rejected batches.

5. Consolidation
- Thorough and uniform; no overworking;
- Internal vibration (vibrator size, frequency, and amplitude; depth of insertion; proper spacing; no movement of concrete by vibrator); and
- External vibration (proper spacing; forms sufficiently rigid to withstand vibration).

6. Finishing of unformed surfaces
- Sufficient layer of surface mortar;
- No finishing while excess bleedwater is on the surface;
- Timing of first floating and final hard troweling;
- No overworking of surface;
- Alignment of surface; and
- Prevention of plastic shrinkage cracks.

7. Finishing of formed surfaces
- Condition of surfaces upon form removal (no honeycombing, peeling, ragged tie holes, or ragged form lines);
- Application of surface treatment;
- Uniformity of surface texture and color; and
- Repair of defects or imperfections.

8. Joints
- Construction and hinged (location, preparation of surfaces, dowels or ties in place and aligned if used);
- Contraction (location, forming, or tooling); and
- Expansion and isolation (location, alignment, joint filler material, and freedom from interference with subsequent movement).

21.6—After concreting (Chapter 10)

1. Supporting forms and shores
- Time of removal;
- Sequence of removal; and
- Location and number of reshores.

2. Protection of concrete from damage
- Impact;
- Overloading;
- Marring of surfaces; and
- Weather extremes (also refer to concreting in cold and hot weather)

3. Curing
- Time curing begins;
- Moist curing (surfaces kept continuously moist);
- Membrane curing (uniform and complete application of curing compound);
- Duration of curing period; and
- Hot and cold weather curing requirements.

4. Joints
- Cleaning and sealing; and
- Timing and alignment of sawn joints.

5. Verification of finished surfaces
- Gap under 10 ft (3 m) straightedge; or
- Measurement of F-numbers for flatness and levelness.

21.7—Special work

21.7.1 *Cold weather concreting*
- Heating of materials and contact surfaces;
- Provisions for protection (enclosures, insulated forms, artificial heating);
- Protection of concrete from drying or carbonation;
- Protection from too-rapid cooling upon removal of enclosures or forms; and
- Plan for cold weather protection.

21.7.2 *Hot weather concreting*
- Limiting combinations of wind, relative humidity, and ambient temperature;
- Cooling of materials;
- Prewetting aggregates and contact surfaces;
- Provisions for protection (windbreaks, fog sprays, plastic sheeting); and
- Plan for hot weather protection.

21.7.3 *Air-entrained concrete*
- Accurate measurement of air-entraining agent;
- Adjustment of mixture to compensate for air content;
- Avoidance of excessive mixing or vibration;
- Tests for air content of concrete (Chapter 19);
- Manner of placement; and
- Finishing (magnesium or aluminum float).

21.7.4 *Two-course floors (Chapter 12)*
- Preparation of base-course surface;
- Mixture proportions and consistency;
- Thickness of top course;
- Consolidation and finishing; and
- Curing.

21.7.5 *Architectural concrete (Chapter 14)*
- Requirements for acceptability;
- Preconstruction samples (mockups);
- Uniformity (materials, workmanship);
- Pigments (use of proper amount; thorough mixing of color with cement);
- Tight joints between form sheathing or lining;
- Form sealers and release agents to avoid sticking or staining;
- No reinforcement too close to surface;
- Vibration to minimize bug holes;
- Form removal to prevent marring;
- Surface treatments (abrasive blasting, bush hammering, grinding, acid etching, tooling);
- Color and texture match mockup;
- Filling of holes left by form ties; and
- Repair of surface defects.

21.7.6 *Tilt-up construction (Chapter 15)*
- Surface of casting platform (no defects; meets wall tolerances);

- Bond prevention (timing and uniformity of curing-compound application);
- Alignment of edge forms;
- Strength of concrete at time of lifting;
- Location of pick-up points;
- Avoidance of excessive pulling, jerking, or jarring during lifting;
- Adequate bracing; and
- Connections to columns (bolting, welding, or keying).

21.7.7 *Preplaced-aggregate concrete (Chapter 15)*
- Gradation and placement of coarse aggregate;
- No aggregate contamination before grouting;
- Void content;
- Grout composition and consistency;
- Condition of grouting equipment;
- Sequence and pressure of grouting operation; and
- Completeness of filling voids.

21.7.8 *Underwater construction (Chapter 15)*
- Avoidance of flowing water;
- Diameter of tremie pipe;
- Placement rate;
- Flowability of mixture; and
- Protection from flowing water for several days.

21.7.9 *Vacuum dewatering (Chapter 15)*
- Condition of vacuum mats;
- Timing and duration of vacuum application;
- Uniformity of processing; and
- Final thickness of slabs.

21.7.10 *Pumped concrete (Chapter 15)*
- Condition of equipment;
- Mixture proportioning to ensure pumpability;
- Pump location;
- Pump lines (minimum bends, firmly supported);
- Angle of boom to minimize freefall of concrete;
- Communication between pump operator and placing crew;
- Point of discharge (no segregation of mixture, displacement of reinforcement, or damage to forms);
- Avoidance of delays to ensure continuous pumping; and
- Complete removal of concrete from pump lines.

21.7.11 *Shotcrete (Chapter 15)*
- Materials (acceptability, proportions);
- Condition of equipment;
- Air pressures (uniform, proper velocity);
- Preparation of surfaces;
- Regulation of water content;
- Mixture uniformity;
- Application (thickness, no sagging, removal of rebound);
- Surface finish; and
- Curing and protection.

21.7.12 *Structural lightweight-aggregate concrete (Chapter 16)*
- Lightweight aggregates (acceptability, prewetting, preventing segregation);
- Mixture proportioning; and
- Consolidation and finishing (no overworking).

21.7.13 *Low-density concrete (Chapter 16)*
- Cellular type (admixtures, timing of operations, mixing processes, foaming agents); and
- Aggregate type (aggregate acceptability, absorption of aggregate, air entrainment).

21.7.14 *High-density concrete (Chapter 16)*
- Heavyweight aggregates (acceptability, specific gravity, gradation);
- No overloading of mixing and conveying equipment; and
- Uniform coarse-aggregate distribution.

21.7.15 *Mass concrete (Chapter 16)*
- Time and rate of placement;
- Avoidance of high or nonuniform temperatures;
- Bonding of lifts;
- Adequate vibration; and
- Prevention of aggregate breakage.

21.7.16 *Shrinkage-compensating concrete (Chapter 16)*
- Acceptability of materials (expansive cement);
- Mixture proportioning and control to ensure required expansion;
- Reinforcement to provide proper restraint;
- Control of slump; and
- Curing and protection.

21.7.17 *Prestressed concrete (Chapter 17)*
- Accurate placement of tendons;
- Secure tying of tendons to prevent displacement;
- Sheathing of tendons (if specified);
- Avoidance of obstruction or excessive friction;
- Strength of concrete at time of prestressing;
- Measurement of tension (by means of jack pressure, lengthening of steel, or both);
- Proper anchoring; and
- Thoroughness of grouting (if specified).

21.7.18 *Grouting under base plates (Chapter 18)*
- Preparation of base;
- Proper mixture (sufficient strength to resist applied loads); and
- Complete filling of voids.

21.7.19 *Pressure grouting (Chapter 18)*
- Drill holes (depth, spacing, freedom from clogging);
- Materials (acceptability, quantities used); and
- Injection (sequence, pressure, completeness of penetration, no damage to structure).

21.7.20 *Mortar and stucco (Chapter 18)*
- Mixture proportions;
- Preparation of backing surface;
- Thickness of each coat;
- Uniform finish; and
- Curing of each coat.

21.7.21 *Tests of concrete (Chapter 19)*

1. Tests of fresh concrete
- Consistency;
- Air content;
- Density;
- Temperature; and
- Uniformity.

2. Strength tests
- Molding specimens;
- Capping cylindrical specimens;
- Field curing of specimens; and

- Shipping specimens to laboratory.
3. Tests of hardened concrete
- Cores for compressive strength;
- Beams for flexural strength; and
- Nondestructive (pulse velocity, impact hammer, penetration probe)
4. Other tests, as needed

21.7.22 *Records and reports (Chapter 20)*
1. Documentation of data and observations pertaining to:
- Ingredients in the concrete mixture;
- Checking of production equipment and procedures;
- Observation of placement and curing;
- Field testing of plastic concrete; and
- Laboratory testing of hardened specimens.

2. Inclusion of supporting data (tables, graphs, sketches, photos)

CHAPTER 22—REFERENCES
22.1—Referenced standards and reports

This chapter lists standards, test methods, and recommended practices of ACI, ASTM International, and other organizations (when the subject is not covered by ACI or ASTM). To obtain copies of the various standards, use the contact information provided. Because these documents are revised frequently, be sure to obtain the most recent revision of a given standard. (The year of adoption typically follows the serial designation.)

Most of these standards, reports, and guides are available as separate documents. It is also possible to purchase and download individual ASTM standards and selected ACI committee documents in portable document format (PDF) from the websites of these organizations.

American Association of State Highway and Transportation Officials (AASHTO)
444 N. Capitol Street, NW, Suite 249
Washington, DC 20001
www.aashto.org

American Concrete Institute (ACI)
P.O. Box 9094
Farmington Hills, MI 48333-9094
www.concrete.org

American Railway Engineering and Maintenance-of-Way Association (AREMA)
8201 Corporate Dr., #1125
Landover, MD 20785
www.arema.org

ASTM International
100 Barr Harbor Dr.
West Conshohocken, PA 19428
www.astm.org

American Welding Society (AWS)
550 N.W. LeJeune Rd.
Miami, FL 33126
www.aws.org

American Concrete Institute

116R	Cement and Concrete Terminology
117	Specifications for Tolerances for Concrete Construction and Materials
201.1R	Guide For Making a Condition Survey of Concrete in Service
201.2R	Guide to Durable Concrete
207.1R	Guide to Mass Concrete
207.3R	Practices for Evaluation of Concrete in Existing Massive Structures for Service Conditions
207.5R	Roller-Compacted Mass Concrete
211.1	Standard Practice for Selecting Proportions for Normal, Heavyweight, and Mass Concrete
211.2	Standard Practice for Selecting Proportions for Structural Lightweight Concrete
211.3R	Guide for Selecting Proportions for No-Slump Concrete
212.3R	Chemical Admixtures for Concrete
212.4R	Guide for the Use of High-Range Water-Reducing Admixtures (Superplasticizers) in Concrete
213R	Guide for Structural Lightweight-Aggregate Concrete
214R	Evaluation of Strength Test Results of Concrete
223	Standard Practice for the Use of Shrinkage-Compensating Concrete
224.1R	Causes, Evaluation, and Repair of Cracks in Concrete Structures
224.3R	Joints in Concrete Construction
228.1R	In-Place Methods to Estimate Concrete Strength
228.2R	Nondestructive Test Methods for Evaluation of Concrete in Structures
232.1R	Use of Raw or Processed Natural Pozzolans in Concrete
232.2R	Use of Fly Ash in Concrete
233R	Slag Cement in Concrete and Mortar
234R	Guide for the Use of Silica Fume in Concrete
301	Specifications for Structural Concrete
302.1R	Guide for Concrete Floor and Slab Construction
302.2R	Guide for Concrete Slabs that Receive Moisture-Sensitive Flooring Materials
303R	Guide to Cast-in-Place Architectural Concrete Practice
304R	Guide for Measuring, Mixing, Transporting, and Placing Concrete
304.1R	Guide for the Use of Preplaced Aggregate Concrete for Structural and Mass Concrete Applications
304.2R	Placing Concrete by Pumping Methods
304.3R	Heavyweight Concrete: Measuring, Mixing, Transporting, and Placing
304.6R	Guide for the Use of Volumetric-Measuring and Continuous-Mixing Concrete Equipment
305R	Hot Weather Concreting
306R	Cold Weather Concreting
308R	Guide for Curing Concrete
309R	Guide for Consolidation of Concrete
311.4R	Guide for Concrete Inspection
311.5	Guide For Concrete Plant Inspection and Field Testing of Ready-Mixed Concrete

318	Building Code Requirements for Structural Concrete
325.6R	Texturing Concrete Pavements
325.9R	Guide for Construction of Concrete Pavements and Concrete Bases
325.12R	Guide for Design of Jointed Concrete Pavements for Streets and Local Roads
336.1	Specification for the Construction of Drilled Piers
336.3R	Design and Construction of Drilled Piers
345R	Guide for Concrete Highway Bridge Deck Construction
346	Specification for Cast-in-Place Concrete Pipe
347	Guide to Formwork for Concrete
349	Code Requirements for Nuclear-Safety-Related Concrete Structures
349.3R	Evaluation of Existing Nuclear Safety-Related Concrete Structures
359	Code for Concrete Containments
363.2R	Guide to Quality Control and Testing of High-Strength Concrete
364.1R	Guide for Evaluation of Concrete Structures before Rehabilitation
423.3R	Recommendations for Concrete Members Prestressed with Unbonded Tendons
437R	Strength Evaluation of Existing Concrete Buildings
503.4	Standard Specification for Repairing Concrete with Epoxy Mortars
504R	Guide to Joint Sealants in Concrete Structures
506R	Guide to Shotcrete
506.2	Specification for Shotcrete
515.1R	Guide for the Use of Waterproofing, Dampproofing, Protective, and Decorative Barrier Systems for Concrete
523.1R	Guide for Cast-in-Place Low-Density Cellular Concrete
523.2R	Guide for Precast Cellular Concrete Floor, Roof, and Wall Units
523.3R	Guide for Cellular Concretes Above 50 pcf and for Aggregate Concrete Above 50 pcf with Compression Strength Less than 2500 psi
524R	Guide to Portland Cement-Based Plaster
546R	Concrete Repair Guide
546.1R	Guide for Repair of Concrete Bridge Superstructures
546.2R	Guide to Underwater Repair of Concrete
548.1R	Guide for Use of Polymers in Concrete
551R	Tilt-Up Concrete Structures

American Railway Engineering and Maintenance-of-Way Association
AREMA Manual of Railway Engineering

ASTM International

A325	Specification for Structural Bolts, Steel, Heat Treated, 120/105 ksi Minimum Tensile Strength
A416/A416M	Specification for Steel Strand, Uncoated Seven-Wire for Prestressed Concrete
A421/A421M	Specification for Uncoated Stress-Relieved Steel Wire for Prestressed Concrete
A490	Specification for Structural Bolts, Alloy Steel, Heat-Treated, 150 ksi Minimum Tensile Strength
A722/A722M	Specification for Uncoated High-Strength Steel Bars for Prestressing Concrete
A775/A775M	Specification for Epoxy-Coated Steel Reinforcing Bars
C29/C29M	Test Method for Bulk Density (Unit Weight) and Voids in Aggregate
C31/C31M	Practice for Making and Curing Concrete Test Specimens in the Field
C33	Specification for Concrete Aggregates
C39/C39M	Test Method for Compressive Strength of Cylindrical Concrete Specimens
C40	Test Method for Organic Impurities in Fine Aggregates for Concrete
C42/C42M	Test Method for Obtaining and Testing Drilled Cores and Sawed Beams of Concrete
C70	Test Method for Surface Moisture in Fine Aggregate
C78	Test Method for Flexural Strength of Concrete (Using Simple Beam with Third-Point Loading)
C87	Test Method for Effect of Organic Impurities in Fine Aggregate on Strength of Mortar
C88	Test Method for Soundness of Aggregates by Use of Sodium Sulfate or Magnesium Sulfate
C94/C94M	Specification for Ready-Mixed Concrete
C109/C109M	Test Method for Compressive Strength of Hydraulic Cement Mortars (Using 2-in. or [50-mm] Cube Specimens)
C117	Test Method for Materials Finer than 75-μm (No. 200) Sieve in Mineral Aggregates by Washing
C123	Test Method for Lightweight Particles in Aggregate
C127	Test Method for Density, Relative Density (Specific Gravity), and Absorption of Coarse Aggregate
C128	Test Method for Density, Relative Density (Specific Gravity), and Absorption of Fine Aggregate
C131	Test Method for Resistance to Degradation of Small-Size Coarse Aggregate by Abrasion and Impact in the Los Angeles Machine
C136	Test Method for Sieve Analysis of Fine and Coarse Aggregates
C138/C138M	Test Method for Density (Unit Weight), Yield, and Air Content (Gravimetric) of Concrete
C142	Test Method for Clay Lumps and Friable Particles in Aggregates

C143/C143M	Test Method for Slump of Hydraulic-Cement Concrete	C494/C494M	Specification for Chemical Admixtures for Concrete
C144	Specification for Aggregate for Masonry Mortar	C495	Test Method for Compressive Strength of Lightweight Insulating Concrete
C150	Specification for Portland Cement	C496/C496M	Test Method for Splitting Tensile Strength of Cylindrical Concrete Specimens
C172	Practice for Sampling Freshly Mixed Concrete	C511	Specification for Mixing Rooms, Moist Cabinets, Moist Rooms, and Water Storage Tanks Used in the Testing of Hydraulic Cements and Concretes
C173/C173M	Test Method for Air Content of Freshly Mixed Concrete by the Volumetric Method		
C177	Test Method for Steady-State Heat Flux Measurements and Thermal Transmission Properties by Means of the Guarded-Hot-Plate Apparatus	C518	Test Method for Steady-State Thermal Transmission Properties by Means of the Heat Flow Meter Apparatus
C183	Practice for Sampling and the Amount of Testing of Hydraulic Cement	C535	Test Method for Resistance to Degradation of Large-Size Coarse Aggregate by Abrasion and Impact in the Los Angeles Machine
C191	Test Method for Time of Setting of Hydraulic Cement by Vicat Needle		
C192/C192M	Practice for Making and Curing Concrete Test Specimens in the Laboratory	C566	Test Method for Total Evaporable Moisture Content of Aggregate by Drying
C227	Test Method for Potential Alkali Reactivity of Cement-Aggregate Combinations (Mortar-Bar Method)	C567	Test Method for Determining Density of Structural Lightweight Concrete
		C586	Test Method for Potential Alkali Reactivity of Carbonate Rocks for Concrete Aggregates (Rock-Cylinder Method)
C231	Test Method for Air Content of Freshly Mixed Concrete by the Pressure Method		
C234	Test Method for Comparing Concretes on the Basis of Bond Developed with Reinforcing Steel (withdrawn 2000)	C595	Specification for Blended Hydraulic Cements
		C597	Test Method for Pulse Velocity Through Concrete
C260	Specification for Air-Entraining Admixtures for Concrete	C617	Practice for Capping Cylindrical Concrete Specimens
C289	Test Method for Potential Alkali-Silica Reactivity of Aggregates (Chemical Method)	C618	Specification for Coal Fly Ash and Raw or Calcined Natural Pozzolan for Use in Concrete
C293	Test Method for Flexural Strength of Concrete (Using Simple Beam with Center-Point Loading)	C666/C666M	Standard Test Method for Resistance of Concrete to Rapid Freezing and Thawing
		C682	Practice for Evaluation of Frost Resistance of Coarse Aggregates in Air-Entrained Concrete by Critical Dilation Procedures (withdrawn 2003)
C295	Guide for Petrographic Examination of Aggregates for Concrete		
C309	Specification for Liquid Membrane-Forming Compounds for Curing Concrete	C684	Test Method for Making, Accelerated Curing, and Testing Concrete Compression Test Specimens
C330	Specification for Lightweight Aggregates for Structural Concrete		
C332	Specification for Lightweight Aggregates for Insulating Concrete	C685/C685M	Specification for Concrete Made by Volumetric Batching and Continuous Mixing
C359	Test Method for Early Stiffening of Hydraulic Cement (Mortar Method)	C702	Practice for Reducing Samples of Aggregate to Testing Size
C403/C403M	Test Method for Time of Setting of Concrete Mixtures by Penetration Resistance	C803/C803M	Standard Test Method for Penetration Resistance of Hardened Concrete
C451	Test Method for Early Stiffening of Hydraulic Cement (Paste Method)	C805	Test Method for Rebound Number of Hardened Concrete
C457	Test Method for Microscopical Determination of Parameters of the Air-Void System in Hardened Concrete	C823	Practice for Examination and Sampling of Hardened Concrete in Constructions
		C845	Specification for Expansive Hydraulic Cement
C469	Test Method for Static Modulus of Elasticity and Poisson's Ratio of Concrete in Compression	C856	Practice for Petrographic Examination of Hardened Concrete
C470/C470M	Specification for Molds for Forming Concrete Test Cylinders Vertically	C876	Test Method for Half-Cell Potentials of Uncoated Reinforcing Steel in Concrete

C878/C878M	Test Method for Restrained Expansion of Shrinkage-Compensating Concrete
C939	Test Method for Flow of Grout for Preplaced-Aggregate Concrete (Flow Cone Method)
C942	Test Method for Compressive Strength of Grouts for Preplaced-Aggregate Concrete in the Laboratory
C989	Specification for Ground Granulated Blast-Furnace Slag for Use in Concrete and Mortars
C1017/C1017M	Specification for Chemical Admixtures for Use in Producing Flowing Concrete
C1064/C1064M	Test Method for Temperature of Freshly Mixed Hydraulic-Cement Concrete
C1074	Practice for Estimating Concrete Strength by the Maturity Method
C1077	Practice for Laboratories Testing Concrete and Concrete Aggregates for Use in Construction and Criteria for Laboratory Evaluation
C1152/C1152M	Test Method for Acid-Soluble Chloride in Mortar and Concrete
C1157	Performance Specification for Hydraulic Cement
C1231/C1231M	Practice for Use of Unbonded Caps in Determination of Compressive Strength of Hardened Concrete Cylinders
C1240	Specification for Silica Fume Used in Cementitious Mixtures
C1260	Test Method for Potential Alkali Reactivity of Aggregates (Mortar-Bar Method)
C1293	Test Method for Determination of Length Change of Concrete Due to Alkali-Silica Reaction
C1383	Test Method for Measuring the P-Wave Speed and the Thickness of Concrete Plates Using the Impact-Echo Method
D75	Practice for Sampling Aggregates
D95	Test Method for Water in Petroleum Products and Bituminous Materials by Distillation
D98	Specification for Calcium Chloride
D448	Classification for Sizes of Aggregate for Road and Bridge Construction
D512	Test Methods for Chloride Ion In Water
D974	Test Method for Acid and Base Number by Color-Indicator Titration
D2419	Test Method for Sand Equivalent Value of Soils and Fine Aggregate
D3665	Practice for Random Sampling of Construction Materials
D4748	Test Method for Determining the Thickness of Bound Pavement Layers Using Short-Pulse Radar
D4788	Test Method for Detecting Delaminations in Bridge Decks Using Infrared Thermography
D4791	Test Method for Flat Particles, Elongated Particles, or Flat and Elongated Particles in Coarse Aggregate
E4	Practices for Force Verification of Testing Machines
E11	Specification for Wire Cloth and Sieves for Testing Purposes
E105	Practice for Probability Sampling of Materials
E329	Specification for Agencies Engaged in Construction Inspection and/or Testing
E1155	Standard Test Method for Determining F_F Floor Flatness and F_L Floor Levelness Numbers

American Welding Society
D1.1	Structural Welding Code—Steel
D1.4	Structural Welding Code—Reinforcing Steel

22.2—Cited references

American Concrete Institute, 1999, *Field Reference Manual: Standard Specifications for Structural Concrete for Building with Selected ACI and ASTM References*, SP-15(99), Farmington Hills, MI, 384 pp.

American Concrete Institute, 2003, *Concrete Repair Manual*, 2nd Edition, Farmington Hills, MI, 2093 pp.

American Institute of Steel Construction, 2006, *Steel Construction Manual*, 13th Edition, AISC, Chicago.

American Society of Civil Engineers, 2000, "Guideline for Structural Condition Assessment of Existing Buildings," ASCE Standard No. 11, Reston, VA, 160 pp.

ASTM International, 1994, *Significance of Tests and Properties of Concrete and Concrete-Making Materials*, STP 169-C, West Conshohocken, PA, 623 pp.

ASTM International, *Annual Book of ASTM Standards*, V. 04.02, Concrete and Aggregates, updated annually.

California Department of Transportation, 1968, "Statistical Quality Control of Highway Construction Materials," *Highway Research Report* No. M&R 631133-9, Final Report, Sacramento, CA, May.

Concrete Reinforcing Steel Institute, 1997a, *Placing Reinforcing Bars*, 7th Edition, Schaumburg, IL, 232 pp.

Concrete Reinforcing Steel Institute, 1997b, *Manual of Standard Practice*, 26th Edition, Schaumburg, IL, 97 pp.

Davis, R. E.; Carlson, R. W.; Kelly, J. W.; and Davis, H. E., 1937, "Properties of Cements and Concretes Containing Fly Ash," ACI JOURNAL, *Proceedings* V. 33, No. 5, May, pp. 577-612.

Dobrowolski, J. A., 1998, *Concrete Construction Handbook*, 4th Edition, McGraw-Hill Book Co., New York, 978 pp.

Feld, J., 1964, *Lessons from Failures of Concrete Structures*, ACI Monograph 1, American Concrete Institute, Farmington Hills, MI, 179 pp.

Hurd, M. K., 2005, *Formwork for Concrete*, SP-4, 7th Edition, American Concrete Institute, Farmington Hills, MI, 500 pp.

Klieger, P., 1952, "Studies of the Effect of Entrained Air on Strength and Durability of Concretes Made with Various

Maximum Sizes of Aggregates," *Research Department Bulletin* RX040, Portland Cement Association, Skokie, IL.

Malhotra, V. M., ed., 1992, *Fly Ash, Silica Fume, Slag, and Natural Pozzolans in Concrete,* Proceedings of the Fourth International Conference, Istanbul, Turkey, SP-132, American Concrete Institute, Farmington Hills, MI, 1692 pp.

McMillan, F. R., and Tuthill, L. H., 1987, *Concrete Primer*, SP-1, 4th Edition, American Concrete Institute, Farmington Hills, MI, 96 pp.

National Ready Mixed Concrete Association, 1992, "Certification of Ready-Mixed Concrete Production Facilities," Silver Spring, MD, 21 pp.

National Ready Mixed Concrete Association, 2002, "Plant Certification Checklist," Silver Spring, MD.

Pennsylvania State University, 1974, "Statistical Quality Control of Highway Construction," V. 1 and 2, College of Engineering, University Park, PA, Dec.

Portland Cement Association, 1995, *Color and Texture in Architectural Concrete*, SP021, Skokie, IL.

Portland Cement Association, 2002, *Design and Control of Concrete Mixtures*, 14th Edition, EB001, Skokie, IL, 212 pp.

Portland Cement Association, 2004, *Finishing Concrete with Color and Texture*, PA 124, Skokie, IL, 72 pp.

Post-Tensioning Institute, 1990, *Post-Tensioning Manual*, 5th Edition, Phoenix, AZ, 406 pp.

Precast/Prestressed Concrete Institute, 1989, *Architectural Precast Concrete*, MNL 122-89, Chicago, 352 pp.

Precast/Prestressed Concrete Institute, 1996, *Manual for Quality Control for Plants and Production of Architectural Precast Concrete Products*, MNL 117-96, Chicago, 226 pp.

Precast/Prestressed Concrete Institute, 1999a, *Erectors Manual*, MNL 127, Chicago, 96 pp.

Precast/Prestressed Concrete Institute, 1999b, *Manual for Quality Control for Plants and Production of Structural Precast Concrete Products*, MNL 116-99, Chicago, 340 pp.

U.S. Army Corps of Engineers, 1994, *Standard Practice for Concrete for Civil Work Structures*, EM 1110-2-2000, Washington, DC, Feb.

U.S. Army Corps of Engineers, 1995, *Evaluation and Repair of Concrete Structures*, EM 1110-2-2002, Washington, DC, June.

U.S. Bureau of Reclamation, 1988, *Concrete Manual*, 8th Edition, Denver, 665 pp.

van Aardt, J.H.P., and Visser, S., "Thaumasite Formation: A Cause of Deterioration of Portland Cement and Related Substances in the Presence of Sulfates," *Cement and Concrete Research*, V. 5, 1975.

INDEX

A

Absolute volume, 48, 49
Absorption, 29
 —Coarse aggregate, 29, 34
 —Fine aggregates, 29, 35
 —Tests, 28
Accelerators, 57
Acceptance
 —Final, 102, 113
 —Inspection, 86
 —Quality, 10
Admixtures, 31, 32, 38, 111
 —High-range water-reducing, 57
 —Measuring, 56
Aggregates, 25, 27, 34
 —Exposed aggregate finish, 112
 —Handling and storage, 36
 —Heavyweight, 127
 —Lightweight, 123, 125
 —Proportioning, 110
 —Sampling, 33
Air content, 61, 145
Air entrainment, 42, 46, 47
Air-free density, 150, 151
Anchorages, 140
Architectural concrete, 107
 —Curing, 112
 —Forms, 109
 —Materials, 110
 —Reinforcement, 110
 —Repairs, 113
 —Uniformity, 109

B

Base plates, grouting, 141
Batching, 53
 —Equipment, 56
 —Operations, 53
Bins, 38, 55
Bond prevention, 115, 116
Bridge decks, 96, 105, 106
Bug hole, 113
Bush hammering, 112

C

Cast-in-place pipe, 115
Cast-in-place prestressed concrete, 139
Cement, 23-25, 110, 111
 —Handling and storage, 36
Cementitious material, 38, 45
Cementitious systems, 142
Checklist, 168
Clay lumps, 34
Coarse aggregate, 33, 34, 47, 150
Completed structures, 152
Compressive strength, 147, 149
 —Normal distribution, 14
Computations, yield, 52
Concrete, 38
 —Fresh, 39
 —Freshly mixed, 145
 —Hardened, 41
 —Mixture proportioning, 43
 —Placing, 71
 —Precast, 132
 —Precast prestressed, 137
 —Preplaced-aggregate, 117
 —Prestressed, 132, 137, 139
 —Shrinkage-compensating, 130
 —Special types of, 123-132
 —Testing, 144
Concrete materials, control charts, 22
Concreting, 64, 71
Consistency, 62, 145
Consolidation, 77
Construction, 94
 —Lift-slab, 116
 —Tilt-up, 115
 —Underwater concrete, 117
Construction joints, 80
Contraction joints, 104
 —Longitudinal, 104
 —Transverse, 103
Control charts, 15-17
 —Concrete materials, 22
Conveying, 72
Curing, 41, 83, 84
 —Accelerated, 84, 149
 —Compounds, 32
 —Membrane, 83
 —Moist, 83
 —Pavements, 102
 —Protection, 82, 84-86

D

Damage, 82
Damp-pack mortar, 141
Dams, 128
Deflection, 138
Density, 125, 146
 —Structural lightweight concrete, 151
 —Tests for, 150, 151
Design reference sample, 108
Detensioning, 138

E

Embedded fixtures, 69
Equipment, 59
 —Batching, 56
 —Mass concrete for dams, 129

—Underwater, 117-119
—Weighing, 54
Excavation, 64
Expansion joints, 104
Exposed aggregate finish, 112

F

Final inspection, 70
Finish demonstration, 109
Finishing, 78
    —Formed surfaces, 80
    —Unformed surfaces, 78
Foams (cellular concrete), 126
Form removal, 81, 110
Form sealers, 110
Forms, 65, 122
    —Architectural concrete, 109
    —Pavements, 96
    —Slipforms, 100
Formwork, 113, 134
Foundations, 64
Friable particles, 34

G

Grading, 96
Gravimetric method, air content, 146
Grinding, 112
Grouting, 117, 140
    —Base plates, 141
    —Machine bases, 141
Grouts, gas-forming, 142

H

Handling and storage, 36, 137
    —Admixtures, 38
    —Aggregates, 36, 37
    —Cement, 36
Hardened concrete, 41
Heat of hydration, 40
Hot weather, 85

I

Inspection, 6, 7, 64, 133, 140
    —Acceptance, 86
    —Checklist, 168
    —Final, 70
    —Placing, 62
    —Plants, 60
Inspector, 6, 8

J

Joint construction, 95
Joint materials, 32
Joint sealing, 105
Joints, 70
    —Construction, 80
        —Longitudinal, 104
        —Transverse, 103
    —Form, 109

L

Lift-slab construction, 116
Lightweight fill concrete, 125
Lightweight insulating concrete, 149
Load test, 153
Longitudinal construction joints, 104
Longitudinal contraction joints, 104

M

Machine bases, grouting, 141
Mass concrete, 129
    —Dams, 128
Membrane curing, 83
Mixture requirements, 92
Mixers, 57-60, 150
Mixing, 53, 57
Mixing plants, inspection, 60
Mixture proportioning, 43, 99
    —Lightweight fill concrete, 125
    —Low-density concrete, 125
    —Mass concrete for dams, 128
    —Shrinkage-compensating concrete, 130
    —Structural lightweight-aggregate concrete, 123
    —Supplementary cementitious materials, 38, 45
Mockup, 108
Moist curing, 83
Mortar, 140, 143

N

Nondestructive tests, 153
Nonshrink grouts, 143
Normal distribution, 13, 14, 20

O

Organic impurities, 28, 34

P

Paste, 47
    —Quantity, 46
Pavements, 96
    —Concrete, 98
    —Forms, 97
    —Foundation, 96
    —Steel reinforcement, 98
Paving, 99, 100
Placing, 70, 71
    —Architectural concrete, 111
    —Special types of concrete, 124
Placing inspection, 62
Planned construction joints, 80
Polymer system, 143
Post-tensioning, 139
Premature traffic, 105
Pressure grouting, 140
Pressure method, air content, 145
Pretensioning, 137
Proportioning, 43-45
Protection, 81-82, 84-85, 112

Q

Quality assurance, 10, 11
Quality control, 10
Quality-control charts, 16, 17, 154

R

Random numbers, 17
Ready mixed concrete, 59
Records, 154
Reinforcement, 67, 98, 110
—Positioning of, 92
—Precast concrete, 135
—Steel, 31, 98, 114
Repairs, 113
—Architectural, 113
—Bug hole, 113
Reports, 154
Reshoring, 81

S

Safety, 10, 122
Samples, 25, 26, 108, 120
—Shipping and handling, 154
Sampling, 18, 25, 144
—Aggregates, 33
—Random, 17
Sand, tests for, 27-29
Sandblasting, 112
Settlement, 40, 66
Shoring, 66
—Reshoring, 81
Shotcrete, 121
Site
—Mixing, 57
Slabs, 92, 95, 96, 116
Slabs-on-ground, 93
—Building, 65
Slipforming
—Cast-in-place pipe, 115
—Vertical structures, 113
Specific gravity
—Coarse aggregate, 34
—Fine aggregates, 35
—Tests for, 30
Specifications, 10, 44
Stabilized base, 97
Standard deviation, 21
Strength
—Specified, 44
—Splitting tensile, 149
Strength tests, 146
Structural slabs, 95
Stucco, 140, 143
Subbase, 93, 96
Subgrade, 93, 96
Surface treatments, 111

T

Temperature control, 61
Tendons, 137-139
Tests
—Aggregate grading, 27
—Aggregates, 27, 28, 54, 153
—Bulk density, 30
—Completed structures, 152
—Concrete, 145
—Fine materials, 27
—Moisture and absorption, 28, 29
—Specific gravity, 30
—Strength, 146, 147
—Uniformity, 150
—Voids, 30
Texturing, 101
Transporting, 38, 59, 111
Tremie concrete, 117
Two-course construction, 94

U

Underwater construction, 65, 117
Unformed surfaces, 78
Uniformity, 59, 109
—Tests, 150
Unplanned construction joints, 81

V

Vertical structures, 113
Vibration, 77, 99
Vibrators, 77, 78
Voids, tests for, 30
Volumetric batching, 60
Volumetric method, air content, 145

W

Water-cementitious material ratio, 39, 45
Weighing equipment, 54
Workability, 39

Y

Yield, 52

ACI 311.4R-05

Guide for Concrete Inspection

Reported by ACI Committee 311

George R. Wargo Chair	Michael T. Russell Secretary

Gordon A. Anderson	John V. Gruber	Venkatesh S. Iyer	Woodward L. Vogt
Joseph F. Artuso	Jimmie D. Hannaman, Jr.	Claude E. Jaycox	Bertold E. Weinberg
Jose Damazo-Juarez	Robert L. Henry	Robert S. Jenkins	Michelle L. Wilson
Mario R. Diaz	Charles J. Hookham	Roger D. Tate	Roger E. Wilson
Donald E. Dixon			

This guide discusses the need for inspection of concrete construction and other related activities, the types of inspection activities involved, and the responsibilities of various individuals and organizations involved in these activities. Field and laboratory testing activities are also considered part of inspection. This guide presents recommendations for inspection plan content and a detailed checklist of inspection attributes that can be adopted for use depending on the scope and needs of individual projects.

Keywords: concrete; construction; inspection; quality assurance; quality control; testing.

CONTENTS
Chapter 1—Introduction, p. 1
1.1—Scope
1.2—Philosophy
1.3—General
1.4—Definitions
1.5—Categories of inspection
1.6—Inspection team

Chapter 2—Responsibilities, p. 3
2.1—Scope
2.2—Owner's responsibilities
2.3—Architect/engineer's inspection responsibilities
2.4—Owner's inspection organization responsibilities
2.5—Contractor's inspection responsibilities
2.6—Manufacturer's or fabricator's inspection responsibilities

Chapter 3—Planning for inspection, p. 5
3.1—Scope
3.2—Written inspection plan
3.3—Building code requirements for special inspections
3.4—Preconstruction conferences
3.5—Meetings
3.6—Qualifications of inspection and testing personnel
3.7—Recommendations for inspection and testing
3.8—Reporting and evaluating inspection and test results

Chapter 4—References, p. 7
4.1—Referenced standards and reports
4.2—Cited references
4.3—Other references

Appendix I—Expanded checklist of inspection attributes, p. 8

Appendix II—Synopsis of ACI 311.5, p. 13

ACI Committee Reports, Guides, and Commentaries are intended for guidance in planning, designing, executing, and inspecting construction. This document is intended for the use of individuals who are competent to evaluate the significance and limitations of its content and recommendations and who will accept responsibility for the application of the material it contains. The American Concrete Institute disclaims any and all responsibility for the stated principles. The Institute shall not be liable for any loss or damage arising therefrom.
Reference to this document shall not be made in contract documents. If items found in this document are desired by the Architect/Engineer to be a part of the contract documents, they shall be restated in mandatory language for incorporation by the Architect/Engineer.

CHAPTER 1—INTRODUCTION
1.1—Scope
This document is primarily intended for guidance in the development of inspection and testing plans that are part of the overall system designed to ensure quality in the finished concrete product. ACI Committee 311 recommends that the owner develop a quality plan, as outlined in ACI 121R, and

ACI 311.4R-05 supersedes ACI 311.4R-00 and became effective October 12, 2005.
Copyright © 2005, American Concrete Institute.
All rights reserved including rights of reproduction and use in any form or by any means, including the making of copies by any photo process, or by electronic or mechanical device, printed, written, or oral, or recording for sound or visual reproduction or for use in any knowledge or retrieval system or device, unless permission in writing is obtained from the copyright proprietors.

that 311.4R be used to develop inspection and testing plans by those organizations assigned by the owner's quality plan to conduct inspections.

1.2—Philosophy

Inspection and testing requirements typically vary, based on the specific scope and needs of construction, and should therefore be tailored to each project individually. The content of an inspection plan is dependent on the type and complexity of the project, special features involved, quality level desired, building code requirements, and the responsibilities of the inspection organization performing the work. Any of these may necessitate the addition of more detailed inspection than conventional or may warrant a reduction from conventional requirements.

1.3—General

Inspection is simply a subsystem of the quality plan. It may be employed by the owner to evaluate future acceptance of the work or by contractors and material producers for quality-control purposes. In addition, inspection may be part of a program of activities performed by government agencies charged with enforcing building codes and other government regulations. The inspection process does not add quality to inspected items. Inspection simply establishes the status of inspected items relative to specified requirements. The information derived from inspections and tests, however, when properly evaluated, and with conclusions and decisions implemented, can result in the improvement of the quality of the product or process. The specified quality is achieved only by implementation of an adequate quality plan. Such a plan affects the entire project, from planning through design and construction to acceptance by the owner. Quality of work during the construction phase is achieved almost entirely by the contractor or producer's quality-control program. This quality-control program involves everyone from management to field supervisors to workers. Quality assurance and quality control should have strong, active support from top management and the active concern and participation of everyone involved in the construction process. Inspection and testing are only a part, though a very important part, of both quality-assurance and quality-control programs.

1.4—Definitions

1.4.1 *Quality assurance (QA)*—A management tool for all planned and systematic actions necessary to ensure that the final product meets the requirements of the contract documents and standards of good practice for the work.

1.4.2 *Quality control (QC)*—Actions taken by a contractor or material producer to provide and document control over what is being done and what is being provided so that the applicable standards of good practice and the contract documents for the work are followed.

1.4.3 *Owner*—The individual or organization having financial and legal responsibility for construction of a project, as well as bearing the ultimate responsibility for the public health, welfare, and safety related to the project. The term "owner" includes those organizations or individuals acting as agents for the owner.

1.4.4 *Architect/engineer (A/E)*—The architect, engineer, architectural firm, engineering firm, or architectural and engineering firm issuing project drawings and specifications, administering the work under contract specifications and drawings, or both.

1.4.5 *Contractor*—The organization responsible for constructing a project according to the project specifications and design drawings. The contractor may also possess the responsibilities of the A/E in designing and building the project and contract execution.

1.4.6 *Construction manager or owner's representative*—The person or management organization responsible to the owner for coordination and review of all contracted work. The person's or organization's role is to coordinate and communicate the entire scope of work to achieve a more efficient construction process.

1.4.7 *Inspection organization*—The organization, agency, or testing laboratory that is responsible for providing inspection and testing for the owner or for providing quality-control inspection and testing for the contractor or producer.

1.4.8 *Inspection*—Visual observations, measurements, and field and laboratory testing of activities, components, and materials to specified requirements along with the recording and evaluation of such data.

1.4.9 *Inspection/test report*—A document that records the results of observations, measurements, and tests as verified by the initials or signature of the individual responsible for the inspection/test activity.

1.4.10 *Material manufacturer or supplier*—The organization responsible for producing or manufacturing a product or material used in the process of construction, or for supplying products or materials to a project, with or without performing additional operations on the product or material.

1.5—Categories of inspection

Inspection activities generally fall into one of the categories described in 1.5.1 through 1.5.4.

1.5.1 *Owner's inspection*—Inspections and tests conducted by or for the owner either by the owner's in-house inspection group or by an independent inspection agency. Owner inspection is a part of the external quality assurance program conducted by the owner. Results of these inspections form the basis of the owner's decision to ultimately accept the work performed by the contractor. Owner-inspection programs should be structured so as to provide the owner with an acceptable degree of assurance that the work of the contractor is in conformance with the contract documents.

1.5.2 *Quality-control inspection: contractor*—A series of formalized activities and procedures that are part of the contractor's operation, providing in-process evaluation of the quality of construction. These activities help to assure the contractor that the finished construction will meet all requirements of the project plans, drawings, and specifications, and will be accepted by the owner.

1.5.3 *Quality-control inspection: producer*—A series of formalized activities and procedures that are part of the fabricating or manufacturing operation of a producer or fabricator of concrete materials, reinforcement, or products who furnishes products to the construction industry or to a specific project. Examples are operations of cement and aggregate producers, concrete producers, precasters, prestressing concrete fabricators, and reinforcing steel mills and fabricators. Production-inspection personnel operate essentially the same way as those described for the contractor. They aid in ensuring that finished products will meet general specifications or those specifications relative to a specific project.

1.5.4 *Compliance inspection*—A series of formalized activities and procedures performed by government agencies charged with the responsibility for enforcing building codes and other regulations. In these cases, compliance inspectors have the responsibility for ensuring that the finished structure conforms to specified codes or regulations. The organization and activities of these inspectors are governed almost entirely by the requirements of building codes or government regulations. An overlap of compliance inspection and owner inspection often occurs when the owner engages the services of a special inspector, as required by some building codes, to oversee and confirm the performance of inspections required by the code. In most cases, the technical requirements of the building code are similar, if not identical, to the requirements given in project specifications and drawings.

1.6—Inspection team

Regardless of classification, an inspection team or group may consist of a number of individuals or, for very small projects, a single individual. Inspection may be performed by a variety of groups, such as:
- Owner's inspection personnel;
- A/E's inspection personnel;
- Laboratory's inspection force;
- Contractor's inspection force; and
- Material manufacturer's and supplier's inspection force.

All inspection force personnel should be qualified and, as applicable, certified to conduct inspections and tests for which they are assigned.

CHAPTER 2—RESPONSIBILITIES
2.1—Scope

This chapter defines the general responsibilities for inspection placed on the owner, A/E, inspection organization contractor, and manufacturer or fabricator in conforming to the recommendations of this guide.

2.2—Owner's responsibilities

2.2.1 The owner should provide for a program of inspection separate and distinct from quality-control inspection conducted by the contractor or by material producers. The A/E should provide the owner with recommendations for the scope and content of inspections and tests to be included in the owner's inspection plan. The owner should review the inspection plan with the A/E and, where appropriate, select the level of inspection required that is consistent with project size, quality, complexity, and the requirements of the local building code.

2.2.2 In conjunction with the A/E, the owner should be responsible for arranging a preconstruction conference that includes all parties involved in the construction project. The conference should include review of the inspection and testing plan(s), and confirm lines of communication, responsibilities, and minimum quality levels for the project. To be effective, the inspection personnel should have the active support of the owner.

2.2.3 The fee for owner inspection should be a separate and distinct item and should be paid by the owner, or by the A/E acting on behalf of the owner, directly to the inspection organization. The owner or A/E should avoid the undesirable practice of arranging payment through the contractor for inspection services intended for use by the owner as a basis of acceptance. Such a practice is not in the owner's interest and may result in a conflict of interests. Impartial service is difficult under such circumstances, and the fees for inspection are eventually paid by the owner in any case.

2.2.4 As a professional service, the selection of the inspection organization/laboratory should be based primarily on qualifications, not on price. It should be done as carefully as the selection of the A/E. The owner should check the physical facilities of the organization/laboratory, review the supervisory program and the qualifications of the supervisory staff, and review accreditations, the latest evaluation, or both, made by the evaluation authority and ensure that any necessary corrective measures have been taken. It should review the organization's ongoing training program of its personnel. The personnel should be certified and meet the qualifications of Section 3.5. The owner should also review the qualifications of all testing and inspection personnel to be assigned to the owner's project. The owner's approval should be required for all personnel before such assignment.

2.2.5 When the project specifications require extensive quality-control inspection by the contractor, the owner should not reduce or eliminate owner inspection. If the contractor's quality-control inspection program becomes the owner's inspection program, the system is nullified. The objections are exactly as stated previously against the practice of having the contractor hire and pay an inspector to perform inspection for the owner. When the owner requires the contractor to have a quality-control inspection program, the owner should still accept responsibility for inspection to provide assurance that the contractor's quality-control program achieves its objectives.

2.3—Architect/engineer's inspection responsibilities

2.3.1 For the protection of the public and the owner, the responsibility for planning and detailing owner inspection should be vested in the A/E as a continuing function of the design responsibility. The A/E should ensure that the program for owner inspection meets all requirements of design specifications and the local building code. The inspection responsibility may be discharged directly, may be conducted by owner personnel, or may be delegated to an independent inspection organization reporting to the A/E.

2.3.2 If the A/E is also responsible for construction, an independent inspection organization should be retained directly by the owner. When the owner provides the A/E service, the owner should also provide inspection or retain an independent inspection organization. Inspection requirements on projects supervised by a construction manager should also be detailed by the A/E and should be carried out by inspection personnel representing the owner.

2.3.3 Some local building codes require that the A/E or another registered professional be engaged by the owner to conduct periodic inspection visits during construction along with an overall review of inspection/testing activities to ensure that work is conducted in compliance with the code.

2.4—Owner's inspection organization responsibilities

2.4.1 The inspection organization/laboratory selected by the owner should perform required inspections in accordance with the contract documents. It is recommended that owner inspections be conducted for items listed in Table 3.1 or as otherwise directed by the owner or A/E. Failure of the owner's inspection to detect defective work or material does not relieve contractors of their responsibility to meet quality requirements for the work. It does not prevent later rejection when such defect is discovered, and does not obligate the owner or A/E for final acceptance.

2.4.2 The inspection organization should be accredited in accordance with ASTM E 329. Testing laboratories should be accredited by ASTM C 1077 and should have its facilities, personnel, and procedures inspected by a qualified evaluation authority at intervals of approximately 24 months.

2.4.3 Owner inspection personnel are responsible for, and can only be involved with, determining that inspected materials, procedures, and final products, as installed, conform to the requirements of the contract documents. The contractor is obligated to meet the requirements of the contract documents. For the inspector to accept less than that required deprives the owner of full value, whereas requiring more than is called for in the contract documents places an unacceptable burden on the contractor. Either action is a misuse of the inspector's authority. The inspection personnel representing the owner have no responsibility or authority to manage the contractor's personnel.

2.4.4 At the conclusion of the project, the owner's inspection organization is often asked to submit a final statement covering the results and final status of inspection/testing activities. Because inspection is typically conducted on only a portion of the total work, the inspection organization is generally not capable of certifying that all work is in conformance with the contract documents (although this type of statement is frequently requested). Properly worded statements usually indicate only that "items documented in inspection and test reports were found to be in conformance…" Should there be items or work that are nonconforming, such as those not corrected by the contractor or accepted by the A/E, the items should be clearly identified in the final statement of inspection.

2.5—Contractor's inspection responsibilities

2.5.1 Coordination and providing scheduling notice to the owner's inspection organization should be included in the contractor's work plans and overall schedule. This will enable timely inspection by the owner's representatives and avoid possible construction delays.

2.5.2 Quality-control inspection, or in-process inspection, is usually performed by contractor personnel or by others, such as independent inspection laboratory personnel, hired by the contractor. To be effective, quality-control inspection results should be reported directly to and have the active support of contractor management. Inspection by or for the contractor, subcontractors, or material suppliers is separate and distinct from inspection conducted for the owner. Qualification and certification requirements for personnel conducting quality-control inspections should be the same as that described previously for owner inspection personnel.

In some construction contracts, the contractor is required to provide a specified amount of inspection as part of a formal quality-control program. When not contractually required, many contractors still maintain a quality-control program that includes inspection personnel separate from the line of supervision, reporting directly to management. The initial cost is often returned many times over through the reduction of mistakes and rejections, resulting in savings in both replacements and repairs. Frequently, this inspection work is an informal and automatic part of the contractor's operations performed by regular production supervisors, although a formal program is more effective.

2.5.3 Inspection performed by, or for, the contractor, particularly when contractually required, will often be much more detailed than is the usual practice for owner inspection. The contractor's personnel should make a more detailed inspection of form alignment, reinforcement positioning, cleanup of forms, and other concrete placement issues. Even if not required by the project specifications, the contractor should use quality-control inspection to reduce the possibility of later rejection by the owner.

2.5.4 Quality-control inspection other than or in addition to that required by the project specifications will be as directed by the contractor's management. These inspection details and criteria will be based on management's judgment as to items and criteria necessary to ensure that all aspects of workmanship and the finished product will meet the requirements of the contract documents and will thus be accepted by the owner.

2.6—Manufacturer's or fabricator's inspection responsibilities

Quality-control inspection by the manufacturer or fabricator should parallel the contractor's programs. Program content depends on contractual requirements, building code requirements, and on the manufacturer's quality-control process. Where facility certifications are available to the industry, they should be required by the owner.

CHAPTER 3—PLANNING FOR INSPECTION

3.1—Scope

This chapter gives specific recommendations to the A/E on inspection-related items that should be included in the contract documents. Recommendations for scheduling meetings or conferences are also discussed due to their direct effect on achieving improved quality in the final work.

3.2—Written inspection plan

All projects can benefit from a written inspection plan. A small project may require only a list of items to be inspected and tests to be conducted for acceptance purposes, but it becomes invaluable in developing adequate communication and understanding among the owner, A/E, contractor, and the inspection organization. All projects should use some form of written plan or checklist. On complex projects, the owner's quality-control plan, as referenced in ACI 121R, detailing the owner's policy statement, quality objectives, scope of work, organizational responsibilities including responsibilities for owner inspection, interface between owner and contractor inspection activities, procedures for documentation of inspections and tests, scheduling and frequency of inspection, reporting of results, handling of nonconformances and design changes, record retention, and auditing of the progress of the work is a necessity. Specifications developed by the A/E should include requirements for the development and submittal of written inspection plans as part of the quality manuals that should be developed by appropriate inspection groups involved with the project.

3.3—Building code requirements for special inspections

Many states and municipalities have adopted the International Building Code (IBC) along with the requirements for special inspections of concrete and other structural materials. Although the manner in which special inspections are carried out can vary, in most cases, a design or engineering professional must be engaged by the owner to oversee and verify that required inspections and tests are performed and that corrective action or engineering evaluation/acceptance is conducted for all items identified as nonconforming to the design specifications or building code requirements. The special inspector provides regular reports during construction to the owner, the A/E, and the local building official, and a final report summarizing and confirming that the work was conducted in accordance with the building code, the contract documents, and the instructions of the A/E. Reference to special inspections should be included in the contract documents where applicable. The recommended outline of concrete inspection activities presented in Table 3.1 of this document was developed to meet or exceed IBC special inspection requirements for activities and materials commonly encountered in concrete construction.

3.4—Preconstruction conferences

A preconstruction conference is recommended for all projects to establish lines of communication at the start of a project. This conference should include all parties involved with the construction. Its main purpose is to identify responsibilities, clarify flow of documents, and establish procedures that will allow construction to proceed in a manner that will ensure the best possible construction quality in the finished work, in accordance with specifications.

Table 3.1—Outline of recommended inspection program activities for concrete construction

I—Concrete production
a. Submittal review of mixture proportions and supporting test data
b. Submittal review of material certifications for concrete materials
c. Preconstruction sampling and testing of concrete materials
d. In-process sampling and testing of concrete materials
e. Inspection of concrete batching and mixing operations
f. Inspection/qualification of concrete mixer trucks
g. Mixer uniformity testing
h. Use of approved mixture proportions in production
II—Inspection and testing of concrete
a. Truck/ticket review and control
b. Field additions of water, admixtures, and other materials
c. Sampling and testing of concrete or grout
d. Initial curing/storage of lab specimens in the field
e. Storage of field-cured specimens
f. Transportation of test specimens to the testing facility
g. Laboratory curing/storage of test specimens
h. Strength testing of specimens
i. Other required tests
III —Preplacement inspection
a. Inspection/testing of subgrade and subbase
b. Inspection of forms and decking
c. Inspection of reinforcing bars
d. Inspection of anchor bolts and embedments
e. Inspection of prestressing strands/bars
f. Inspection of pretensioning
g. Inspection of welding of reinforcing steel
h. Inspection of shear stud installation
IV—Placement inspection
a. Inspection of concrete transportation and handling
b. Inspection of concrete placement and consolidation
c. Inspection of leveling and screeding operations
d. Inspection of finishing operations
e. Inspection of surface treatment applications
f. Inspection of initial curing and protection
V—Postplacement inspection
a. Inspection of concrete moist curing
b. Inspection of concrete protection and monitoring of curing temperatures
c. Verification of in-place strength before removal of shoring or loading of the structure
d. Inspection of post-tensioning
e. Inspection for damage following form removal or loading
f. Inspection of surface treatments and surface repairs
g. Inspection of repairs to A/E instructions
h. Examination of in-place concrete strength by testing of cores or nondestructive methods
VI—Precast erection
a. Inspection of transportation and lifting of elements
b. Inspection of bearing and alignment
c. Inspection of grouted connections
d. Inspection of welded connections

3.5—Meetings

Regular meetings of the contractor, A/E, concrete producer, inspection organization, and testing laboratory are recommended. The frequency of meetings is contingent on the size, schedule, complexity of the project, and on the existence of unidentified problems. These meetings allow for open dialogue and should be used to identify potential problem areas before they develop; meetings can also provide a platform to resolve noncompliant construction. The agenda should allow for review of the past period's activity and a schedule of activities for current and future periods.

3.6—Qualifications of inspection and testing personnel

The qualifications of personnel conducting inspections and tests are critical to attaining the desired level of quality. Nonstandard testing practices and ineffective inspection can miss critical items and result in an inaccurate assessment of the work examined and are of no benefit to the project. Erroneous results of tests and inspections can also cause costly actions that are unwarranted.

The ACI certification program currently outlines training programs and certification of inspection personnel in the following categories (ACI certification program [CP] study guides in parentheses):
- Concrete field testing technician—Grade I (CP 1);
- Concrete laboratory testing technician—Grades I and II (CP 16 and 17);
- Concrete strength testing technician (CP 19);
- Aggregate field testing technician (CP 40);
- Aggregate laboratory testing technician (CP 41);
- Concrete construction special inspector (CP 21);
- Concrete transportation construction inspector (CP 31); and
- Concrete flatwork technician (CP 10).

All personnel performing concrete inspection and testing work should be certified and demonstrate a knowledge and ability to perform the necessary inspection and testing procedures equivalent to the minimum guidelines set forth for certification by ACI for the appropriate category listed. Personnel may be certified by ACI or by organizations whose programs provide for written and performance examinations considered equivalent to the corresponding ACI program.

3.7—Recommendations for inspection and testing

3.7.1 *General*—Owner inspection need only be detailed enough to permit adequate evaluation of the product or process. The contractor and concrete producer should be encouraged to provide their own formalized quality-control programs.

In many instances, there is a tendency for contractors and concrete producers to rely on the activities of the owner's inspection to control quality instead of conducting their own quality-control inspection. Specifications that provide for a program of inspection on behalf of the owner should include a statement indicating that the activities of the owner's inspection group do not relieve the contractor of his ultimate responsibility for quality of the work. Additionally, failure of the owner's inspection activities to identify nonconforming conditions does not prevent rejection at a later date and does not obligate the owner or A/E for final acceptance.

If there is concern by the owner or A/E about the adequacy of the quality control planned by the contractor, project specifications can direct the contractor to provide specific testing and inspection activities as part of the quality-control program, with results transmitted to the owner and A/E. When this is done, owner inspection should not be eliminated, but its scope may be adjusted, as deemed appropriate, to satisfy quality-assurance concerns.

Should a concern about the adequacy of quality control develop during the course of the project, the scope and frequency of owner inspection should necessarily be increased until the contractor's quality-control performance removes reason for concern.

3.7.2 *Owner's inspection*—The A/E should evaluate the necessity of conducting prequalification tests of the materials to be used in the project. If materials with past service records are to be used, earlier qualification tests may be relied on, or satisfactory performance in a similar environment may be used as the basis for acceptance. If prequalification tests are to be conducted, the A/E should specify the tests and the acceptance criteria.

The approval of concrete mixture proportions to be used should be based on reliable historical performance data or trial mixture results that should accompany the submittals. Methods of proportioning and other criteria established by ACI 318 and ACI 301 should be followed.

Based on the project's size and complexity, evaluate the need for certification of batch plants before concrete production and consider a qualification program for truck mixers, including mixer uniformity tests. National Ready Mixed Concrete Association certification procedures are recommended.

Sampling and testing concrete materials at established intervals during construction is usually required, and some properties will need to be monitored on a daily, weekly, or monthly basis as established by the A/E. Generally, qualification tests will not need to be repeated during construction, but new qualification tests should be performed whenever there is a change in material or material source. Material test reports for cement, admixtures, and reinforcing steel can usually be relied on for acceptance of these materials as delivered from the material manufacturer. To ensure more reliability, manufacturers' quality-assurance and quality-control programs should be formulated in accordance with ACI 121R.

Daily inspection of batching may be needed, depending on the level of plant automation, concrete strength, and quality level required as established by the A/E. Regular checks for yield and aggregate moisture content are desirable. ACI 311.5 provides recommendations for this type of inspection.

Inspection of forming, preplacement, placement, and post-placement of concrete activities should be part of the acceptance process for most projects, and special precautions should be considered during hot- and cold-weather concreting.

When the time frame for form removal or loading of the structure is dependent on the structural strength and stability of the concrete, monitoring of in-place concrete strength by testing of field-cured cylinders or by using some form of nondestructive testing is required. Procedures and criteria established by ACI 305R, 306R, and 347 should be followed.

Daily strength tests of concrete to confirm concrete production quality and design assumptions are almost always required.

3.7.3 *Quality-control inspection*—Quality-control inspection is a functional responsibility of the contractor. As a minimum, the contractor has direct quality-control responsibility for all preplacement, placement, and postplacement activities. Contractual relationships will determine whether the contractor or the concrete producer will be directly responsible for concrete quality control.

Contractors purchasing concrete from an independent concrete producer usually rely on the producer's quality control and seldom get directly involved in the production process. The contractor, however, should monitor the quality-control reports of the concrete producer and the properties of plastic concrete delivered to the project.

Contractors operating their own concrete production facilities should assume direct responsibility for all quality-control activities.

3.7.4 *Recommended inspection activities*—Table 3.1 presents an outline of inspection activities that should be implemented for owner inspection of concrete structures where the safety of the public must be considered. When not otherwise restricted by building code requirements, the A/E may reduce the scope of inspection activities presented in Table 3.1 as deemed appropriate for the size and nature of the concrete work. Some or all of the activities noted in the table may be adopted by contractors or material suppliers as part of their quality-control programs for inspection.

Appendix I presents an expanded checklist of inspection attributes that are intended for use in the development of detailed inspection checklists for each inspection activity listed in Table 3.1. Examples of inspection checklists and reports can be found in the *ACI Manual of Concrete Inspection* (SP-2).

3.7.4.1 *Specialty work*—Some construction projects may require items of inspection not listed in Table 3.1 or Appendix I. Such items can be added by the A/E to ensure adequate conformance to quality requirements. For this reason, the inspection items listed are intended to cover only those construction activities and materials most commonly encountered in concrete construction. Inspection items for specialty work, such as pressure grouting, shotcrete, high-performance concrete, self-consolidating concrete, two-course floors, super-flat floors, terrazzo, stucco, masonry, cast stone, tile, architectural concrete, painting, preplaced-aggregate concrete, tilt-up construction, underwater construction, vacuum concrete, and slipform construction are intentionally omitted from Table 3.1 and Appendix I. It is intended that the A/E will develop inspection criteria for specialty work that is appropriate to the specific needs of these activities.

3.7.4.2 Appendix II is a synopsis of ACI 311.5, which has been developed as a separate document. ACI 311.5 is intended to be used on projects where the A/E needs specific guidance on items to include in a batch-plant inspection program or in a program of testing for fresh and hardened properties of ready mixed concrete. Instructions should be modified as necessary to meet the individual needs and requirements of each project.

3.7.4.3 *High-strength concrete*—The use of high-strength concrete (8000 psi [55 MPa] or greater) requires more testing and inspection because a high degree of confidence of quality is required. Recommendations from ACI 363R and 363.2R should be followed, as this report does not address any special requirements for high-strength concrete.

3.8—Reporting and evaluating inspection and test results

Results of all inspections and tests conducted by the owner's inspection personnel should be documented in inspection (test) reports and promptly transmitted or communicated to the A/E. Distribution of these reports to the owner, the contractor, and, in some instances, to the local building official, is also frequently required. Contract documents typically require that results of contractor quality-control inspections and tests also be transmitted to the A/E and others for review.

Results of inspections and tests, specifically those results that fail to meet requirements, need to be evaluated and dispositioned by the A/E. The A/E's disposition of a nonconforming condition generally falls into one of the following categories:
- Accept as is;
- Rework/repair and reinspect; or
- Reject (remove or replace).

Follow-up inspection and reporting of repair/rework activities shall be conducted by the responsible inspection group in accordance with the disposition instructions provided by the A/E.

CHAPTER 4—REFERENCES
4.1—Referenced standards and reports

The standards and reports listed below were the latest editions at the time this document was prepared. Because these documents are revised frequently, the reader is advised to contact the proper sponsoring group if it is desired to refer to the latest version.

American Concrete Institute
116R	Cement and Concrete Terminology
121R	Quality Management System for Concrete Construction
301	Specifications for Structural Concrete
305R	Hot Weather Concreting
306R	Cold Weather Concreting
311.5	Guide for Concrete Plant Inspection and Field Testing of Ready-Mixed Concrete
318	Building Code Requirements for Structural Concrete
347	Guide to Formwork for Concrete
363R	Report on High-Strength Concrete
363.2R	Guide to Quality Control and Testing of High-Strength Concrete

CP 1   Concrete Field Testing Technician—Grade I
CP 10   Concrete Flatwork Technician Finisher
CP 16   Concrete Laboratory Testing Technician—Grade I
CP 17   Concrete Laboratory Testing Technician—Grade II
CP 19   Concrete Strength Testing Technician
CP 40   Aggregate Field Testing Technician
CP 41   Aggregate Laboratory Testing Technician
CP 21   Concrete Construction Special Inspector
CP 31   Concrete Transportation Construction Inspector

ASTM International
C 1077 Standard Practice for Laboratories Testing Concrete and Concrete Aggregates for Use in Construction and Criteria for Laboratory Evaluation
E 329   Standard Specification for Agencies Engaged in the Testing and/or Inspection of Materials Used in Construction

The preceding publications may be obtained from the following organizations:

American Concrete Institute
P.O. Box 9094
Farmington Hills, MI 48333-9094
www.concrete.org

ASTM International
100 Barr Harbor Dr.
West Conshohocken, PA 19428-2959
www.astm.org

4.2—Cited references
International Code Council, Inc., 2003, "International Building Code," Country Club Hills, Ill., 655 pp.

National Ready Mixed Concrete Association, 2005, "Quality Control Manual," 9th Revision, Nov., Silver Spring, Md.

4.3—Other references
Abdun-Nur, E. A., 1981, "Contractual Relationships, An Essential Ingredient of the Quality Assurance System," *Transportation Research Record* No. 792, Transportation Research Board, pp. 1-2.

Abdun-Nur, E. A., 1982a, "Inspection and Quality Assurance," *Concrete International,* V. 4, No. 9, Sept., pp. 58-62.

Abdun-Nur, E. A., 1982b, "Incentive Specifications for Concrete," *Concrete International,* V. 4, No. 9, Sept., pp. 20-24.

ACI Manual of Concrete Practice, 2005, American Concrete Institute, Farmington Hills, Mich.

ACI Committee 311, 1999, *ACI Manual of Concrete Inspection* (SP-2), 9th Edition, American Concrete Institute, Farmington Hills, Mich., 220 pp.

Annual Book of ASTM Standards, V. 04.01, Cement, Lime, and Gypsum, and V. 04.02, Concrete and Mineral Aggregates, ASTM International, West Conshohocken, Pa.

Dixon, D. E., 1982, "Guidance in the Training and Qualification of Inspection Personnel," *Concrete International*, V. 4, No. 9, Sept., pp. 84-87.

Henry, R. L., 1982, "Quality Control and Acceptance Inspection as Viewed by the Testing Laboratory," *Concrete International*, V. 4, No. 9, Sept., pp. 75-78.

Jaycox, C. E., 1982, "Guidance in the Establishment of an Inspection Program," *Concrete International*, V. 4, No. 9, Sept., pp. 79-83.

Keifer, O., Jr., 1981, "Control Charts Catch Changes, Can Cut Costs," *Concrete International*, V. 3, No. 11, Nov., pp. 12-16.

Mayer, C. W., 1982, "Quality Control by the Contractor," *Concrete International*, V. 4, No. 9, Sept., pp. 72-74.

Post-Tensioning Institute, 1990, *Post-Tensioning Manual*, 5th Edition, Phoenix, Ariz., 406 pp.

Post-Tensioning Institute, 2004, *Design of Post-Tensioned Slabs on Ground*, 3rd Edition, Phoenix, Ariz., 90 pp.

Prestera, J. R., 1982, "Quality Control Inspection by the Ready-Mixed Concrete Producer," *Concrete International*, V. 4, No. 9, Sept., pp. 67-71.

Weinberg, B. E., 1982, "Product Control and Acceptance Inspection as Viewed by the Owner and Designer," *Concrete International*, V. 4, No. 9, Sept., pp. 62-66.

APPENDIX I—EXPANDED CHECKLIST OF INSPECTION ATTRIBUTES
Preconstruction testing of materials
1. *Coarse and fine aggregate properties*
   a. Grading and fineness modulus, ASTM C 136;
   b. Amount of material finer than 75 μm (No. 200), ASTM C 117;
   c. Soundness, ASTM C 88;
   d. Lightweight particles, ASTM C 123;
   e. Specific gravity and absorption, ASTM C 127 or C 128;
   f. Water-soluble chlorides, ASTM C 1218;
   g. Reactivity of aggregate, ASTM C 227, C 289, C 342, C 586, C 1260, and C 1293;
   h. Bulk unit weight, ASTM C 29; and
   i. Petrographic examination, ASTM C 295.

2. *Fine aggregate properties*
   a. Organic impurities, ASTM C 40; and
   b. Effect of organic impurities on strength, ASTM C 87.

3. *Coarse aggregate properties*
   a. Abrasion, ASTM C 131 or C 535;
   b. Flat or elongated particles, ASTM D 4791; and
   c. Friable particles, ASTM C 142.

4. *Cementitious materials*
   a. Physical properties, ASTM C 150, C 595, C 845, or C 1157;
   b. Chemical properties, ASTM C 150 or C 595;
   c. Physical/chemical properties of silica fume, ASTM C 1240;
   d. Physical/chemical properties of fly ash and natural pozzolans, ASTM C 618; and
   e. Physical/chemical properties of ground-granulated blast-furnace slag, ASTM C 989.

5. *Water*
   a. Strength versus control, ASTM C 109;

b. Time of set versus control, ASTM C 191;
c. Total solids content, ASTM D 1888;
d. Total chlorides, ASTM D 512; and
e. Potable water (local health standards).

6. *Admixtures*
   a. Air-entraining admixtures, ASTM C 260 as required;
   b. Water-reducing admixtures, ASTM C 494 as required;
   c. Admixtures for flowing concrete, ASTM C 1017 as required; and
   d. Review of test documentation and warnings.

7. *Reinforcing steel—ASTM A 615, A 617, A 706, A 767, A 775, A 934, A 955M, or A 996*
   a. Deformations: spacing, height, and gap;
   b. Weight per linear ft;
   c. Bending properties;
   d. Tensile properties: yield, strength, tensile strength, and percentage of elongation; and
   e. Chemical properties.

8. *Prestressing steel—ASTM A 416, A 421, A 722, A 779, A 866, or A 882*
   a. Quantity (ft/lb);
   b. Diameter of strand;
   c. Grade of strand;
   d. Packaging;
   e. Special requirements; and
   f. Item 7 requirements for bars.

9. *Concrete*
   Freezing-and-thawing resistance, ASTM C 666

Mixture proportion qualification and approval
- As defined by ACI 318 and ACI 301.

Certification of batch plants and truck mixers
- Certification to National Ready-Mixed Concrete Association requirements before construction.
- Certification of ready-mix operations to ASTM C 94 requirements.

Inspection of batch plants and truck mixers before or during construction—ACI 304R

1. *Aggregate storage areas*
   a. Cleanliness;
   b. Separation of materials;
   c. Handling of materials;
   d. Aggregate spray system and drainage;
   e. Approved sources;
   f. Cold-weather provisions (heat, cover); and
   g. Hot-weather provisions (cool, cover).

2. *Cement silo storage*
   a. Weathertight;
   b. Temperature of shipment;
   c. Mill certification with bulk shipment; and
   d. Retesting (if longer than 6 months of storage by manufacturer or vendor).

3. *Cement bag storage*
   a. Storage on pallets;
   b. Identification of type, brand, and manufacturer;
   c. Protection from moisture;
   d. Mill certification with bag shipment; and
   e. Retesting (if longer than 3 months of local storage).

4. *Admixture storage and usage*
   a. Temperature control;
   b. Contamination control;
   c. Agitation;
   d. Retesting (if longer than 6 months of storage by manufacturer or vender); and
   e. Identification of type, brand, and manufacturer.

5. *Batching equipment*
   a. Check of scales and measuring devices every 90 days;
   b. Dial and balance scales accurate within ±0.20% of scale capacity;
   c. Digital scales accurate within ±0.25% of scale capacity;
   d. Return to zero indication;
   e. Adequate separation of bins;
   f. Free discharge of materials with tight closing gates;
   g. Weighing hoppers freely suspended;
   h. Conditions of fulcrum and pivot points;
   i. Water delivery system leak-free;
   j. Measurement of water accurate to ±1%;
   k. Separate dispensers for each admixture;
   l. Admixture dispensing system leak free and accurate to ±3%; and
   m. Indicating devices in full view of operator.

6. *Batching operations—ASTM C 94, C 685*
   a. Cement and supplementary cementitious materials measured within ±1% of desired weight;
   b. Aggregates measured within ±2% of desired weight (±1% when a cumulative weight is taken);
   c. Allowable water adjusted for free moisture in aggregates;
   d. Admixtures discharged separately within a volumetric tolerance of ±3% using a method that does not allow concentrated admixtures to contact each other; and
   e. Verification of batch ticket information to requirements of ASTM C 94.

7. *Mixing operation and qualification of mixers*
   a. Mixer blades free of buildup;
   b. Inspection of blades for holes or cracks;
   c. Height of mixer blades measured for wear;
   d. Mixture uniformity tests for stationary or truck mixtures, ASTM C 94;
   e. Truck mixing 70 to 100 revs;
   f. Central mixing—a minimum of 1 min for first m^3 (yd^3) + 15 s for each additional m^3 (yd^3); and
   g. Truck water dispensers accurate to within ±1%.

8. *Sampling and testing aggregates during construction*
   a. Sampling, ASTM D 75;
   b. Grading and fineness modulus, ASTM C 136;
   c. Amount of material finer than 75 μm (No. 200), ASTM C 117;
   d. Friable particles, ASTM C 142;
   e. Coal and lignite, ASTM C 123;
   f. Specific gravity and absorption, ASTM C 127 or C 128; and
   g. Organic impurities, ASTM C 40.

9. *Preplacement inspection*
   a. Lines and grades;
   b. Location;
   c. Elevation;
   d. Dimensions;
   e. Shape;
   f. Drainage;
   g. Preparation of surface; and
   h. Bearing.

10. *Forms and decking*
    a. Specified type;
    b. Location;
    c. Dimensions;
    d. Tolerances;
    e. Alignment;
    f. Stability (bearing, shores, tees, and spacers);
    g. Surface preparation;
    h. Tightness;
    i. Chamfer strips;
    j. Inspection openings;
    k. Cleanliness;
    l. Temperature;
    m. Accessories (such as ties, cones, and clamps); and
    n. Metal deck installation (puddle welds, tek screws, crimping, and overlaps).

11. *Reinforcing steel*
    a. Size (diameter, length, bends, and anchorage);
    b. Grade;
    c. Location (number of bars, spacing, and cover);
    d. Splicing (lap length, mechanical joint, weld joint, welder qualifications, and welding procedures);
    e. Placement (wire tying, bar supports, and side-form spacers);
    f. Cleanliness (no loose rust, oil, paint, or dried mortar);
    g. Protective coating; and
    h. Shear stud installation (size, location, and spacing).

12. *Prestressing steel (pretensioned and post-tensioned)*
    a. Strand, wire, or bar placement (wire tying, bar supports, and side-form spacers);
    b. Size;
    c. Location;
    d. Type;
    e. Draping;
    f. Anchorage;
    g. Tensioning sequence;
    h. Loading and elongation measurements;
    i. Concrete stressing strength verification;
    j. Cleanliness;
    k. Condition of sheathing and protective coating;
    l. Grouting of post-tensioned tendons; and
    m. Sealing of end anchors.

13. *Embedments*
    a. Location;
    b. Size; and
    c. Condition.

14. *Blockouts*
    a. Location;
    b. Size; and
    c. Condition.

15. *Joints*
    a. Type;
    b. Location; and
    c. Filler material.

Placement inspection

1. *Conditions*
   a. Coordination of concrete delivery;
   b. Protection against sun, rain, hot or cold weather conditions; and
   c. Lighting and power.

2. *Field inspection and testing of concrete*
   a. Use of specified mixture;
   b. Field water additions (minimum 30 drum revs), verify w/cm;
   c. Sampling freshly mixed concrete, ASTM C 172;
   d. Slump, ASTM C 143;
   e. Temperature of freshly mixed concrete and ASTM C 1064 (maximum and minimum as specified);
   f. Air content (pressure or volumetric), ASTM C 231 and C 173;
   g. Density (unit weight), ASTM C 138;
   h. Yield, ASTM C 138;
   i. Cylinder specimens, ASTM C 31 (identification, mixture, location, and date);
   j. Discharge of ready-mixed concrete truckload before 300 revs or 90 min, ASTM C 94; and
   k. Initial curing of cylinder specimens, ASTM C 31.

3. *Conveyance of concrete*
   a. Nonreactive materials;
   b. Prevention of segregation and loss of materials;
   c. Prevention of contamination;
   d. Condition of conveying equipment (smooth surfaces, no holes, and cleanliness); and
   e. Use of drop-chutes or funnel hoses to contain freefall.

4. *Placement and consolidation of concrete*
   a. Precautions taken for hot or cold weather conditions;

b. Preparation of contact surfaces;
c. Ability of conveying method to place concrete in all areas of placement;
d. Mortar bedding (use of starter mixture);
e. Prevention of segregation (no chuting or dropping against forms or reinforcement);
f. Depth of layer (maximum limit);
g. External vibration (spacing to prevent dead spots);
h. Internal vibration (depth of insertion, spacing, time, vertical insertion, no movement of concrete by vibration). Vibrators to be equipped with rubber heads when consolidating concrete around epoxy-coated reinforcing bars;
i. Even layering around openings and embedments;
j. Removal of bleed water; and
k. Removal of temporary ties and spacers.

Post-placement inspection and tests
1. *Finishing, curing, and formwork and shore removal*
   a. Specified finish;
   b. Protection of surfaces from cracking due to rapid drying (avoid direct heat);
   c. Proper curing temperature;
   d. Form removal (field-cured cylinder tests or other approved tests);
   e. Curing compound (conformance to ASTM C 309, application); and
   f. Finish of formed surfaces (patching and repairs where necessary).

2. *Shoring and reshore removal*
   a. Verification of in-place strength;
   b. Location;
   c. Number;
   d. Time of removal; and
   e. Sequence of removal.

3. *Tests of hardened concrete*
   a. Curing of specimens, ASTM C 31;
   b. Preparation of concrete cores, ASTM C 42;
   c. Capping, ASTM C 617;
   d. Tests for compressive strength, ASTM C 39;
   e. Tests for split tensile strength, ASTM C 496;
   f. Equilibrium unit weight of lightweight concrete, ASTM C 567;
   g. Flexural strength, ASTM C 293 and ASTM C 78;
   h. Density, absorption, and voids, ASTM C 642
   i. First crack strength and toughness (fiber-reinforced), ASTM C 1018;
   j. Nondestructive tests, ASTM C 597, C 803, C 805, C 900, and C 1074;
   k. Petrographic analysis, ASTM C 856; and
   l. Restrained expansion of shrinkage-compensating concrete, ASTM C 878.

Precast erection
a. Number and location of lifting lugs;
b. Blocking during storage and transportation;
c. Placement and condition of bearing pads;
d. Alignment of connection pockets;
e. Sampling and testing of grout, ASTM C 109;
f. Alignment of embedment plates and weld clips;
g. Welding procedures and qualification of welders; and
h. Inspection for damage and visual appearance.

REFERENCED STANDARDS AND REPORTS FOR APPENDIX I

The standards and reports listed below were the latest editions at the time this document was prepared. Because these documents are revised frequently, the reader is advised to contact the proper sponsoring group if it is desired to refer to the latest version.

American Concrete Institute
121R	Quality Management System for Concrete Construction
301	Specifications for Structural Concrete
304R	Guide for Measuring, Mixing, Transporting, and Placing Concrete
318	Building Code Requirements for Structural Concrete

ASTM International
A 416	Standard Specification for Steel Strand, Uncoated Seven-Wire Stress-Relieved for Prestressed Concrete
A 421	Standard Specification for Uncoated Stress-Relieved Steel Wire for Prestressed Concrete
A 615	Standard Specification for Deformed and Plain Carbon-Steel Bars for Concrete Reinforcement
A 617	Specification for Axle-Steel Deformed and Plain Bars for Concrete Reinforcement
A 706	Standard Specification for Low-Alloy Steel Deformed and Plain Bars for Concrete Reinforcement
A 722	Standard Specification for Uncoated High-Strength Steel Bar for Prestressing Concrete
A 767	Standard Specification for Zinc-Coated (Galvanized) Steel Bars for Concrete Reinforcement
A 775	Standard Specification for Epoxy-Coated Steel Reinforcing Steel Bars
A 779	Standard Specification for Steel Strand, Seven-Wire, Uncoated, Compacted, Stress-Relieved for Prestressed Concrete
A 866	Standard Specification for Medium Carbon Anti-Friction Bearing Steel
A 882	Standard Specification for Filled Epoxy-Coated Seven-Wire Prestressing Steel Strand
A 934	Standard Specification for Epoxy-Coated Prefabricated Steel Reinforcing Bars
A 955M	Standard Specification for Deformed and Plain Stainless-Steel Bars for Concrete Reinforcement [Metric]
A 996	Standard Specification for Rail Steel and Axle Steel Deformed Bars for Concrete Reinforcement
C 29	Standard Test Method for Bulk Density (Unit Weight) and Voids in Aggregate
C 31	Standard Practice for Making and Curing Concrete Test Specimens in the Field
C 39	Standard Test Method for Compressive Strength of Cylindrical Concrete Specimens

C 40	Standard Test Method for Organic Impurities in Fine Aggregates for Concrete	C 309	Standard Specification for Liquid Membrane-Forming Compounds for Curing Concrete
C 42	Standard Test Method for Obtaining and Testing Drilled Cores and Sawed Beams of Concrete	C 494	Standard Specification for Chemical Admixtures for Concrete
C 78	Standard Test Method for Flexural Strength of Concrete (Using Simple Beam with Third-Point Loading)	C 496	Standard Test Method for Splitting Tensile Strength of Cylindrical Concrete Specimens
C 87	Standard Test Method for Effect of Organic Impurities in Fine Aggregate on Strength of Mortar	C 535	Standard Test Method for Resistance to Degradation of Large-Size Coarse Aggregate by Abrasion and Impact in the Los Angeles Machine
C 88	Standard Test Method for Soundness of Aggregates by Use of Sodium Sulfate or Magnesium Sulfate	C 567	Standard Test Method for Unit Weight of Structural Lightweight Concrete
C 94	Standard Specification for Ready-Mixed Concrete	C 586	Standard Test Method for Potential Alkali Reactivity of Carbonate Rocks for Concrete Aggregate (Rock-Cylinder Method)
C 109	Standard Test Method for Compressive Strength of Hydraulic Cement Mortars (Using 2 in. or [50 mm] Cube Specimens)	C 595	Standard Specification for Blended Hydraulic Cements
C 117	Standard Test Method for Materials Finer than 75-μm (No. 200) Sieve in Mineral Aggregates by Washing	C 597	Standard Test Method for Pulse Velocity Through Concrete
		C 617	Standard Practice for Capping Cylindrical Concrete Specimens
C 123	Standard Test Method for Lightweight Particles in Aggregate	C 618	Standard Specification for Coal Fly Ash and Raw or Calcined Natural Pozzolan for Use in Concrete
C 127	Standard Test Method for Density, Relative Density (Specific Gravity), and Absorption of Coarse Aggregate	C 642	Standard Test Method for Density, Absorption, and Voids in Hardened Concrete
C 128	Standard Test Method for Density, Relative Density (Specific Gravity), and Absorption of Fine Aggregate	C 666	Standard Test Method for Resistance of Concrete to Rapid Freezing and Thawing
		C 685	Standard Specification for Concrete Made by Volumetric Batching and Continuous Mixing
C 131	Standard Test Method for Resistance to Degradation of Small-Size Coarse Aggregate by Abrasion and Impact in the Los Angeles Machine	C 803	Standard Test Method for Penetration Resistance of Hardened Concrete
C 136	Standard Test Method for Sieve Analysis of Fine and Coarse Aggregates	C 805	Standard Test Method for Rebound Number of Hardened Concrete
C 138	Standard Test Method for Density (Unit Weight), Yield, and Air Content (Gravimetric) of Concrete	C 845	Standard Specification for Expansive Hydraulic Cement
		C 856	Standard Practice for Petrographic Examination of Hardened Concrete
C 142	Standard Test Method for Clay Lumps and Friable Particles in Aggregates	C 878	Standard Test Method for Restrained Expansion of Shrinkage-Compensating Concrete
C 143	Standard Test Method for Slump of Hydraulic Cement Concrete	C 900	Standard Test Method for Pullout Strength of Hardened Concrete
C 150	Standard Specification for Portland Cement		
C 172	Standard Practice for Sampling Freshly Mixed Concrete	C 989	Standard Specification for Ground Granulated Blast-Furnace Slag for Use in Concrete and Mortars
C 173	Standard Test Method for Air Content of Freshly Mixed Concrete by the Volumetric Method	C 1017	Standard Specification for Chemical Admixtures for Use in Producing Flowing Concrete
C 227	Standard Test Method for Potential Alkali Reactivity of Cement-Aggregate Combinations (Mortar-Bar Method)	C 1018	Standard Test Method for Flexural Toughness and First-Crack Strength of Fiber-Reinforced Concrete (Using Beam with Third-Point Loading)
C 231	Standard Test Method for Air Content of Freshly Mixed Concrete by the Pressure Method	C 1064	Standard Test Method for Temperature of Freshly Mixed Hydraulic-Cement Concrete
C 260	Standard Specification for Air-Entraining Admixtures for Concrete	C 1074	Standard Practice for Estimating Concrete Strength by the Maturity Method
C 289	Standard Test Method for Potential Alkali-Silica Reactivity of Aggregates (Chemical Method)	C 1077	Standard Practice for Laboratories Testing Concrete and Concrete Aggregates for Use in Construction and Criteria for Laboratory Evaluation
C 293	Standard Test Method for Flexural Strength of Concrete (Using Simple Beam with Center-Point Loading)	C 1157	Standard Performance Specification for Hydraulic Cement
C 295	Standard Guide for Petrographic Examination of Aggregates for Concrete	C 1202	Standard Test Method for Electrical Indication of Concrete's Ability to Resist Chloride Ion Penetration

C 1218   Standard Test Method for Water-Soluble Chloride in Mortar and Concrete
C 1240   Standard Specification for Silica Fume Used in Cementitious Mixtures
C 1260   Standard Test Method for Potential Alkali Reactivity of Aggregates (Mortar-Bar Method)
D 75   Standard Practice for Sampling Aggregates
D 512   Standard Test Methods for Chloride Ion in Water
D 1411   Standard Test Methods for Water-Soluble Chlorides Present as Admixtures in Graded Aggregate Road Mixes
D 4791   Standard Test Method for Flat Particles, Elongated Particles, or Flat and Elongated Particles in Coarse Aggregate

The preceding publications may be obtained from the following organizations:

American Concrete Institute
P.O. Box 9094
Farmington Hills, MI 48333-9094
www.concrete.org

ASTM International
100 Barr Harbor Dr.
West Conshohocken, PA 19428-2959
www.astm.org

APPENDIX II—SYNOPSIS OF ACI 311.5
Guide for Concrete Plant Inspection and Testing of Ready Mixed Concrete
Reported by ACI Committee 311

This guide is intended for use in establishing basic duties and reports required of inspection personnel. It can be used for all types and sizes of projects but should be supplemented with additional inspection requirements when the complexity of the project so dictates. Refer to ACI 311.4R for guidance on additional requirements and to SP-2 for more detailed information on concrete production practices and inspection and testing of concrete.

This guide recommends minimum requirements for inspection at the concrete plant when such inspections are required by specifications or the owner. It also recommends minimum requirements for field and laboratory testing of concrete. It is intended for use by specifiers, architects, engineers, owners, contractors, or other groups needing to monitor the ready-mixed concrete producers' activities at the concrete plant and concreting activities at the project site through the use of an independent inspection agency or in-house inspection organization.

This guide also recommends minimum testing laboratory qualifications, minimum inspector qualifications, duties, and reports.

ACI 311.6-18

Specification for Testing Ready Mixed Concrete

An ACI Standard

Reported by ACI Committee 311

Michael C. Jaycox,* Chair
Tracy Grover, Secretary

Joseph W. Clendenen
Mario R. Diaz
Robert L. Henry

Jose Damazo Juarez
Venkatesh S. Iyer
Claude E. Jaycox*

Robert S. Jenkins
David Savage
Jose Damazo Juarez

Woodward L. Vogt
George R. Wargo
Michelle L. Wilson

Consulting Members

Joseph F. Artuso
Ann Balogh
Roger E. Wilson

*Task group preparing this specification.

This Reference Specification covers Testing Agency requirements for field and laboratory testing of ready mixed concrete delivered to the project. It is intended for use by specifiers, Architects, Engineers, Owners, and other groups interested in monitoring the quality of concrete used in project construction. This Reference Specification can be made applicable to any construction project by citing it in a Contract. The specifier supplements the provisions of this Reference Specification as needed by specifying individual project requirements in the Contract.

Keywords: air content; compressive strength; laboratory; ready mixed concrete; sampling; slump; Specification; technician; temperature; testing; unit weight.

(mandatory portion follows)

CONTENTS

SECTION 1—GENERAL REQUIREMENTS, p. 1
1.1—Scope, p. 1
1.2—Qualifications, p. 2
1.3—Definitions, p. 2
1.4—Referenced standards, p. 2
1.5—Certification and accreditation organizations, p. 2
1.6—Units of measurement, p. 3

SECTION 2—TESTING READY MIXED CONCRETE, p. 3
2.1—Sampling, p. 3
2.2—Frequency of sampling and testing, p. 3
2.3—Tests, p. 3
2.4—Number of strength test specimens, p. 3
2.5—Curing of strength test specimens, p. 3
2.6—Testing for strength, p. 3

SECTION 3—TRANSMITTALS, p. 3
3.1—Scope, p. 3
3.2—Reports, p. 3
3.3—Report information, p. 3

(nonmandatory portion follows)

NOTES TO SPECIFIERS, p. 4
General notes, p. 4

FOREWORD TO CHECKLISTS, p. 4

MANDATORY REQUIREMENTS CHECKLIST, p. 5

OPTIONAL REQUIREMENTS CHECKLIST, p. 5

SUBMITTALS CHECKLIST, p. 5

(mandatory portion follows)

SECTION 1—GENERAL REQUIREMENTS

1.1—Scope

1.1.1 *Work specified*—This Specification sets the minimum requirements for testing of ready mixed concrete

---

ACI 311.6-18 supersedes ACI 311.6-09, became effective July 30, 2018, and was published August 2018.

Copyright © 2018, American Concrete Institute.

All rights reserved including rights of reproduction and use in any form or by any means, including the making of copies by any photo process, or by electronic or mechanical device, printed, written, or oral, or recording for sound or visual reproduction or for use in any knowledge or retrieval system or device, unless permission in writing is obtained from the copyright proprietors.

at the project site when specific field tests and specimens prepared for laboratory tests properties including strength are used as a basis for acceptance of concrete as delivered to the site. It includes requirements for making and curing test specimens, performing field and laboratory tests, and qualifying personnel and laboratories performing these tests.

1.1.2 *Work not specified*

1.1.2.1 This Specification does not apply to testing requirements and responsibilities required of the Construction Contractor. ACI 301 identifies the testing responsibilities required of the Construction Contractor.

1.1.2.2 This Specification does not apply to other duties, such as inspection, monitoring of concrete production, mixing, and construction; or conformity to project specifications and drawings. ACI 311.7 addresses these responsibilities of the inspector. The person testing the concrete in 1.1.1 is not required to be trained or qualified to perform duties other than those specified in 2.3 (tests).

1.1.3 *Test Reports*—The Testing Agency shall provide the testing services and reports of tests as required by the Project Specifications.

1.2—Qualifications

1.2.1 The Testing Agency shall submit qualifications of field technicians and laboratory testing technicians, as defined in 1.2.1.1 to 1.2.1.2, to Owner or Owner's representative.

1.2.1.1 *Field technician*—Technicians conducting field tests of concrete shall be certified as ACI Concrete Field Testing Technician – Grade I, unless otherwise specified.

1.2.1.2 *Laboratory technician*—Technicians conducting laboratory testing shall be certified as ACI Concrete Laboratory Testing Technician – Level 1 or ACI Concrete Strength Testing Technician, unless otherwise specified.

1.2.2 *Testing Agency*—The Owner or the Owner's representative shall not delegate these services to the Construction Contractor. The Testing Agency shall meet the requirements of ASTM E329 and be accredited in accordance with the requirements of ASTM C1077. Unless otherwise specified, the Testing Agency shall be accredited by one or more of the following: the AASHTO Accreditation Program (AAP), the National Voluntary Laboratory Accreditation Program (NVLAP), the American Association for Laboratory Accreditation (A2LA), the International Accreditation Service (IAS), the Construction Materials Engineering Council (CMEC), or the Washington Area Council of Engineering Laboratories (WACEL).

1.3—Definitions

accreditation—the act or result of having a Testing Agency's quality procedures, equipment calibration, and personnel reviewed and approved by an independent third-party agency.

certification—the act of having an individual's skills and knowledge of a certain subject tested and verified by an independent third-party agency.

Contract—the professional agreement between the Owner and Testing Agency for providing testing services.

Contract Documents—a set of documents supplied by Owner to Contractor as the basis for testing; these documents contain contract forms, contract conditions, Specifications, drawings, addenda, and contract changes.

Contractor—as used in this document, the person, firm, or entity under contract.

Owner—the corporation, association, partnership, individual, public body, or authority for whom the testing services are performed.

Project Specification—the written document that details requirements for the Work in accordance with service parameters and other prescriptive or performance criteria.

Testing Agency—the firm or entity under contract for providing testing services.

Work—the entire construction or separately identifiable parts thereof required to be furnished as described in the Contract Documents.

1.4—Referenced standards

Standards of ACI and ASTM cited in this Specification are listed by name and designation, including year.

1.4.1.1 *American Concrete Institute*

ACI 301-16—Specifications for Structural Concrete

ACI 311.7-18—Specification for Inspection of Concrete Construction

1.4.1.2 *ASTM International*

ASTM C31/C31M-18—Standard Practice for Making and Curing Concrete Test Specimens in the Field

ASTM C39/C39M-18—Standard Test Method for Compressive Strength of Cylindrical Concrete Specimens

ASTM C138/C138M-17—Standard Test Method for Density (Unit Weight), Yield, and Air Content (Gravimetric) of Concrete

ASTM C143/C143M-15—Standard Test Method for Slump of Hydraulic Cement Concrete

ASTM C172/C172M-17—Standard Practice for Sampling Freshly Mixed Concrete

ASTM C173/C173M-16—Standard Test Method for Air Content of Freshly Mixed Concrete by the Volumetric Method

ASTM C231/C231M-17—Standard Test Method for Air Content of Freshly Mixed Concrete by the Pressure Method

ASTM C1064/C1064M-12—Standard Test Method for Temperature of Freshly Mixed Hydraulic-Cement Concrete

ASTM C1077-17—Standard Practice for Agencies Testing Concrete and Concrete Aggregates for Use in Construction and Criteria for Testing Agency Evaluation

ASTM E329-18—Standard Specification for Agencies Engaged in Construction Inspection, Testing, or Special Inspection

1.5—Certification and accreditation organizations

1.5.1 *Certification organization*—The following certification organization referred to in this Specification is listed.

American Concrete Institute

1.5.2 *Accreditation organizations*—The following accreditation organizations referred to in this Specification are listed.

AASHTO Accreditation Program (AAP)

AASHTO Materials Reference Laboratory (AMRL)
The American Association for Laboratory Accreditation (A2LA)
Construction Materials Engineering Council (CMEC)
International Accreditation Service (IAS)
National Voluntary Laboratory Accreditation Program (NVLAP), Standards Services Division
Washington Area Council of Engineering Laboratories, Inc.

1.6—Units of measurement
Values in this Specification are stated in inch-pound units. A companion Specification in SI units is also available.

SECTION 2—TESTING READY MIXED CONCRETE

2.1—Sampling
Sample the ready mixed concrete for testing required in this Specification in accordance with ASTM C172/C172M, unless otherwise specified.

2.2—Frequency of sampling and testing
2.2.1 Tests listed in 2.3.1 through 2.3.4 shall be performed and strength test specimens shall be obtained per requirements of 2.3.6 and tested according to requirements of 2.3.5, at least once for each 150 yd^3, or fraction thereof, for each concrete mixture placed in any one day, unless otherwise specified.

2.2.2 Tests 2.3.1 through 2.3.4 shall be performed on the first delivery of each concrete mixture delivered to the project each day, and if a concrete mixture is adjusted, on the first delivery of the adjusted mixture, and then randomly throughout the placement if needed, unless otherwise specified.

2.3—Tests
2.3.1 *Slump test*—ASTM C143/C143M
2.3.2 *Temperature test*—ASTM C1064/C1064M
2.3.3 *Air content test*—ASTM C173/C173M or C231/C231M
2.3.4 *Density (unit weight) test*—ASTM C138/C138M
2.3.5 *Making and curing compressive strength specimens*—ASTM C31/C31M
2.3.6 *Compressive strength test*—ASTM C39/C39M

2.4—Number of strength test specimens
An acceptance strength test set consists of a minimum of two 6 x 12 in. specimens or three 4 x 8 in. specimens to be tested at the age of 28 days, unless otherwise specified.

2.5—Curing of strength test specimens
2.5.1 *Initial curing*—Owner or Owner's representative will provide and maintain adequate facilities on the project site for initial storage and curing of the concrete specimens, unless otherwise specified. Specimens shall be stored under conditions that meet the requirements of ASTM C31/C31M and shall be verified by Testing Agency. Such storage shall have temperature controls to maintain ASTM C31/C31M temperature requirements. Calibrated temperature recording devices shall be used to record daily maximum and minimum temperatures of the initial curing environment.

2.5.2 *Transportation*—Testing Agency will recover and transport concrete specimens in accordance with ASTM C31/C31M.

2.5.3 *Final curing*—Final curing of strength test specimens shall be done in accordance with ASTM C31/C31M.

2.6—Testing for strength
Determine compressive strengths of cylinders in accordance with 2.3.6.

SECTION 3—TRANSMITTALS

3.1—Scope
The Testing Agency's scope of authority, the requirements for transmittal of reports such as timelines, methods of delivery and distribution list, and the procedure for notification of deficiencies such as, methods of delivery, timelines and distribution list, and required observations shall be defined in the Contract with the Owner before the start of the project.

3.2—Reports
3.2.1 *Daily report*—The Testing Agency shall, daily, submit reports that document the results of field concrete tests performed in accordance with 2.2 and distributed in accordance with 3.1, unless otherwise specified.

3.2.2 *Nonconformance report*—Deficient items shall be communicated in a nonconformance report to the distribution list established in accordance with 3.1. Unless otherwise stated in the Contract Documents, a nonconformance report shall be issued within 24 hours for any concrete found deficient, unless otherwise specified.

3.2.3 *Final report*—A final field report and a laboratory report per ASTM test standards shall be provided to the distribution list within a time established in accordance with 3.1.

3.3—Report information
Reports shall include items from 3.3.1 through 3.3.12, and information required by ASTM standards referenced in 2.3:
3.3.1 Project name
3.3.2 Client name
3.3.3 Concrete supplier
3.3.4 Date and time of sampling and field testing
3.3.5 Dates that strength test specimens will be tested
3.3.6 Name of field and laboratory technicians and certification numbers
3.3.7 Delivery truck number, ticket, mixture designation, and locations of sampling
3.3.8 Results of air content, temperature, slump, and density (unit weight) tests
3.3.9 Specified compressive strength of concrete and the designated test age
3.3.10 Location of placement represented by the strength test specimens
3.3.11 Location of sampled concrete within the placement
3.3.12 Report curing method and maximum and minimum temperatures of the curing environment during the initial curing period

(nonmandatory portion follows)

NOTES TO SPECIFIERS

General notes

G1. ACI Specification 311.6-18 is to be used by reference or incorporation in its entirety in the Owners Contract for testing services. Do not copy individual sections, parts, articles, or paragraphs into a Contract, because taking them out of context may change their meaning.

G2. If Sections or Parts of ACI Specification 311.6-18 are copied into a Contract or any other document, do not refer to them as an ACI Specification, because the Specification has been altered.

G3. A statement such as the following will serve to make ACI Specification 311.6-18 a part of a Contract:

> "Testing services for (this Contract) shall conform to all requirements of ACI 311.6-18 published by the American Concrete Institute, Farmington Hills, Michigan, except as modified by this Contract."

G4. Each technical Section of ACI Specification 311.6-18 is written in the three-part section format of the Construction Specifications Institute, as adapted for ACI requirements. The language is imperative and terse.

G5. ACI Specification 311.6-18 is written to the Testing Agency and is to be used by the Owner or the Owner's representative to specify the concrete testing on a project. When a provision of this Specification requires action on the Testing Agency's part, the verb "shall" is used. If the Testing Agency is allowed to exercise an option when limited alternatives are available, the phrasing "either... or..." is used. Statements provided in the Specification as information to the Testing Agency use the verbs "may" or "will." Informational statements typically identify activities or options that "will be taken" or "may be taken" by the Owner or Owner's representative.

FOREWORD TO CHECKLISTS

F1. This foreword is included for explanatory purposes only; it does not form a part of ACI Specification 311.6-18.

F2. ACI Specification 311.6-18 may be referenced by the specifier in the Contract for any building project, together with supplementary requirements for the specific project. Responsibilities for project participants must be defined in the Contract. ACI Specification 311.6-18 cannot and does not address responsibilities for any project participant other than the Testing Agency.

F3. Checklists do not form a part of ACI Specification 311.6-18. Checklists assist the specifier in selecting and specifying project requirements in the Contract.

F4. The Mandatory Requirements Checklist indicates work requirements regarding specific qualities, procedures, materials, and performance criteria that are not defined in ACI Specification 311.6-18. The specifier must include these requirements in the Contract.

F5. The Optional Requirements Checklist identifies specifier choices and alternatives. The checklist identifies the sections, parts, and articles of the ACI Reference Specification 311.6-18 and the action required or available to the specifier. The specifier should review each of the items in the checklist and make adjustments to the needs of a particular project by including those selected alternatives as mandatory requirements in the Contract.

F6. The Submittals Checklist identifies information or data to be provided by the Testing Agency before, during, or after construction.

F7. *Recommended references*—Documents and publications that are referenced in the checklists of ACI Specification 311.6-18 are listed below. These references provide guidance to the specifier and are not considered to be part of ACI Specification 311.6-18.

American Concrete Institute
CP-1—Technician Workbook for ACI Certification of Concrete Field Testing Technician – Grade I
CP-19—Technician Workbook for ACI Certification of Concrete Strength Testing Technician

MANDATORY REQUIREMENTS CHECKLIST

Section/Part/Article	Notes to Specifiers
2.4	Specify cylinder specimen size.
3.1	Specify Testing Agency's scope of authority, the requirements for submittal of reports (timeliness, methods of delivery, and distribution list), procedures for notification of deficiencies, and required observations.
3.2.1	Specify if daily reports are not required and if summaries are required and their frequencies.

OPTIONAL REQUIREMENTS CHECKLIST

Section/Part/Article	Notes to Specifiers
1.2.1.1	Other field technician certifications equivalent to ACI Concrete Field Testing Technician – Grade I may be specified. Equivalent certification should include written and performance examinations to relevant standards as accepted by the accreditation agency.
1.2.1.2	Other laboratory technician certifications equivalent to ACI Concrete Laboratory Technician – Level 1 or ACI Concrete Strength Testing Technician may be specified. Equivalent certification should include written and performance examinations to relevant standards as accepted by the accreditation agency.
1.2.2	Laboratory accreditations equivalent may be specified.
2.1	Specify actual sampling requirements.
2.2.1	Specify when sampling frequency must be increased or decreased due to the specific scope of the project.
2.2.2	Specify if these tests of fresh concrete need only be performed when concrete is sampled for making cylinders for strength tests or if they need to be performed at each delivery.
2.4	Specify number and age of test for acceptance cylinders. Specify what additional test cylinders and ages of tests are needed for early strength indication, formwork removal, field curing, or others.
2.5.1	Specify if the Construction Contractor will be responsible for the facilities construction and maintenance. The Testing Agency is responsible *only* for verification of proper facilities and temperatures.
3.2.2	Other timelines may be specified.

SUBMITTALS CHECKLIST

Section/Part/Article	Submittal items and notes to Specifier
1.2.1.1	Documentation of current field technician certification for required testing.
1.2.1.2	Documentation of current laboratory technician certification for required testing.
1.2.2	Documentation of current Inspection Agency/Laboratories Accreditation Certifications.
3.2.1	Daily field reports.
3.2.2	Nonconformance reports.
3.2.3	Final reports.

ACI 311.7-18

Specification for Inspection of Concrete Construction

An ACI Standard

Reported by ACI Committee 311

Michael C. Jaycox,* Chair Tracy Grover, Secretary

Joseph W. Clendenen	Jose Damazo Juarez	Robert S. Jenkins	George R. Wargo
Mario R. Diaz	Venkatesh S. Iyer	David Savage	Michelle L. Wilson
Robert L. Henry	Claude E. Jaycox*	Woodward L. Vogt	

Consulting Members

Joseph F. Artuso Ann Balogh Roger E. Wilson

*Task group preparing this specification.

This reference specification covers quality assurance inspection services for preplacement, placement, and post-placement of concrete construction. This reference specification can be made applicable to a particular construction project by citing it in the inspection services contract. The specifier shall supplement the provisions of this reference specification as needed by specifying individual project requirements in the inspection services contract. The materials, processes, quality control measures, and inspections described in this specification should be tested, monitored, or performed as applicable only by individuals holding the appropriate ACI Certifications or equivalent.

Keywords: accreditation; certification; concrete construction; inspection agency; placement; post-placement; preplacement; special inspection; specification.

(mandatory portion follows)

CONTENTS

SECTION 1—GENERAL REQUIREMENTS, p. 2
1.1—Scope, p. 2
1.2—Units of measurement, p. 2
1.3—Referenced standards, p. 2
1.4—Definitions, p. 2
1.5—Reporting requirements, p. 3
1.6—Qualifications, p. 3
1.7—Items requiring inspection, p. 3

SECTION 2—PREPLACEMENT INSPECTION, p. 3
2.1—General requirements, p. 3
2.2—Formwork, shoring, and embedments, p. 3
2.3—Mild steel reinforcement, p. 4
2.4—Bonded post-tensioned steel reinforcement, p. 4
2.5—Unbonded post-tensioned steel reinforcement, p. 5
2.6—Stressing and grouting, p. 5

SECTION 3—PLACEMENT INSPECTION, p. 5
3.1—Concrete quality, p. 5
3.2—Conveyance and placement, p. 5
3.3—Consolidation, p. 5
3.4—Finishing, p. 5

SECTION 4—POST-PLACEMENT INSPECTION, p. 5
4.1—Protection and curing, p. 5
4.2—Forms, shore removal, and reshoring, p. 5
4.3—Strength test specimens, p. 6

SECTION 5—CONCRETE ANCHORS, p. 6
5.1—General, p. 6

ACI 311.7-18 supersedes ACI 311.7-14, became effective March 16, 2018, and was published April 2018.
Copyright © 2018, American Concrete Institute
All rights reserved including rights of reproduction and use in any form or by any means, including the making of copies by any photo process, or by electronic or mechanical device, printed, written, or oral, or recording for sound or visual reproduction or for use in any knowledge or retrieval system or device, unless permission in writing is obtained from the copyright proprietors.

SECTION 6—ERECTION OF PRECAST CONCRETE, p. 6

SECTION 7—SHOTCRETE, p. 6

(nonmandatory portion follows)

NOTES TO SPECIFIERS, p. 6
General notes, p. 6
Foreword to checklists, p. 6

(mandatory portion follows)

SECTION 1—GENERAL REQUIREMENTS

1.1—Scope
1.1.1 *Services specified*—This specification sets the minimum requirements for quality assurance inspection of concrete construction whenever inspection services are required by the local building codes or the Contract Documents. The services covered are directed to Special Inspections covered in the current edition of the IBC. This specification is also valid for other projects at the option of the user. It includes inspection requirements for preplacement (reinforcement, forms, and embedments), placement (conveyance and consolidation), and post-placement (finishing, curing, stressing, grouting, and formwork removal) of concrete construction, as well as requirements for personnel qualification and Inspection Agency accreditation. All inspection services required in this specification shall be performed by an Inspection Agency engaged by the Owner.

1.1.2 *Independence*—An Inspection Agency shall be objective, competent, and independent from the Contractor responsible for the Work being inspected. The Owner or the Owner's Representative shall not delegate these services to the Contractor, Construction Manager, Construction Manager at Risk, or any other party responsible for constructing the Work.

1.1.3 *Coordination with Contractor*—The Owner shall instruct the Contractor to coordinate with the Inspection Agency, provide access to the Work, and provide notification when the Work is ready for inspection.

1.2—Units of measurement
1.2.1 *Units*—Inch-pound is the standard units of measurement in this document.

1.3—Referenced standards
Standards cited in this specification are listed by name and designation, including year, unless otherwise specified.

1.3.1 *American Concrete Institute*
ACI 301-16—Specifications for Structural Concrete
ACI 311.6-18—Specification for Ready Mixed Concrete Testing Services
ACI 318-19—Building Code Requirements for Structural Concrete
ACI 355.4-11—Qualification of Post-Installed Adhesive Anchors in Concrete and Commentary

1.3.2 *ASTM International*
ASTM A706/A706M-15—Standard Specification for Deformed and Plain Low-Alloy Steel Bars for Concrete Reinforcement
ASTM E329-14—Standard Specification for Agencies Engaged in Construction Inspection, Testing, or Special Inspection

1.3.3 *American Welding Society*
AWS D1.4/D1.4M: 2011—Structural Welding Code—Reinforcing Steel

1.3.4 *International Code Council*
IBC 2015

1.4—Definitions
The following definitions govern in this specification. For definitions of other terms, refer to "ACI Concrete Terminology (ACI CT)."

acceptance—acknowledgement by Architect/Engineer that submittal or completed work is acceptable.

accreditation—act or result of having an Inspection Agency's quality system, including equipment calibration and personnel certifications, reviewed and accredited by a qualified third-party Accreditation Agency.

Architect/Engineer)—architect, engineer, architectural firm, or engineering firm developing Contract Documents or administering the Work under construction Contract Documents, or both.

certified—having met the requirements of a program that determines and attests that an individual is qualified to perform a specific service.

Contractor—person, firm, or entity under contract for the construction of the Work.

continuous inspection—full-time observation of Work requiring inspection by a Special Inspector who is present in the specific area where the Work is being performed.

discrepancy—an item that does not conform to the contract documents and is reported to the contractor for correction.

inspection agency—person, firm, or entity under contract for providing inspection services.

inspection services—periodic or continuous Special Inspections as required in this specification and Inspection Services Contract between the Owner and Inspection Agency.

Inspection Services Contract—set of documents supplied by the Owner to the Inspection Agency that serves as the basis for the agreement to perform inspection services; these documents can contain contract forms, contract conditions, Statement of Special Inspections, specifications, drawings, addenda, and contract changes. A different set of Contract Documents will exist between the Owner and construction Contractor.

Non-conformance—discrepancy that is not corrected prior to completion of that phase of the work and is reported to the building official and RDPRC so the RDPRC can determine the action to be taken.

Owner—corporation, association, partnership, individual, or public body or authority for whom the Work is constructed.

periodic inspection—part-time or intermittent observation of Work requiring Special Inspection by a Special Inspector who is present in areas where the work is being performed or at the completion of the Work. All items requiring inspection must be inspected before closing or before the Work is no longer accessible to inspect.

Project Specification—written document that details requirements for the Work in accordance with service parameters and other specific criteria.

Registered Design Professional in Responsible Charge—used in place of Architect/Engineer in the IBC.

Statement of Special Inspection—a list of the required special inspections and code references that is based on the project contract documents and prepared by the RDPRC.

Special Inspection—inspection of construction requiring the expertise of a Special Inspector to ensure compliance with the approved Contract Documents.

Special Inspector—employee of the Special Inspection Agency or an individual that is qualified and certified to perform the inspections described in this specification

testing agency—person, firm, or entity under contract for monitoring the quality of the materials as specified in the Contact Documents.

verify—act of reviewing, inspecting, testing, and reporting that Work inspected is or is not in conformance with Contract Documents

Work—entire construction or separately identifiable parts thereof required to be furnished under the Contract Documents.

1.5—Reporting requirements

1.5.1 Comply with special inspection requirements of the applicable Building Code and with the additional requirements of the Contract Documents.

1.5.2 Notify the Contractor of discrepancies from approved Contract Documents daily, usually as soon as detecting the discrepancy. If uncorrected prior to the completion or covering of the Work, before further Work occurs, notify the Contractor, Owner, Architect/Engineer, and building official with copies of the Non-Conformance Report (NCR).

1.5.3 Compile an NCR log (NCL) or similar document noting construction items not conforming to the Contract Documents. The NCL shall describe the nonconformance, date noticed, and individuals or organizations informed. Upon completion of the corrective action, update the NCR and NCL to indicate the action and date of acceptance by the Architect/Engineer, Building Official, or both, as appropriate.

1.5.4 Submit progress reports to the Owner, Architect/Engineer, and the Building Official, describing inspections and tests that were performed, nonconformances, and compliance of Work with the Contract Documents.

1.5.5 Submit a final summary report stating whether Work requiring inspection was, to the best of the inspector's knowledge, in conformance with the approved Contract Documents and applicable provisions of the building code.

1.6—Qualifications

1.6.1 *Inspection agency*—Unless otherwise specified, Inspection Agency shall submit to Owner its qualifications and the qualifications of its inspectors as required in 1.6.1.1 and 1.6.1.2.

1.6.1.1 *Special Inspectors*—Unless otherwise specified, Inspectors conducting Special Inspections of concrete preplacement (reinforcement, embedments, and forms), placement (conveyance and consolidation), and post-placement (finishing, curing, stressing, grouting, and form and shore removal) as defined in the local building code, shall be currently certified by ACI as a Concrete Construction Special Inspector (CCSI) or Concrete Transportation Construction Inspector (CTCI) or as acceptable to the local building official.

1.6.1.2 *Inspection Agency*—Unless otherwise specified, The Inspection Agency shall conform to the requirements of ASTM E329 as it pertains to concrete inspection and local building code requirements. The Inspection Agency shall be accredited by The American Association for Laboratory Accreditation (A2LA) or The International Accreditation Service (IAS) or as acceptable to the local Building Official.

1.7—Items requiring inspection

1.7.1 Unless otherwise specified in the local building code or Contract Documents, items requiring verification and inspection shall be continuously or periodically inspected in accordance with Table 1.7.1. Any item not listed in Table 1.7.1 shall be deemed requiring periodic inspection.

SECTION 2—PREPLACEMENT INSPECTION

All work listed in Section 2 shall be verified in accordance with Table 1.7.1 and ACI 301 Section 4, and as specified in the Contract Documents.

2.1—General requirements

2.1.1 Review the Contract Documents and required local Building Codes for inspection requirements.

2.1.2 Review structural and material submittals accepted by the Architect/Engineer or RDPRC.

2.1.3 Take part in preconstruction meetings to discuss project inspection requirements and proper sequencing of required inspection services, and the duties of all parties concerned.

2.1.4 Before concrete construction, verify that the substrate condition (soil subgrade, decking, or other materials beneath the concrete construction) has been inspected and approved for concrete placement by the Architect/Engineer or RDPRC.

2.1.5 Verify use of required design mixture per Table 1.7.1 Item 5.

2.2—Formwork, shoring, and embedments

All work listed in 2.2 shall be verified per Table 1.7.1 Item 12; ACI 301, 2.3; and as specified in the Contract Documents.

Table 1.7.1—Required Special Inspections of concrete construction

Item	Continuous special inspection	Periodic special inspection	Referenced standard and codes
1. Inspect reinforcement, including prestressing tendons, and verify placement.	—	X	ACI 318, Ch. 20, 25.2, 25.3, and 26.6.1 through 26.6.4
2. Reinforcing bar welding: 　a) Verify weldability of reinforcing bars other than ASTM A706 　b) Inspect single-pass fillet welds, maximum 5/16 in. 　c) All other welds	— — X	X X —	AWS D1.4; ACI 318, 26.6.4
3. Inspect anchors cast in concrete.	—	X	ACI 318, 26.7, 26.13
4. Inspect anchors post-installed in hardened concrete members. 　a) Adhesive anchors installed in horizontally or upwardly inclined orientations to resist sustained tension loads. 　b) Mechanical anchors and adhesive anchors not defined in 4a.	X —	— X	ACI 318, 26.7, 26.13; ACI 355.4 ACI 318, 26.7, 26.13
5. Verify use of required design mixture.	—	X	ACI 318, Ch. 19 and 26.4.4
6. During concrete and shotcrete placement, verify fabrication of strength tests specimens and perform slump, air content tests, and determined temperature of the concrete.	X	—	ACI 311.6; ACI 318, 26.12
7. Inspect concrete placement for proper application techniques.	X	—	ACI 318, 26.5.2
8. Verify maintenance of specified curing temperature and techniques.	—	X	ACI 318, 26.5.3 through 26.5.5
9. Inspect prestressed concrete for: 　a) Application of prestressing forces 　b) Grouting of bonded prestressing tendons.	X X	—	ACI 318, 26.10.1 and 26.13.3.2
10. Inspect erection of precast concrete members.	—	X	ACI 318, 26.9
11. Verify in-place concrete strength prior to stressing of tendons in post-tensioned concrete and prior to removal of shores and forms from beams and structural slabs.	—	X	ACI 318, 26.11.2.1 and 26.13.3.3f
12. Inspect formwork for shape, location, and dimensions of the concrete member being formed.	—	X	ACI 318, 26.11.1.2(b)
13. Inspect shotcrete placement for proper application techniques, materials, and testing	X	—	IBC 1908

2.2.1 Verify the formwork complies with requirements for cleanliness and will provide concrete elements of the specified size and shape.

2.2.2 Verify that the location and preparation of joints comply.

2.2.3 Verify that the type, quantity, size, spacing, and location of embedments are as specified and are supported and secured with materials approved by the Architect/Engineer or RDPRC, to prevent displacement during concrete placement.

2.2.4 Verify that block-outs are correct for location, size, and condition.

2.3—Mild steel reinforcement

All work listed in 2.3 shall be verified per Table 1.7.1 Items 1 and 2; ACI 301, 3.3; and as specified in the Contract Documents and approved placing drawings.

2.3.1 Verify that reinforcing steel is of the type, grade, and size specified.

2.3.2 Verify that reinforcing steel is free of oil, dirt, and loose rust and that steel is acceptably coated, if specified.

2.3.3 Verify that reinforcing steel is located in accordance with the Contract Documents and is adequately supported and secured with materials approved by the Architect/Engineer or RDPRC, to prevent displacement during concrete placement.

2.3.4 Verify that minimum concrete cover will be provided based on formwork configuration.

2.3.5 Verify that placement of reinforcing steel complies with required spacing, profile, and quantity requirements, as indicated in the Contract Documents.

2.3.6 Verify that hooks, bends, ties, stirrups, and supplemental reinforcements are fabricated and properly placed.

2.3.7 Verify that required lap lengths, stagger, and offsets are provided.

2.3.8 Verify proper installation of approved mechanical connections in accordance with the manufacturer's instruction, evaluation reports, or both.

2.3.9 Verify that welds of reinforcing steel and other elements are as specified and have been inspected and approved by a certified welding inspector.

2.4—Bonded post-tensioned steel reinforcement

All work listed in 2.4, 2.5, and 2.6 shall be verified per Table 1.7.1 Items 7, 9, and 11; ACI 301, 13.2; and as specified in the Contract Documents.

2.4.1 Verify that the handling and storage of tendon strands on-site did not contaminate or damage strands.

2.4.2 Verify placement and condition of tendons and anchorages.

2.4.3 Verify that ducts, including inlets and outlets, are of the required size, mortar tight, and are in specified locations.

2.4.4 Verify that the pump and jack system to be used for tendon tensioning has been calibrated within the prior 6 months.

2.4.5 Verify that the prestressing force is applied within the allowable tolerance and is recorded accurately.

2.4.6 Verify that the stressing sequence is followed, if specified.

2.4.7 Verify that the elongations are within the required tolerance and are recorded accurately.

2.5—Unbonded post-tensioned steel reinforcement

2.5.1 Verify that the handling and storage of tendon strands on-site did not contaminate or damage strands.

2.5.2 Verify placement location of tendons and anchorages and that sheathing is intact.

2.5.3 Verify that procedure approved by the Architect/Engineer or RDPRC is followed for any repair of sheathing.

2.5.4 Verify location and profile of tendons and ducts.

2.5.5 Verify that tie bars, chairs, and bolsters are the specified type and are secured to the tendon to prevent displacement of tendon during concrete placement.

2.6—Stressing and grouting

2.6.1 Verify that the pump and jack system to be used for tendon tensioning have been calibrated within the prior 6 months.

2.6.2 Verify that the stressing sequence is followed, if specified.

2.6.3 Verify that the stressing forces and tendon elongations are within the required tolerances and are recorded accurately.

2.6.4 Verify that the in-place concrete strength is achieved before transferring stressing force or stressing.

2.6.5 Verify that the grout mixture for tendon grouting is approved by Architect/Engineer or RDPRC, the grout is applied continuously until the vents discharge a steady stream of grout, and the grouting pressure is attained after the vents are closed.

SECTION 3—PLACEMENT INSPECTION

All work listed in Section 3 shall be verified per Table 1.7.1 Items 6 and 7; ACI 301, 2.3 and 5.3; tested per ACI 311.6; and as specified in the Contract Documents.

3.1—Concrete quality

3.1.1 Verify that all individual batch tickets indicate deliveries of the mixture are as specified.

3.1.2 Verify time limits of mixing, total water added, drum revolution limits and specified consistency, and workability for the placements are as specified.

3.1.3 Verify that the required type, quantity, and frequency of tests to be performed on the fresh concrete are performed by a certified ACI Concrete Field Testing Technician, as specified in ACI 311.6.

3.1.4 Observe sampling of concrete, field testing of fresh concrete, and making of test specimens.

3.1.5 Verify slump, temperature, and air content.

3.2—Conveyance and placement

3.2.1 Verify acceptable condition of the place of deposit before the concrete is placed. Place of deposit shall be free of snow, ice, mud, water, and loose material.

3.2.2 Verify that methods of conveying and depositing concrete avoid contamination and segregation of the mixture.

3.2.3 Verify use of drop chutes or funnel hoses to contain free fall and that free fall does not exceed requirements.

3.2.4 Verify that grout used to lubricate the pump hose is not incorporated into the placement.

3.2.5 Report any unexpected delays that may have caused a cold joint.

3.2.6 Verify that the depths of concrete layers are as specified.

3.3—Consolidation

3.3.1 Verify that the vibrator is inserted through the full depth of the layer and 6 in. into the previous layer and that the vibrator is not used to move concrete horizontally.

3.3.2 Verify that the spacing and duration of external vibration are consolidating the concrete without segregation.

3.3.3 Verify that revibration is performed if required.

3.3.4 Verify that formwork is tight and appears stable.

3.4—Finishing

3.4.1 Verify that screeding, floating, edging, troweling, and texturing are completed.

3.4.2 Verify that saw-cut joints are cut within the specified time and within tolerances for depth, width, location, and alignment.

SECTION 4—POST-PLACEMENT INSPECTION

All work listed in Section 4 shall be verified per Table 1.7.1 Item 8; ACI 301, 5.3.6; and as specified in the Contract Documents.

4.1—Protection and curing

4.1.1 Verify that concrete is protected from temperature extremes and from sunlight and wind that could detrimentally decreases the quality of the hardened concrete or produce plastic shrinkage cracking due to rapid drying, as in accordance with the Contract Documents.

4.1.2 Verify that the curing compound or method approved by the Architect/Engineer or RDPRC, if specified, is applied at the specified time and rate of application.

4.1.3 Verify that the specified curing is applied and for the specified duration, per the Contract Documents.

4.2—Forms, shore removal, and reshoring

4.2.1 Verify that form removal, shore removal, and reshoring are performed per the Contract Documents.

4.2.2 Verify that the concrete has attained the minimum required in-place strength before formwork removal by reviewing the strengths of field-cured specimens or the results of in-place test obtained by testing agency.

4.2.3 Observe and document the condition of the concrete surface after form removal and report to the Architect/Engineer or RDPRC on the possible need for repairs.

4.3—Strength test specimens

4.3.1 Verify that proper test specimen identification is used and that the correct numbers of specimens are cast.

4.3.2 Verify the initial site storage, curing, and protection of strength test specimens are per contact documents.

4.3.3 Verify that specimens are protected against damage while stored, and loaded into transportation vehicle to the Testing Laboratory.

4.3.4 Verify the storage conditions and locations of field-cured test specimens used to confirm in-place concrete strength before form and shoring removal and tendon stressing or transfer of tendon stress.

4.3.5 Verify that the strength of test specimens is achieved in accordance with the Contract Documents by review of the testing agency reports.

4.3.6 Verify the testing agency conforms to ACI 311.6, unless otherwise specified.

SECTION 5—CONCRETE ANCHORS

All work listed in Section 5 shall be verified per Table 1.7.1 Items 3 and 4; ACI 355.4; and as specified in the Contract Documents.

5.1—General

5.1.1 Verify concrete has reached an age of 21 days before installation of anchors.

5.1.2 Verify that the installer for adhesive anchors is a certified ACI Adhesive Anchor Installer, unless otherwise specified.

5.1.3 Verify installer has the proper manufacturer's product installation instructions (MPII).

5.1.4 Verify the installer has a copy of the design requirement for the field condition of the concrete, including minimum age of concrete; concrete temperature range; moisture condition of concrete at time of installation; type of lightweight concrete, if applicable; and requirements for hole drilling and preparation.

SECTION 6—ERECTION OF PRECAST CONCRETE

All work shall be verified per Table 1.7.1 Item 10; ACI 301, 13.3; and as specified in the Contract Documents.

SECTION 7—SHOTCRETE

All work shall be verified per Table 1.7.1 Item 13, and as specified in the Contract Documents.

(nonmandatory section follows)

NOTES TO SPECIFIERS

General notes

G1. ACI Specification 311.7-16 is to be used by reference or incorporation in its entirety in the Project Specification. Do not copy individual parts, sections, articles, or paragraphs into the project specification because taking them out of context may change their meaning.

G2. If sections or parts of ACI Specification 311.7-16 are copied into the Project Specification or any other document, do not refer to them as an ACI Specification, because the specification has been altered.

G3. A statement such as the following will serve to make ACI Specification 311.7-16 a part of the Project Specification:

> "Work on (Project Title) shall conform to all requirements of ACI Specification 311.7-16, 'Specification for Inspection of Concrete Construction,' published by the American Concrete Institute, Farmington Hills, MI, except as modified by these contract documents."

G4. Each technical Section of ACI Specification 311.7-16 is written in the three-part Section format of the Construction Specifications Institute, as adapted for ACI requirements. The language is imperative and terse.

G5. If ACI Specification 311.7-16 is used with another ACI specification that contains overlapping provisions, identify which requirements are in conflict and state in the contract documents which requirements control.

Foreword to checklists

F1. This foreword is included for explanatory purposes only; it is not a part of ACI Specification 311.7-16.

F2. ACI Specification 311.7-16 may be referenced by the specifier in the inspection services contract for any building project, together with supplementary requirements for the specific project. Responsibilities for project participants must be defined in the inspection services contract. ACI Specification 311.7-16 cannot and does not address responsibilities for any project participant other than inspection agency.

F3. Checklists do not form a part of ACI Specification 311.7-16. Checklists assist the specifier in selecting and specifying project requirements in the inspection services contract.

F4. The Optional Requirements Checklist identifies specifier choices and alternatives. The checklist identifies the sections, parts, and articles of ACI Specification 311.7-16 and the action required or available to the specifier. The specifier should review each of the items in the checklist and make adjustments to the needs of a particular project by including those selected alternatives as mandatory requirements in the inspection services contract.

OPTIONAL REQUIREMENTS CHECKLIST

Section/Part/Article	Notes to Specifiers
1.3	References other than the latest edition of the included documents may be specified.
1.6.1 and 1.6.1.1	Other special inspectors' certifications equivalent to ACI CCSI or CTCI may be specified if allowed by the building code official. Equivalent certification should include written and performance examinations to relevant codes and specifications as accepted by the Accreditation Agency.
1.6.1.2	Inspection agency accreditations equivalents may be specified.
1.7.1	Inspection may be changed from continuously or periodically, by the local building codes.
4.2.6	Contractors' testing agency accreditations' equivalents may be specified.
5.1.2	Anchor installer certification equivalents may be specified.

ACI Certification Programs

The American Concrete Institute's 25+ certification programs provide workers with the credentials to build the best concrete structures in the world. Ensuring crew members are certified gives contractors a leading edge, as many local, national, and international building codes require ACI Certified personnel on the jobsite. ACI Certification can help build a solid reputation with owners and specifiers, giving contractors the potential to become eligible for more projects.

Individuals and agencies looking to verify active certifications held by specific individuals should visit *www.concrete.org/verify* for access to the ACI Certified Personnel Directory.

ACI Certification programs typically fall into three main categories: Construction/Specialist Programs, Testing Programs, Inspection Programs, and. A partial list follows. For the full scope and requirements of each program, visit *www.concrete.org/certification*.

INSPECTION PROGRAMS
Individuals who inspect various concrete-related installation and construction practices.

Adhesive Anchor Installation

Adhesive Anchor Installation Inspector
- An individual who has demonstrated the knowledge required to properly inspect the installation of adhesive anchors in concrete. The Inspector must understand the responsibilities and qualification requirements of the Installer, inspection requirements as cited in the documents governing the construction of the project, and the process governing the qualification of anchor systems.

Concrete Construction Inspection

Concrete Construction Special Inspector
- A person qualified to inspect and record the results of concrete construction inspection based on codes and job specifications. The program covers inspection during preplacement, placement, and post-placement operations.

Concrete Transportation Construction Inspector
- A person who has demonstrated proficiency in concrete inspection methods for transportation projects, including preplacement, placement, and post-placement operations.

CSA-Based Concrete Construction Special Inspector (Canada Only)
- A person qualified to inspect and record the results of concrete construction inspection based on Canadian and U.S. codes and job specifications. The program covers inspection during preplacement, placement, and post-placement operations.

TESTING PROGRAMS
Individuals who perform, record, and report the results of materials field/laboratory tests

Aggregate Testing

Aggregate Testing Technician – Level 1
- An individual who has demonstrated the knowledge and ability to properly perform, record, and report the results of basic field and laboratory procedures for aggregates.

Aggregate Testing Technician – Level 2
- An individual who has demonstrated the knowledge and ability to properly perform, record, and report the results of advanced laboratory procedures for aggregates.

Aggregate/Soils Base Testing Technician
- An individual who has demonstrated the knowledge and ability to properly perform, record, and report the results of basic field and laboratory procedures for aggregates and soils.

Cement Testing

Cement Physical Tester
- An individual who has demonstrated the knowledge and ability to properly perform, record, and report the results of specific ASTM tests on the physical properties of cementitious materials.

Field Concrete Testing

Concrete Field Testing Technician – Grade I
- An individual who has demonstrated the knowledge and ability to properly perform and record the results of seven basic field tests on freshly mixed concrete.

CSA-Based Concrete Field Testing Technician – Grade I (Canada Only)
- An individual who has demonstrated the knowledge and ability to properly perform and record the results of seven basic field test methods and standard practices on freshly-mixed concrete as prescribed by Canadian Standards Association methods.

Self-Consolidating Concrete Testing Technician
- An individual who has demonstrated the knowledge of and ability to properly perform five standard test methods and practices on self-consolidating concrete.

Laboratory Concrete Testing

Concrete Strength Testing Technician
- An individual who has demonstrated the knowledge and ability to properly perform, record, and report the results of four basic laboratory procedures related to the determination of concrete compressive and flexural strength.

Concrete Laboratory Testing Technician – Level 1
- An individual who has obtained concurrent certifications as both ACI Concrete Strength Testing Technician AND ACI Aggregate Testing Technician—Level 1.

Concrete Laboratory Testing Technician – Level 2
- An individual who has demonstrated the knowledge and ability to properly perform, record, and report the results of advanced laboratory procedures for aggregates and concrete.

Masonry Testing

Masonry Field Testing Technician
- An individual who has demonstrated the knowledge and ability to perform sample preparation and testing of masonry construction-related materials—including materials for brick, structural clay tile, concrete masonry units, prisms, mortars, and grouts—in the field.

Masonry Laboratory Testing Technician
- An individual who has demonstrated the knowledge and ability to perform laboratory procedures required for preparation and testing of masonry construction-related materials including brick, structural clay tile, concrete masonry units, prisms, mortars, and grouts.

CONSTRUCTION/SPECIALIST PROGRAMS
Individuals who carry out concrete construction practices in various specialty applications.

Adhesive Anchor Installation

Adhesive Anchor Installer
- An individual who has demonstrated the ability to read, comprehend, and execute instructions to properly install adhesive anchors in concrete. The Adhesive Anchor Installer must also demonstrate possession of the knowledge to properly assess ambient conditions, concrete condition, materials, equipment, and tools for

installing adhesive anchors and determine when it is appropriate to proceed with installation of an adhesive anchor or when additional guidance from a supervisor/foreman/project engineer is needed.

Concrete Flatwork Finishing

Specialty Commercial/Industrial Flatwork Finisher and Technician
- A person who has demonstrated a basic knowledge of finishing procedures for Specialty Commercial/Industrial Concrete Flatwork Finishing, including high tolerance floor construction, application of surface treatments, and silica fume concrete, and who has demonstrated skill in operating mechanized finishing equipment.

Concrete Flatwork Finisher
- A Concrete Flatwork Finisher is a craftsman who has demonstrated the physical skills and ability needed to place, consolidate, finish, edge, joint, cure and protect concrete flatwork.

Decorative Concrete Flatwork Finisher
- A Decorative Concrete Flatwork Finisher is a craftsman who has demonstrated knowledge about and the ability to place, finish, cure, and protect decorative concrete flatwork.

Concrete Foundation

Residential Concrete Foundation Technician
- An individual who has demonstrated knowledge of residential concrete foundation construction and the codes and standards that apply to this segment of the concrete construction industry.

Concrete Quality Management

Concrete Quality Technical Manager
- An individual who has demonstrated the knowledge and experience necessary to supervise an effective concrete quality control/quality assurance program and perform duties on behalf of the Architect/Engineer in technical matters pertaining to the concrete used in a project.

Shotcrete Construction

Shotcrete Nozzleman (Dry-Mix Process)
- An ACI-certified Shotcrete Nozzleman (Dry-Mix Process) is an individual who has demonstrated knowledge of the dry-mix shotcrete process (via written examination) and the ability to properly place dry-mix shotcrete (via performance examination shooting a limited-size test panel). The Shotcrete Nozzleman (Dry-Mix Process) also possesses substantially more shotcrete work experience in actual dry-mix shotcrete field applications than a Shotcrete-Nozzleman-in-Training (Dry-Mix Process).

Shotcrete Nozzleman (Wet-Mix Process)
- An ACI-certified Shotcrete Nozzleman (Wet-Mix Process) is an individual who has demonstrated knowledge of the wet-mix shotcrete process (via written examination) and the ability to properly place wet-mix shotcrete (via performance examination shooting a limited-size test panel). The Shotcrete Nozzleman (Wet-Mix Process) also possesses substantially more shotcrete work experience in actual wet-mix shotcrete field applications than a Shotcrete-Nozzleman-in-Training (Wet-Mix Process).

Tilt-Up Concrete Construction

Tilt-Up Supervisor and Technician
- A Tilt-Up Supervisor is a person who has demonstrated proficiency in and an understanding of overall on-site administrative and technical management for producing tilt-up projects by passing the ACI written examination and meeting work experience requirements.
- A Tilt-Up Technician is a person who has an understanding of overall on-site administrative and technical management for producing tilt-up projects by passing the ACI written examination, but who lacks sufficient work experience to qualify as a tilt-up supervisor.